Global Health Care Equivalency in the Age of Nanotechnology, Nanomedicine and Artificial Intelligence

This groundbreaking volume explores the vision of a Global Health Care Equivalency (GHCE) system powered by artificial intelligence and quantum computing technologies, operating on secure quantum-encrypted blockchains to create an equitable and decentralized healthcare framework accessible to all people worldwide.

Readers will gain profound insights into how emerging technologies can transform global healthcare delivery through detailed analyses of quantum encryption, blockchain implementation, and AI-driven diagnostics. The book presents a comprehensive roadmap for achieving healthcare equity, offering practical frameworks for technological integration, ethical considerations for implementation, and strategies for overcoming geopolitical barriers to create a truly universal healthcare system.

This visionary work is essential reading for healthcare policy makers, medical technology researchers, AI and quantum computing specialists, global health advocates, and graduate students in healthcare administration, biomedical engineering, and international development. It will particularly benefit professionals working at the intersection of healthcare and emerging technologies who seek to address global health disparities.

Global Health Care Equivalency in the Age of Nanotechnology, Nanomedicine and Artificial Intelligence

Edited by Frank J. Boehm and Raj Bawa

CRC Press
Taylor & Francis Group
Boca Raton London New York

CRC Press is an imprint of the
Taylor & Francis Group, an **informa** business

Designed cover image: Frank J. Boehm

First edition published 2026
by CRC Press
2385 NW Executive Center Drive, Suite 320, Boca Raton FL 33431

and by CRC Press
4 Park Square, Milton Park, Abingdon, Oxon, OX14 4RN

CRC Press is an imprint of Taylor & Francis Group, LLC

ISBN: 978-1-138-63199-1 (hbk)
ISBN: 978-1-041-15560-7 (pbk)
ISBN: 978-1-003-68009-3 (ebk)

DOI: 10.1201/9781003680093

Typeset in Times
by KnowledgeWorks Global Ltd.

I lovingly dedicate this book to the memory of my father, Josef Boehm, who ignited in me the flame of imagination by example of his unquenchable curiosity, quest for knowledge, and fascination with life and the universe; to my mother, Charlotte Boehm, whose amazing fortitude, generosity, and contagious enthusiasm for life has inspired all around her to never give up on their dreams and to always look up (we recently celebrated her 99th Birthday!); and to my sweeties, Liz Balfour and Jazmyn Balfour-Boehm, whose unconditional love and support has allowed me to realize my dreams. Further, to my amazing grandsons, Skye (SkyeGuy) Barney and Linden Barney (beautiful sons of Jazmyn and Steve Barney) who are incarnations of pure love and joy, for whom I deeply hope and trust that the world they inherit will be one that is far more unified, loving, equitable, wise, and compassionate.

I also dedicate this book to several passed colleagues who I will always consider as "forces of nature", whose presence, passion, diligence, leadership, and kindness will be sorely missed - Veer Endra, Ioan Opris, Ned Seeman, Selwyn Super, and Jack Taunton.

Contents

Preface

Since the publication of my first book (*Nanomedical Device and Systems Design: Challenges, Possibilities, Visions* – CRC Press, 2013), it became increasingly evident that synergies between the concurrently rapid advances in nanotechnology and nanomedicine in conjunction with the recent exponential progress in artificial intelligence (AI) would likely facilitate the capacity to actualize three envisaged global scale paradigm shifting technology platforms. Thus, I endeavored to explore these possibilities through the simultaneous generation of what I consider to be a set of three companion books. The rationale is that I believe these applications will be inextricably linked.

This volume, *Global Health Care Equivalency in the Age of Nanotechnology, Nanomedicine and Artificial Intelligence*, delves into the potential emergence of an envisaged equitable, cost-effective, non-monopolizable, globally distributed/decentralized AI/QC-driven healthcare system, which would operate on safe/secure nested quantum encrypted blockchains. With the attainment of mature Global Health Care Equivalency (GHCE), which may well be significantly expedited when advanced AI (and soon enough the emergence of Artificial General Intelligence (AGI) and Artificial Superintelligence (ASI)) are coupled with quantum computation (QC):

> we can envisage a future world where any individual on the planet has access to the same advanced and cost effective nanomedical diagnostic and therapeutic technologies, no matter how wealthy or impoverished they are, no matter where they happen to reside, or under what conditions they live. Progress toward this goal will be incremental, with each successive wave of nanomedical technologies being more advanced than the previous wave.
>
> *NanoApps Medical, Inc., 2024*

Multiple nascent iterations, operational/efficacy trials, and the beta testing of GHCE will be driven by further advances in 3D printing. Increasingly robust telemedical infrastructures will begin to progressively transition many of the common health concerns of patients (as well as healthy individuals) to be resolved by various nanomedical means, in their own homes. These early developmental stages might be guided by readily accessible top-tier online physicians on demand, and/or potentially, formally certified medical AI agents that have been granted access (by patients and/or their caregivers) to comprehensive knowledge of their medical histories, data, and specific needs. These medical personnel would initially be approved to prescribe only certain medicines at specific dosages, which could be safely fabricated in a patient's dwelling using envisaged countertop microwave-sized appliances referred to as Factory at Home (F@H) systems.

The feedstocks for F@H systems may initially be comprised of the full range of chemical constituents required to synthesize basic medicines, as well as amino acids for the fabrication of therapeutic peptides. With further technological advances, standardized nanoparticles of various species might serve as F@H feedstocks, which are then custom functionalized, replete with targeting molecules, to generate a dynamic range of high precision diagnostic, therapeutic, and theranostic nanoparticles to address myriad conditions (from colds to cancers). More advanced iterations of F@H would rely on molecular and atomic feedstocks to enable the fabrication of molecular/atomic resolution (passive and semi-autonomous) nanodevices that integrate nanoelectronics, sensing, communications, and computational capacities. Finally, the feedstocks of mature F@H systems would be comprised of pristine atomic species that enable the molecular manufacturing (MM) of atomic resolution autonomous nanomedical robots (~Ø1 μm), endowed with further capabilities (e.g., propulsion, navigation, and a full armamentarium of cell repair instrumentation).

A robust, massively decentralized healthcare system would, for the most part, render hospitals and clinics obsolete. Exceptions would, of course, be those required to accommodate maternity and pediatric needs, address emergency trauma cases, pandemic responses, large-scale mass casualties from natural disasters, extreme weather events, or due to anthropogenic conflicts. Broken bones, serious injuries, burns, etc. would still require triaged care. That said, most common ailments, and even more serious diseases, such as cancers, might be eradicated via envisioned domestically (F@H) fabricated autonomous *sentinel* class nanomedical devices (Domschke and Boehm, 2018), which would significantly augment the human immune system against all "non-self" bacterial and viral pathogens, as well as toxic agents. This would significantly reduce the need to ever visit a hospital or clinic for treatment (except for emergencies … albeit even many of these might also be rapidly addressed in the future by sufficiently sophisticated legions of autonomous nanorobots that continuously reside within the human body (in dwell mode) until instantaneously activated via a perceived inflicted trauma). Additionally, most age-correlated ailments might be negated, as aging itself would be addressed by specialized classes of autonomous nanomedical devices that continually monitor the body for/manage atherosclerosis, osteoporosis, as well as any inflammation, or myopathy.

The prerequisite technology platform and tipping point toward mature GHCE will arrive with the emergence of hypothetical (for now) molecular manufacturing (MM); the topic of my second book, *Molecular Manufacturing: Emergence of the Grand Equalizer* (Boehm, CRC Press, 2025). It is envisaged that MM may enable the cost-effective domestic fabrication of the advanced autonomous nanomedical devices described above, exemplars of which I articulate in my first book.

The third book in this series, *Nanomedical Brain/Cloud Interface: Explorations and Implications* (Boehm, Institute of Physics Publishing (IOPP), 2025), explores how a hypothetical (for now) nanomedically enabled brain/cloud interface (B/CI) may endow us with capacities to instantaneously access the full breadth of human knowledge, significantly enhance our ability to learn, and enable ultrahigh resolution fully sensorial immersive experiences (including episodic consensual inhabitation within any other willing participant on the planet), while empowering us to keep pace/integrate with AI. Consequently, this technology platform may, in effect, facilitate even further enhancements/efficacies for MM and GHCE, which from the author's perspective would close the loop. This, as each of these three technology platforms would likely synergistically cross-pollinate; all with the innate capacity to reciprocate. Further, the broad tech-trusted acceptance and ubiquitous implementation of a mature B/CI may be nothing less than the next evolutionary step for humanity, one where we seamlessly merge with our technologies.

To be realistic and clear, since GHCE would constitute such a truly massive and multifaceted paradigm shifting undertaking, it could not be expected to suddenly magically supplant current (2025) medical and healthcare paradigms and infrastructures. Rather, it might inspire and herald the emergence of a significant evolutionary leap, or indeed, revolution in medicine and healthcare, wherein medical personnel, as well as medical device and pharmaceutical infrastructures, remain engaged, albeit playing different roles. This, with the aim of significantly enhancing the quality and personalization of healthcare for all of humanity, which would concomitantly be of immense benefit planetwide.

It may be the case that GHCE rolls out via disparate scenarios, or the combination of multiples. On one hand, since big pharma (for instance) and other large corporations already have enormous manufacturing infrastructures in place, they might initially be repurposed to generate and supply the molecular/atomic feedstocks required to synthesize myriad diagnostic and therapeutic compounds and functionalized nanoparticles via F@H systems (which themselves would need to be manufactured), not only for their existing customer bases worldwide, but also to serve practically everyone on the planet. They may also be engaged with and have a stake in the initial development of various classes of nanomedical devices (passive, semi-autonomous, and fully autonomous), which would subsequently be fabricated domestically in a massively distributed/decentralized manner.

Due to the globally decentralized molecular manufacturing (MM) (via F@H systems) of nanomedicines, and further downstream, autonomous nanomedical robotics, the centralized control and monopolies over drugs and healthcare would eventually fade to obsolescence. Ultimately, Recykl@ Home (R@H) systems would emerge with the capacity to disassemble all manner of household wastes down to their atomic constituents, which would then be recycled/repackaged as pristine atomic feedstocks to continuously resupply the F@H/molecular manufacturing units. This would close the loop.

Some may hold the view that big pharma, health insurance, huge medical device and medical equipment conglomerates, hospitals, etc. primarily operate on the premise of maximizing bottom line profits, or institute cultural paradigms of profit above all else, which appear unlikely to change. We shall see. Things inexorably do transform, perhaps induced by the unseen energies of quantum entangled alchemy, changes of heart, of conscience, or otherwise.

GHCE might well initially roll out in the real world (spanning a certain temporal domain) as a further extension and coalescing of/advancement in home healthcare; expanding on conventional telemedicine, domestic healthcare robotics, 3D printing, etc. Once the advantages of increasingly distributed/decentralized and personalized GHCE become more and more evident and relevant/ practical for the benefit of patients, communities, and finally entire populations who become more deeply engaged with it, a critical mass threshold is likely be attained and exceeded, after which all manner of (perceived as outdated) paradigms will begin to fall away.

> Although the attainment of GHCE would indeed be a truly extraordinary achievement, we must keenly recognize that in reality, such a massive undertaking on a global scale will require intense and sustained high spirited collaboration between like minded individuals who have widely varied backgrounds within diverse areas of expertise. Highly specialized laboratories, dynamic administrative institutions, regulatory bodies, and government agencies worldwide will also be vital prerequisites for its realization, as will significant resources and time. We are all amazing individuals, each with our own unique talents and expertise that might contribute and be applied to this noble enterprise. Our working together might culminate in a positive critical mass on a global scale toward the actual realization of GHCE, which would ultimately be of benefit for each and every one of us, as we are finally "all" supported in our natural human right of optimal health for life.
>
> *Frank J. Boehm, NanoApps Medical, 2025*

It is hoped that the prospective concepts herein may inspire others downstream to cumulatively create a critical mass of focused intention toward the potential realization of Global Health Care Equivalency for the benefit of humankind and the planet.

Domschke A, Boehm FJ. Chapter 19: "The "Sentinel": A Conceptual Nanomedical Strategy for the Enhancement of the Human Immune System. In *Immune Aspects of Biopharmaceuticals and Nanomedicines*. Bawa R., Szebeni J., Webster TJ, Audette GF. (eds.), Singapore: Pan Stanford. 2018. https://www.crcpress.com/Immune-Effects-of-Biopharmaceuticals-and-Nanomedicines/Bawa-Szebeni-Webster-Audette/p/book/9789814774529

Acknowledgments

I am indeed deeply grateful to all the contributing authors and co-authors for their incredible efforts toward the creation and infusion of their amazing and inspiring intellects and life forces into this book, who are listed in the Contributors section. I would be surely be remiss for not acknowledging K. Eric Drexler for presenting to the world his original vision of the boundless possibilities of nanotechnology, molecular manufacturing, and nanomedicine as articulated in his highly inspirational book *Engines of Creation* (1987). Further, my true appreciation to Robert A. Freitas Jr. for further enlightening me as to the virtually limitless and exciting potential of nanomedicine through his acute insights and most excellent and groundbreaking book *Nanomedicine, Volume I: Basic Capabilities* (1999) and *Nanomedicine* book series. Also, my sincere gratitude to Richard Satava (former DARPA Program Manager and Professor of Surgery at the University of Washington-Seattle) for initially inspiring me to articulate my nanomedical concepts on paper, while advising me to always endeavor to amply support (via the scientific literature) their potential viability and promise. Finally, my special thanks and deep gratitude go to Amanda Scott (CEO NanoApps Medical, Inc., Creative Director Alias Studio) for her amazing and tireless efforts in support and promotion of my nanomedical concepts and publications as well as the latest progress in nanotechnology, nanomedicine, and AI through the creation and curation of our excellent NanoApps Medical, Inc. and NanoApps Athletics, Inc. sites, and social media. I also give my sincere thanks and kudos to Heinz Hoenen for his exceptional efforts and contributions to our international social media presence.

I would like to express my deep appreciation to the following individuals for providing information, insights, and other forms of support in facilitating the realization of this project: Amara Angelica, Elizabeth Balfour, Jazmyn Balfour-Boehm, Steve Barney, Raj Bawa, C. JoyBell C, Aicheng Chen, Han Chen, Simon Corrie, Pieter Cullis, Ernest Daddey, Aubrey de Grey, Angelika Domschke, Ted Duke, Godfrey Gandawa, Claude Giroux, Urs Hafeli, David Harakal, Robert R. Hieronimus, Bruce Johnson, Jonathan Kitzen, Ingemar Patrick Linden, Haijin Liu, Adriana Marias, Kannan Mavila, Margaret Morris, Sabino Padilla, Raphael Rakowski, Jeffery Rosenfeld, Mark Schultz, Shannon Smith, Frank Soda, Melanie Swan, Bryan Sykes, Jack Taunton, Ingrid Vasiliu-Feltes, Natasha Vita-More, Julia Walker, and Douglas Westcott.

I sincerely apologize to any individuals whom I may have inadvertently overlooked.

My sincere thanks and gratitude go out to my publisher, Taylor & Francis Group/CRC Press, for having the confidence and trust in me to see this project through, and allowing me the complete freedom to explore some of the exciting possibilities of an equitable, economical, non-monopolizable, tech-trusted, massively decentralized healthcare system I refer to as Global Health Care Equivalency (GHCE). In particular, I wish to convey my appreciation to Michael Slaughter, Michele Smith, and Marc Gutierrez for their incredible patience and support throughout the writing process. In addition, I very much appreciate the efforts of Jennifer Petruzziello and the staff at Taylor & Francis Group/CRC Press, as well as Mayank Sharma and the staff at KnowledgeWorks Global Ltd. in the production of this book.

Lastly, I will forever be grateful to my truly inspiring and wise father, Josef Boehm; my ever sweet and generous mother, Charlotte Boehm; my loving sister, Renata Swanson, and brother, John Boehm; Elizabeth Balfour, and our incredible daughter Jazmyn Balfour-Boehm and her husband Steve Barney, who blessed us with our beautiful and amazing grandsons Skye (SkyeGuy) Barney, and Linden Barney, for their unconditional love, encouragement, and unwavering support without which this book could not have been possible.

About the Editors

Frank J Boehm comes from a diverse background that includes a 12-year stint as a full time rock/pop musician in various bands that toured North America (1975–1987), and 17 years as a Senior Mechanical Designer involved in the design and development of dedicated robotic tooling for complex automated assembly systems (1987–2004). He has been further engaged as a Nanotechnology Consultant, Technology Scout, Product Designer, Research Associate, and Academic Writer/Editor for several international corporations and university laboratories (since 1996). Frank has been involved with nanotechnology and especially nanomedicine since 1996 (autodidactically trained), which inspired the development of numerous concepts and designs for advanced nanomedical diagnostic and therapeutic components, devices, and systems to potentially address myriad disease states. His aim is to develop and transform these concepts into real-world applications for global benefit. Frank serendipitously encountered the concept of nanotechnology on the Internet and immediately became fascinated with its virtually limitless potential, particularly as it relates to the field of medicine. He passionately proceeded to evolve and textually articulate various advanced near-term and longer-term nanomedical concepts and designs. Concomitantly, he initiated correspondence with numerous nanotechnology and nanomedicine research scientists and thought leaders from across the globe and founded a startup company, NanoApps Medical, Inc. (2009) with the aim of investigating and developing advanced, innovative, and cost-effective nanomedical diagnostic and therapeutic devices and systems for the benefit of humanity. Upon recognizing the immense potential of nanomedicine to impart positive paradigm shifts across the medical domain (e.g., precisely targeted drug delivery, vascular/neurological/cellular plaque removal, eradication of cancers via hypothermic nanoparticles, completely non-invasive surgical procedures, enhancement of physiological systems, and extended longevity), Frank was deeply motivated to write more extensively on the topic. This culminated in the authoring/publication of his first book, *Nanomedical Device and Systems Design: Challenges, Possibilities, Visions* (CRC Press, 2013). Subsequently, he authored/co-authored numerous book chapters and journal articles, including an in-depth paper titled "Human Brain/Cloud Interface" (*Frontiers in Neuroscience*, 2019), and co-founded NanoApps Athletics, Inc. in the same year. Frank then set his focus on a far more expansive investigation into the possibilities of nanotechnology/nanomedicine coupled with AI/AGI/ASI and quantum computation via the simultaneous compilation of a set of three companion books (with multiple chapter contributors for each) that explore a trifecta of envisaged global-scale paradigm-shifting technology platforms (*Nanotechnology, Nanomedicine, and AI: Toward the Dream of Global Health Care Equivalency* (CRC Press, 2025), *Molecular Manufacturing: Emergence of the Grand Equalizer* (CRC Press, 2025), and *Nanomedical Brain/Cloud Interface: Explorations and Implications* (IOP Publishing, 2025).

Raj Bawa, MS, PhD, MD, is president of Bawa Biotech LLC (founded 2002), a biotech/pharma consultancy and patent law firm based in Ashburn, Virginia, USA. Trained as a microbiologist and biochemist, he is an inventor, author, entrepreneur, professor, and registered patent agent (since 2002) licensed to practice before the US Patent & Trademark Office. He is currently a scientific advisor to Teva Pharmaceutical Industries, Israel (since 2010), professor (adj) at Northern Virginia Community College, Annandale, Virginia (since 2004), and vice president/chief IP officer at Guanine, Inc., Rensselaer, New York (since 2016). He has served as a principal investigator of various research grants, most recently as a principal investigator of a CDC grant to develop an assay for carbapenemase-resistant bacteria. He was an adjunct professor at Rensselaer Polytechnic Institute, Troy, NY, from 1998 to 2018. His education includes a BSc (Honors School) in microbiology, an MS in cancer biology, a PhD in biophysics/biochemistry, and an MD. In the 1990s, Dr. Bawa held various positions at the US Patent & Trademark Office, including primary examiner from 1996 to 2002. Currently, he is a life member of Sigma Xi, co-chair of the nanotech and precision medicine committees of the American Bar Association, and founding director of the American Society for Nanomedicine (established 2008). He has authored over 100 publications, edited 10 texts, and is on the editorial boards of numerous peer-reviewed journals, including serving as an associate editor of *Nanomedicine* (Elsevier). Some of Dr. Bawa's awards include the Innovations Prize from the Institution of Mechanical Engineers, London, UK (2008); Appreciation Award from the Undersecretary of Commerce, Washington, DC (2001); the Key Award from Rensselaer's Office of Alumni Relations (2005); and Lifetime Achievement Award from the American Society for Nanomedicine (2014).

Contributors

Keshav Alagarsamy
Nanoworld Laboratory
University of Cincinnati
Cincinnati, Ohio

Frank J. Boehm
NanoApps Medical Inc.
NanoApps Athletics Inc.
NanoApps Consulting
Thunder Bay, Ontario, Canada

C. JoyBell C.
International Journal of History and Philosophy of Medicine, United States
Studia Humanitatis Mrongoviensis: Polish Scientific Journal, Mragowo, Poland
New York, New York

Manila, Philippines
Studia Humanitatis Mrongoviensis: Polish Scientific Journal
Mrągowo, Poland

Marc Cahay
University of Cincinnati
Cincinnati, Ohio

Sumeet Chaudhary
Nanoworld Laboratory
University of Cincinnati
Cincinnati, Ohio

Simon R. Corrie
Department of Chemical and Biological Engineering
Monash University
Clayton, Victoria, Australia

Zhongyun Dong
College of Medicine
University of Cincinnati
Cincinnati, Ohio

Godfrey Gandawa
Springfield Research University
Ezulwini, Eswatini (Swaziland)

Sook Kuan Goh
Nanoworld Laboratory
University of Cincinnati
Cincinnati, Ohio

Kevin J. Haworth
College of Medicine
University of Cincinnati
Cincinnati, Ohio

Nicholas Lalley
Nanoworld Laboratory
University of Cincinnati
Cincinnati, Ohio

Ingemar Patrick Linden
Geneva College of Longevity Science
Brooklyn, New York

Adriana Marais
Proudly Human
Foundation for Space Development Africa
Cape Town, South Africa

Raphael Rakowski
Medically Home
Neurofinity
Sedona, Arizona

Jeffrey V. Rosenfeld
Department of Neurosurgery, Alfred Hospital
Melbourne, Victoria, Australia
Department of Surgery, Monash University
Melbourne, Victoria, Australia

Sarah Pixley
College of Medicine
University of Cincinnati
Cincinnati, Ohio

Maham Rahimi
College of Medicine
University of Cincinnati
Cincinnati, Ohio

Mark Schulz
Nanoworld Laboratory
University of Cincinnati
Cincinnati, Ohio

Melanie Swan
UCL Centre for Blockchain Technologies
DIYgenomics
New York, New York

Bryan Sykes
Public Health Advocate Research
Black Men's Health Alliance (BMH)
DiaBeatIt
South Africa

Ingrid Vasiliu-Feltes
University of Miami
Miami, Florida

Natasha Vita-More
Geneva College of Longevity Science
Human+
Digital University
Center for the Future Mind (Florida Atlantic University)
The Institute for Ethics and Emerging Technologies
Longevity Training LIT
Scottsdale, Arizona

Griffin Wagner
Nanoworld Laboratory
University of Cincinnati
Cincinnati, Ohio

Julia Walker
Department of Chemical Engineering
Monash University
Clayton, Victoria, Australia

Douglas Westcott
Independent MedTech Research
Vancouver, British Columbia, Canada

John Yin
Nanoworld Laboratory
University of Cincinnati
Cincinnati, Ohio

1 From Newton to Nano

Beyond the Boundaries of the Possible

Douglas Westcott

CONTENTS

1.1 INTRODUCTION

We live in an age that too often seems more comfortable building walls that separate, instead of constructing stronger bonds of partnership, an age that stresses difference versus integration, contestation over collaboration and wider cooperation. As well-known science presenter Bill Nye, with his customary wry wit, remarked in a radio interview on the Canadian Broadcasting Corporation, "If you like to worry, you were born at the right time" (Nye, 2017). The second decade of the 21st century in which we now live has been described recurrently in the media and academia as the "age of disruption." *The Economist* magazine notes that "the notion of disruption, with its promise to destroy the status quo and then renew it, is the most fashionable idea in global business since the craze of emerging markets over a decade ago." The magazine, however, goes on to point out a discrepancy; namely, that "there is a puzzle at the heart of the orthodoxy."

"Few bosses, in public or in private, expect their own firms to decline …" "The tech revolution, it seems, will be momentous, but harmless, with no corporate victims. Something does not add up" (Schumpeter, 2017). Something, indeed, does not "add up," and it is not just limited to the board-rooms of big business and the risk and rewards of disruption on their business models. The history of science is an ongoing story of "things not adding up." The essence of the scientific method can

DOI: 10.1201/9781003680093-1

be characterized as a systematic response to two plus two failing to equal four, wherever it occurs in the progression from hypothesis to experiment. Scientists rigorously reassess the theoretical premises of whatever is being observed, measured, and tested. All potential inadequacies in the methodological "food chain" must be thoroughly re-examined, re-evaluated, and, if necessary, initial hypotheses reformulated.

1.2 MAX PLANCK AND THE BLACKBODY PROBLEM

Something definitely was not adding up for physicists studying electromagnetic radiation at the end of the 19th century. There was a big problem, namely, that it was a prediction of classical physics that an ideal blackbody (i.e., a piece of material that radiates corresponding to its temperature) at thermal equilibrium will emit radiation in all frequency ranges, emitting more energy as the frequency increases. Quite simply, by calculating the total amount of radiated energy, it could be shown that a blackbody would release an *infinite amount of energy*, which contradicted the principles of the conservation of energy, the first law of thermodynamics (Figure 1.1).

A new model for the behavior of blackbodies was required. When the 40-year-old German theoretical physicist Max Planck turned his attention to the problem in 1898, the question had been explored experimentally, but no theoretical treatment agreed with experimental values. After many now well-known attempts to solve this problem, including those by fellow German physicist Wilhelm Wein, as well as English physicists Lord Rayleigh and James Jeans, it was Planck who, in 1900, came up with the solution to the "ultraviolet catastrophe" as it was later to become known. Whereas Wien's formula and the Rayleigh-Jeans Law could not explain the full spectrum of a blackbody, Max Planck's famous equation solved the problem by assuming that light was discrete; that is, light was made of "packets" of energy, which he famously christened quanta. In the field of thermodynamics, two plus two, indeed, finally added up to four.

FIGURE 1.1 A blackbody comprises a hypothetically perfect absorber of all incident radiation, which subsequently emits radiation as a sole function of its temperature. The blackbody shown (NASA) is employed to calibrate instrumentation for the Earth Radiation Budget Experiment (ERBE) and Clouds and the Earth's Radiant Energy System (CERES). (Image from Wikimedia Commons.)

In order to make his ultimate discovery, however, Planck was, as he later recounted, literally "ready to sacrifice my previous convictions about physics" (Planck, 1969). These "convictions" were those he had shared with his peers who were contemptuous of atomic theory, particularly the new field of statistical mechanics introduced by Ludwig Boltzmann. So disturbing were the ramifications of what Planck had discovered that he himself stated, "my unavailing attempts to somehow reintegrate the action quantum into classical theory extended over several years and caused me much trouble" (Turok, 2013). But in spite of his serious reservations, as Niels Bohr would later say, "his belief in the compelling force of logical reasoning from facts was so strong that he did not flinch from announcing the most revolutionary idea which has ever shaken physics" (Born, 1947).

It is upon this unflinching foundation of conducting himself like a superb detective, dogged in the pursuit of the truth irrespective of where the trail of clues may lead, that Planck's everlasting fame and reputation rest. It is worth noting in this context that a contemporary of Planck, a fellow shifter of paradigms, would receive, at long last, his Nobel prize in 1921, not for his astonishing theory of relativity, but, rather, for the "photoelectric effect." His name was Albert Einstein. Relativity, like atomic theory that preceded it in the 19th century, was largely "unproven" as it was argued by the scientific establishment of the day. This led Planck to write many years later, reflecting back on his early years, that, "a new scientific truth does not triumph by convincing its opponents and making them see the light, but rather because its opponents eventually die, and a new generation grows up that is familiar with it" (Planck, 1949).

1.3 THE MAX PLANCK CENTER

If we fast-forward just over a century, we see today Max Planck's revolutionary idea, nature's rulebook of quantum mechanics, at work in evolutionary, tangible fashion in nanotechnological research centers, university laboratories, and private and public enterprises around the globe. The Max Planck Center-EPFL Center for Molecular Nanoscience and Technology in Lausanne, Switzerland, symbolizes the incredible journey from Planck's quantum reservations to tangible, quantum-based applications. The Max Planck Center is representative of the field of nanoscience aggregating physics, chemistry, biology, and engineering in an interdisciplinary approach to manipulate and gain control of the molecular and atomic domains. The intention is to design systems and functionalities which approach that of living matter. Researchers are learning from nature's rulebook, "borrowing," as Ted Sargent describes it in his book, *The Dance of the Molecules: How Nanotechnology Is Changing Our Lives*, "from the genius of her methods" (Sargent, 2006).

By bringing the nanoscale and life sciences under one umbrella, the Max Planck Center is no less enthusiastic about the "exciting possibilities" in, as their website puts it, "establishing Nanoscale Science and Technology-based methodologies in life science-related fields." The Center goes on to say that these "hold great promise to revolutionize research in supramolecular chemistry, biology, biotechnology, and medicine. This is expected to result in a new era in healthcare, with novel strategies for the treatment and management of a range of diseases" (The Max Planck Center, 2025).

Ian Goldin and Kris Kutarna, in their book *Age of Discovery: Navigating the Risks and Rewards of Our New Renaissance*, remark that, "medicine is gaining nature's power to design life; engineering the power to design matter. This is not ordinary progress, but the revolution that comes when we hit the outer limits of our present paradigms and adopt new ones" (Goldin and Kutarna, 2016). They go on to point out, however, that amidst the excitement of hitting the "outer limits of present paradigms," let alone proceeding to adopt "new ones," another constant, perhaps as irreducible as Planck's constant itself, abides in history. This is, namely, that "a flourishing of genius does not just happen. It arises out of social and intellectual conditions that enable creativity to shine forth" (Goldin and Kutarna, 2016). In many ways, it is ironic that such a revolutionary technological advancement as nanoscience is set against a background of weakening social cohesion today. Thus, in such a turbulent societal environment, the authors note, "the same connective and developmental forces that fuel human imagination also breed complexity and concentrate our activities in dangerous ways" (Goldin and Kutarna, 2016).

1.4 ANTHROPOGENIC COGNITIVE "BLIND SPOT"

In such an environment, where the velocity of these "connective and developmental forces" in global trade, finance, population migration, and technology entangle the globe at seemingly ever-increasing speed, how do people respond? In their book, *The Tree of Knowledge: The Biological Roots of Human Understanding*, biologists Humberto Maturana and Francisco Varela begin their study of human cognition and perception by discussing what they call the "temptation of certainty." They draw our attention to a human cognitive bias that "we tend to live in a world of certainty, of undoubted, rock-ribbed perceptions: our convictions prove that things are the way we see them and there is no alternative to what we hold as true. This is our daily situation, our cultural condition, our common way of being human" (Maturana and Varela, 2008, p. 18).

The authors proceed to demonstrate how the apparent firmness of our experiential world "wavers," as they describe it, via two examples, the first being the famous phenomenon of the visual blind spot which most of us have tried in grade school science class. The commonly accepted explanation for the blind spot is that the image of the dot on a page they use for their demonstration, in one specific position in front of our eyes (with one being closed), falls into the area of the retina where the optic nerve emerges; hence, it is not sensitive to light and is thus termed the "blind spot." The authors immediately posed an important question: "What is rarely stressed in giving this explanation, however, is: 'How come we don't go around with a hole that size all the time?'" The simple answer is that our visual experience is a "continuous space." We do not normally – unless, of course, we do the sort of deliberate observational manipulations they perform in their book or we do in science class – perceive the discontinuity which is always there.

As the authors succinctly state, "we do not see, what we do not see" (Maturana and Varela, 2008, p. 19). Maturana and Varela use the blind spot as a key foundational metaphor for beginning their discussion of the phenomenon of cognition and the concept of our essential "doing" in "bringing forth of a world." So-called optical illusions, such as the blind spot or colored shadows, are not errors, annoying residues, or obstacles to be eradicated in our understanding of how we bring forth knowledge. Quite the contrary, they are an essential characteristic of our way of knowing. This phenomenon is not a superfluous mental feature, but rather,

> It is associated with the deepest roots of our cognitive being, however strong our experience may be. And because these roots go to the very biologic base … this bringing forth of a world manifests itself in all our actions. Certainly, it manifests itself in all those actions of human social life where it is often evident, as in the case of values and preferences.
>
> *Maturana and Varela, 2008, p. 27*

What then becomes of these "values and preferences" when the larger macro environment in which we all dwell becomes increasingly complex, as it has over the last 100 years, along with greater velocity of change in the first two decades of the 21st century? For their part, Goldin and Kutarna explored this issue in great depth by noting that, "the more complex our interactions become, the harder it is for us to see relationships of cause and effect. We develop cognitive 'blind spots' in our vision of the events around us. How can we make good decisions when we can't see their consequences?" (Goldin and Kutarna, 2016, p. 176).

There is no argument that both Planck and Einstein were two of the greatest minds in the history of science. The former, in the words of Carlo Rovelli, in his *Seven Brief Lessons of Physics*, was the "father" of quantum mechanics and the latter, the "parent who nurtured it" (Rovelli, 2015, p. 16). Yet, significantly, both of these titans of science, both the literal and spiritual fathers of quantum theory, had serious difficulties with the ramifications and consequences of what Niels Bohr and Werner Heisenberg, working alongside each other in Copenhagen, were proposing as they refined and, by 1925, provided the necessary equations to systematize quantum mechanics. So great was the cognitive dissonance engendered by his own theory that Planck, for his part, was actually hoping that wave mechanics would eventually supplant quantum mechanics. Einstein was similarly disturbed by the thought of such a random basis to our world, and he entered into the famous "light box" series of debates with Niels Bohr to defend and explain his objections to the Copenhagen interpretation of quantum mechanics. As Rovelli described it,

> The young lions of the Copenhagen group were dismayed: how was it possible that *Einstein* should think this? Their spiritual father, the man who had shown the courage to think the unthinkable, now retreated and was afraid of this new leap into the unknown that he himself had triggered. The same Einstein that had shown that time is not universal, and that space is curved was now saying that the world cannot be *this* strange.
>
> *Rovelli, 2015, p. 18*

Thus, even two of the greatest thinker's history has produced, it would appear, can have "blind spots" when confronted with disturbing facts, when the "relationship between cause and effect" becomes too cognitively "complex" as they intrude upon deeply held value and belief systems. Once again, we are reminded of Planck's previously quoted sardonic remark from his autobiography that a new scientific truth does not triumph by convincing its opponents and making them see the light, but rather because its opponents eventually die, and a new generation grows up that is familiar with it. This would, in later years, become known famously as "Planck's principle" (Brown, 2015) and it is somewhat ironic in this context that Planck was not referring to his disquiet with the theoretical implications of what he had himself "fathered."

Rather, it was his experience as a much younger theoretical scientist, whose paradigm-shifting resolution to the ultraviolet catastrophe ran into the same sort of cognitive dissonance from the 19th century scientific establishment of the day. History, and not just the scientific kind, provides us with numerous examples of individuals responding to social complexity overloading cognitive capacity by "filling in" the blind spot gap and, in this manner, making the uncomfortably discontinuous more comfortably continuous. This provides, needless to say, a rich and fertile terrain for superstition, magical thinking, ideological certitude, and stereotypical bias to thrive.

The above list is, given the innate human hard wiring at work, far from exhaustive. Various and many are human cognitive biases, and Nobel Prizes have been awarded in recent years to academics like Daniel Kahneman, Robert Tversky, and most recently, economist Richard Thaler for their groundbreaking work demonstrating that homo sapiens is a preponderantly cause-and-effect seeking species. Human beings are pattern seekers, and while this was, from an evolutionary standpoint, a most handy survival mechanism to have had in our arsenal while avoiding a hungry predator on the African savannah, we also "see" or fill in patterns when, as often as not, they aren't really there.

1.5 COGNITIVE CAPACITY AND SOCIAL COMPLEXITY

People can certainly "fill in" the gaps between our cognitive capacities and the ever-increasing magnitude of social complexities in a number of ways. Ted Sargent aptly refers to this phenomenon at the close of his book, *The Dance of the Molecules*. Anyone observing the lightning-fast developments in a field of "complex interactions" such as nanotechnology – from the molecular up to the macro human scale – provides fertile ground for individuals to oscillate between "irrational exuberance or disproportionate despondency" (Sargent, 2006, p. 211). Even the most perfunctory review of 21st century revolutionary technologies must have quantum computing, AI, and nanotechnology, as well as the complex social ramifications accompanying them, at the epicenter of any such discussion.

In April 2012, *The Economist* magazine ran a special report titled, "The Third Industrial Revolution," which surveyed the present and projected future state of manufacturing worldwide. The report highlighted the emerging "manufacturing revolution" in additive manufacturing (3D printing), and how this might realistically be scaled up by continuous production processes to make mass market, mass customization possible. It also discussed the effects of increased online accessibility for the crowdsourcing of networked manufactured products, as well as issues such as merging materials research with manufacturing technologies more seamlessly and efficiently. The report surveyed a number of innovations, including a special casting system for Rolls-Royce turbine blades, as well as the use of carbon fiber, recycled plastic waste, and new battery technologies.

With respect to the last of these, the writer did give a nod to nanotechnology by stating, "increasingly, product engineering will begin at the nanoscale. Nanotechnology is already used to enhance some products." Indeed, this may be abundantly evidenced today by perusing the Nanotechnology Products Database, which lists 8878 products that involve nanotechnology (Nanotechnology Products Database, 2019). A brief discussion of the use of hydrophilic titanium dioxide to produce self-cleaning glass in buildings is followed and contrasted by hydrophobic coatings, which would greatly improve the efficiency and durability of machines like steam turbines and desalination plants. The well-known biomimetic work of Andrea Belcher was also mentioned, which describes the use of viruses to grow improved materials such as semiconductors, transparent coatings for touch screens, and catalysts for turning natural gas into oil and plastics (*The Economist*, 2012).

1.6 THE FOURTH INDUSTRIAL REVOLUTION

Fast-forward to 2016, when the World Economic Forum released a report titled, *The Fourth Industrial Revolution: What it means, how to respond*, authored by the organization's founder and executive chairman, Klaus Schwab. During the 20th century, centralized electricity, the telephone, radio and television, cheap oil, and internal combustion vehicles on national road systems converged to create an infrastructure for the Second Industrial Revolution. It took about a century to go from this to the introduction of the microchip, personal computing, and the Internet, better known as the digital revolution, at the end of the 20th century. From the late-stage advancements of the Third Industrial Revolution, as described in *The Economist* Special Report in 2012, to the pronouncement by the World Economic Forum's founder of a Fourth Industrial Revolution took only five years!

Schwab sets out his stall at the outset of his paper in no uncertain terms. He observed that,

> There are three reasons why today's transformations represent not merely a prolongation of the Third Industrial Revolution but rather the arrival of a Fourth and distinct one: velocity, scope, and systems impact. The speed of current breakthroughs has no historical precedent. When compared with previous industrial revolutions, the Fourth is evolving at an exponential rather than a linear pace. Moreover, it is disrupting almost every industry in every country. And the breadth and depth of these changes herald the transformation of entire systems of production, management, and governance.
>
> *Schwab, 2016*

Schwab goes on to discuss the transformational impacts on business of emerging technology breakthroughs in fields such as artificial intelligence (AI), robotics, the Internet of Things (IoT), autonomous vehicles, 3D printing, nanotechnology, biotechnology, materials science, energy storage, and quantum computing.

Few would suggest that Karl Schwab is indulging in "magical thinking" in suggesting that a new industrial revolution is already upon us characterized by "velocity, scope, and systems impact," that "the speed of current breakthroughs has no historical precedent," or that "today's transformations" are "evolving at an exponential rather than a linear pace." Certainly, the World Economic Forum is not known for expressing spasms of "irrational exuberance" on global issues of economic interest to its powerful, elite member audience. Quite to the contrary, Schwab is defining something we can all *feel*, if not always so accurately define, all around us in today's disruptive social environment. Schwab goes on, in his conversations with global CEOs and senior business executives, to pinpoint an important "underlying theme," namely, that "the acceleration of innovation and the velocity of disruption are hard to comprehend or anticipate and that these drivers constitute a source of constant surprise, even for the best connected and most well informed" (Schwab, 2016).

The velocity of change apparently has most powerful, wealthy, "best-connected and most well-informed" CEOs of the world's largest corporations – many with annual revenues larger than the annual GDP of many countries in the world – feeling like the Red Queen from *Alice's Adventures in Wonderland*: they have to run forever faster in place just to maintain their present position. It really should not come as too much of a "surprise" that one doesn't have to drill down too far beneath the early 21st century surface to discover reductive solutions to cognitive blind spots arising from our increasingly complex, entangled, and disrupted social reality.

We live in an era of seemingly unending technological "breakthroughs," "game changers," "next big thing," and "paradigm shifts" themselves giving way to "seismic shifts" and "tsunamis" of transformation. Is this all just rhetorical hyperbole to attract listeners and viewers, more "eyeballs" producing more "clicks" to provide more revenue to the bottom line? Perhaps, but only in part. No matter how such claims might cue an eye-roll from a seasoned participant in the lab or clinic, or the most cynical observer from the media, none of us, including the most "well informed" can ever be sure in an age of maximum disruption, how to "anticipate," where the signal from the noise truly resides.

Furthermore, when things constantly do not quite "add up," it is human nature, even for true geniuses like Planck and Einstein, to become apprehensive about broader implications. If we re-examine the worldview these scientists were instrumental in overthrowing, we might gain some needed perspective to better glean the signal from the noise in our own cognitive perceptions regarding the real risks and challenges that face nanotechnology in general, and nanomedicine, in particular. Let us pause and cast a brief backward glance at scientific history.

1.7 PARADIGM SHIFTERS: COPERNICUS, GALILEO, AND DESCARTES

Prior to the mid-15th century, the predominant worldview could be summed up in one word, namely, faith. Historians Will and Ariel Durant, in Volume IV of their *Story of Civilization*, "The Age of Faith," observe that "the preeminent bequest of the Age of Faith was religion" (Durant and Durant, 1950). They affirm Western society's wholesale reliance on faith as this period's most salient feature. Medieval science in the West during this era was a discipline that rested, in no uncertain terms, on the twin pillars of authority: Aristotle and the Church. Thomas Aquinas adroitly – perhaps too adroitly for his critics – synthesized Aristotle's natural philosophy with Christian doctrine into a worldview where reason and faith were handmaids before the altar of a higher, omnipresent, and omnipotent power. Soon, however, these twin pillars would seem much less solid and unassailable than they appeared during the medieval period.

The onset of the 16th century begins with Copernicus overthrowing the Ptolemaic geocentric model of the solar system and science, which gradually led to a new scientific method to formulate valid criteria for explanations based on data extracted from experimental observations, gathering momentum in the 16th and 17th centuries. This culminated with the achievements of Kepler, Galileo, Bacon, and Newton in the fields of astronomy, physics, and natural science. Johannes Kepler (1571–1630) painstakingly formulated empirical laws for planetary motion which provided further evidentiary support to the Copernican theory. His contemporary, Galileo Galilei, "father of the scientific method," as Einstein called him, proceeded to analyze the "laws of nature" in mathematical terms.

The language of the universe was "written in mathematics and its characters are triangles, circles, and other geometric figures." The well-chronicled reward for this great polymath who provided

FIGURE 1.2 The Italian polymath Galileo Galilei was best known for his contributions to astronomy, physics, engineering, philosophy, and mathematics and has been referred to as the "father of observational astronomy." (Image from Wikimedia Commons, public domain, portrait by Justus Sustermans (1597–1681), cropped.)

a pivotal step in the transition away, and final separation from a world viewed through the lens of faith and religiosity, was a meeting with the Inquisition in 1633. Never were the difficulties that the human race faces in changing deep-seated belief preferences more highlighted than during the trial of Galileo (Figure 1.2). Faced with a new astronomical theory that completely disrupted a secure and ideologically fixed worldview, as fixed and stationary as the Earth was regarded to be under the geocentric theory, the authorities of the time, the Church of Rome, were merciless in their persecution of the unfortunate Galileo, and the "heresy" of his heliocentrism.

Galileo's astronomical observations, as disruptive as they were, however, could not provide complete and definitive proof at that time that the Earth was not the center of the cosmos. Even the much-improved telescope that Galileo invented and employed in 1610 could not observe a stellar parallax that would have provided the definitive proof he required. Thus, it was convenient for the Church to be able to rely on the current scientific disagreement and use scientist Tycho Brahe's compromise system as arguments against Galileo's evidence. It is a solecism to suggest that the conventional wisdom of the time should have seen the "light" of the new observational science that provided compelling proof for a paradigm shift. And when compelling proof is in fact provided, even the scientists who provided the proof may not want to be convinced if the results disrupt their entrenched belief systems too radically.

Arthur Koestler, in his book, *Act of Creation*, used the term "snow blindness," referring "to that remarkable form of blindness which often prevents the original thinker from perceiving the meaning and significance of *his own* discovery. Jealousy apart, the antibody reaction directed against new ideas seems to be much the same whether the idea was let loose by others–or oneself" (Koestler, 2014). When social complexity overwhelms our cognitive capacity to comprehend unsettling change, a pervasive sort of snow blindness can affect fields of vision, even, as we have witnessed, for the very scientists whose theories engender this belief dislocation in the first place.

René Descartes arrived upon the scientific scene at the end of the 16th century, and this brilliant mathematician and first modern philosopher may have indeed doubted that all (except the fact that he *could* doubt) had absolutely no doubt about his own genius. What Descartes *was* certain about was scientific mathematical certainty, and his most celebrated discovery of analytic geometry was motivated in large measure by his desire to describe nature in purely mathematical terms. He was able to reduce all physical phenomena to exact mathematical relationships. As the great philosopher and mathematician put it succinctly himself, "my entire physics is nothing other than geometry." Descartes's radical doubt and analytic method may have been one of his great bequests to science; however, his separation of reality into mind, or *res cogitans* ("thinking thing"), and matter, *res extensa* ("extended thing"), has had a deep and lasting influence on our Western worldview ever since.

For Descartes, the world was, as Fritjof Capra, in his book *The Systems View of Life: A Unifying Vision*, remarks, "a machine and nothing but a machine." Capra goes on to describe what this view of the natural world entailed, namely,

> Nature worked according to mechanical laws, and everything in the material world could be explained in terms of the arrangement and movement of its parts. This mechanical picture of nature became the dominant paradigm of science in this period following Descartes. It guided all scientific observation and the formulation of all theories of natural phenomena until twentieth century physics brought about a radical change. The elaboration of mechanistic science in the seventeenth, eighteenth, and nineteenth centuries, including Newton's grand synthesis, was but the development of the Cartesian idea. Descartes gave scientific thought its general framework – the view of nature as a perfect machine, governed by exact mathematical laws.
>
> *Capra and Luisi, 2016*

Indeed "Newton's grand synthesis" cemented a mechanical picture of nature that became the "dominant paradigm," and this was as firmly embedded for the following three centuries as surely as the faith-based geocentric view of the universe had prevailed in the medieval period up to the late Renaissance.

Living as we do today in a Fourth Industrial Revolution, where the "velocity, scope, and systems impact" of today's transformations are exponential rather than linear, our "belief perseverance," as the field of decision theory refers to it, is being tested and challenged as never before. Nobel-prize winning psychologist Daniel Kahneman, in *Thinking, Fast and Slow*, notes that, "sustaining doubt is harder than sliding into certainty" (Kahneman, 2011). And his foundational psychological principle of "cognitive ease," whereby human beings use judgment heuristics to simplify decisions made under radical uncertainty, echoes Maturana and Varela's theory of cognition, when he remarks that, "we are prone to exaggerate the consistency and coherence of what we see" (Kahneman, 2011).

1.8 PEERING THROUGH RHETORICAL PRISMS AND BREAKTHROUGH HYPE

Thus, in our age of extreme disruption, where forces of innovation battle the forces of reaction in a climate of radical uncertainty, frequently extreme positions are the consequence. Nanotechnology, in its earlier and more formative phase over two decades ago, underwent a wave of excitement that could be said to reflect and even symbolize the period in general. Occurring concurrently in the years just prior to and following the dotcom bubble and bust at the close of the 20th century, nanotechnology was swept up in the general feeling of what Federal Reserve Chairman Alan Greenspan famously termed "irrational exuberance," the phrase by which he used to characterize the skyrocketing stock market of the roaring 1990s.

Nanotechnology made the cover of *Red Herring* magazine in June 2001. The special issue was titled, "Nanotechnology: This Could Change Everything" (Schwartz, 2001). This was in the period following the "tech wreck", which had wrought its devastation on Silicon Valley (in the process of losing over 450,000 jobs), and an almost wishful plea for a new, sustainable technological breakthrough to take the innovative "torch" from the explosive information superhighway in the 1990s. In fact, by the end of the next year, *Red Herring* magazine would itself be defunct and add its name to the pervasive bloodbath in the technology space.

For its part, nanomedicine today is emerging from the lab in an environment that is certainly not in the sort of irrational exuberance of the dotcom bubble era but, nevertheless, one in which medical "breakthroughs" seem exponential indeed. The recent and rapid development of personalized, precision medicine arose in large part from the ambitious and successful mapping of the complete genetic sequence of the human species by the Human Genome Project in 2003. As Dr. W. Gregory Feero, Associate Editor of the *Journal of the American Medical Association*, stated in the May 9, 2017, edition of *JAMA*, the U.S. Healthcare system, despite being in a period of "turbulence," has been "undergoing a deeper transformation." He explained that, "this shift is inexorably moving medicine from an endeavor in which care for individual patients is driven by trial and error informed by studies designed to measure population outcomes to one in which care is selected based on a deep understanding of health and disease attributes unique to each individual" (Feero, 2017). It appears evident that medicine's more complete understanding of how disease operates at the molecular level is beginning to yield tangible results.

Nothing perhaps better typifies this growing excitement than *The Economist* magazine's front cover of its September 16, 2017, issue titled, "Closing in on Cancer," with a graphic of a shooting target riddled with bullet holes represented as getting closer and closer to the bull's-eye. The special report states, reassuringly, that,

> New research tools, such as easily generated antibodies, rapid gene sequencing and ever easier genetic engineering, have revolutionized biologists' understanding of cancer. This understanding has allowed more specific approaches to the disease to be developed, and the trend will continue. What is more, the tools of molecular biology have moved out of the lab and into the clinic. Genetic tests are used to find the precise vulnerabilities of a particular patient's cancer. Antibodies attack the specific molecules that have gone haywire. The cells of patients with cancer are engineered to better fight the disease.
>
> *The Economist, 2017*

If we add to the above, the rapid developments going on in immunotherapy, liberating the immune system's response to cancer, it is no wonder that, as the report notes, "the excitement at oncology conferences is palpable." Perhaps we should call this reaction more of a rational than an irrational type of exuberance. The special report goes on to underline that, "although immunotherapy is still new, it has already radically shifted the treatment and research landscape" (*The Economist*, 2017).

For its part, the pharmaceutical industry (as reported by Evaluate Group in their *EvaluatePharma – World Preview 2017, Outlook to 2022* release), the total number of new drugs approved and launched globally from 2013 to 2015 set new records – 165 in total, before recording a slower pace or trend in 2016 (EvaluatePharma, 2017). Prescription drug sales are forecast to grow at a healthy annual compound rate of 6.5%, eventually hitting $1.06 trillion worldwide in 2022. Much of this growth is set to come from some of the industry's hottest therapy areas, including new cancer immunotherapies like Opdivo and Keytruda.

It is worth noting that biologics are set to contribute 52% of the top 100 product sales by 2022. In a report discussing "the beginning of the end of blood cancer," Jim Greenwood at *BIOtech Now* (October 23, 2017) reported that, "In his inaugural address this June as chairman of the world's largest biotechnology industry group, CEO John Maraganore of Alnylam Pharmaceuticals dropped more than a few jaws with his prediction that we can 'expect that all hematological malignancies will be cured in the next decade.'" Greenwood stated emphatically, "It was a 'wow' moment: envisioning an end to blood cancer *that* soon" (Greenwood, 2017).

The basis for this enthusiasm was the U.S. Food and Drug Administration (FDA) approval of Novartis' "living drug," Kymriah, better known as CAR T-cell therapy, that genetically modifies a patient's own blood cells to attack an untreatable form of leukemia in children and young adults. On August 30, 2017, the FDA called the approval "historic." And, of course, it may well turn out to be just such a "game-changer." But let us take a step back before champagne corks are popped and victory declared by the members of the biotech, pharma, and medical communities over this undeniably significant approval. In fact, the champagne wasn't even "flat" before the second "living drug" or biologic to get approval followed seven weeks later with the announcement that the FDA sanctioned the second-ever gene therapy for sale in the United States. Gilead Sciences will offer a new CAR-T therapy with the potential to send cancer into remission for thousands of patients with untreatable lymphoma.

At the same time as these major drug approval developments were taking place, the medical science reporter Kelley Crowe of the Canadian Broadcasting Corporation, along with other news organizations, was reporting that "when researchers in the U.K. took a close look at the effectiveness of new cancer drugs approved by the European Medicines Agency between 2009 and 2013, they discovered that only about half showed 'a significant improvement in survival or quality of life.'" The findings that Crowe was referring to appeared in two of the world's most prestigious medical journals: the first, titled "Assessment of Overall Survival, Quality of Life, and Safety Benefits Associated With New Cancer Medicines," appeared in *JAMA* in March 2016 (Salas-Vega et al., 2017). This was followed by a second study titled, "Availability of evidence of benefits on overall survival and quality of life of cancer drugs approved by European Medicines Agency: retrospective cohort study of drug approvals 2009-13," published in the *British Medical Journal (BMJ)* in October 2017 (Davis et al., 2017).

Earlier in 2017, Liz Szabo of *Kaiser Health News* provided similar research in a detailed report titled, "Widespread Hype Gives False Hope To Many Cancer Patients." Szabo reported that patients and families are "bombarded with the news that the country is winning the war against cancer. The news media hypes research results to attract readers. Drug companies promise 'a chance to live longer' to boost sales. Hospitals woo paying customers with ads that appeal to patients' fears and hopes." Szabo quotes Dr. Otis Brawley, chief medical officer at the American Cancer Society, stating bluntly that, "we're starting to believe our own bullshit" (Szabo, 2017).

For his part, Dr. Walid Geilad, co-director of the Center for Pharmaceutical Policy and Prescribing at the University of Pittsburgh, remarked, a little more guardedly, that, "It's in the interest of almost every stakeholder in the health system to be optimistic about these therapies" (Szabo, 2017). And make no mistake – there are huge financial rewards for the participants in the health system going forward as the baby boom cohort ages over the next 20 years. In this regard, *The Economist* confirmed that, "the pipeline of oncology drugs in clinical development has grown 45% in the past decade. There are currently about 600 drugs under development at biotech and pharma companies" (*The Economist*, 2017).

The United States alone spent nearly $88 billion treating cancer in 2014, and given that this disease is integrally related to aging, spending is projected according to the National Cancer Institute to explode by over 27% by 2020. A spokesperson for the Pharmaceutical Research and Manufacturers of America, an industry group, said in a statement, "in the last decade we've seen a number of scientific advances transform the landscape of many cancers" (Szabo, 2017). This declaration seems intuitively obvious enough at first glance, after all there is that word "transform" again, and do we not live in a radically transformative age? But do the numbers actually back up this confident picture? When we consult the previously quoted *JAMA* and *BMJ* research on these claims, we find a somewhat different picture. The *JAMA* report states in its findings the following survivorship statistics in the data:

> An analysis of health technology assessment reports found that new cancer drugs were associated with increased overall survival by an average of 3.43 months between 2003 and 2013, with 43% increased overall survival by 3 months or longer, 11% by less than 3 months and 30% were not associated with an increase in overall survival. Most new cancer drugs improved quality of life and were associated with reduced patient safety.
>
> *Salas-Vega et al., 2017*

Much of the problem with the efficacy of new cancer medicines occurred, the report further emphasizes, because of "non-validated evidence." In their study of the European Medicines Agency (EMA) drug approvals for the period 2009–2013, the *BMJ* found in the segment with the most significant prolongation of survival (35%) that, "the magnitude of benefit on overall survival ranged from 1.0 to 5.8 months (median 2.7 months)." Unlike the more restrained conclusions of the *JAMA* report, the *BMJ* pulled no punches. It concluded that,

> Although the goal of cancer treatment is to improve the quantity and quality of life, clinical trials designed to gain regulatory approval for new drugs often evaluate indirect or "surrogate" measures of drug efficacy. These endpoints show that an agent has biological activity, but they are not reliable surrogates for improved survival or quality of life in all settings, and two recent systematic reviews suggest that the strength of association between surrogates in cancer clinical trials and life extension is generally low.
>
> *Davis et al., 2017*

The problem noted by the *BMJ* scientists was that trials far too often used these "surrogate endpoints" that measured whether the cancer progressed, or how many patients experienced tumor shrinkage, rather than actually measuring whether patients lived longer. This so-called "progression-free survival" has a very low confidence association between clinical trials and, as the *BMJ* puts it, overall "life extension." Kelly Crowe, in her radio health report for the CBC, asked the central question here: ultimately, "what does this mean for patients?" Queen's University oncologist and researcher Dr. Christopher Booth answered, "that's the million-dollar question. We've always thought in medicine that any medical intervention should help people live longer and better lives, which means any new treatment in a clinical trial should be shown to increase survival and improve quality of life" (Crowe, 2017).

But surely the results of two of the world's newest and biggest selling immunotherapy, PD 1 checkpoint inhibitors, blockbuster drugs, *Opdivo* and *Keytruda*, showed far better survival outcomes than earlier drugs on the market. Merck's *Keytruda*, in its television ads, stated that 71% of patients given the drug were alive "at the time of patient follow-up" compared with 58% of those who received chemotherapy. Szabo, however, noted that "what the ad fails to mention is that the 'time of follow-up' was 11 months" (Szabo, 2017). As for Bristol-Meyers Squibb's *Opdivo* (total 2016 worldwide sales of $4.73 billion), its television ads projected the words "a chance to live longer" on the sides of skyscrapers as amazed crowds look up. How much longer we may well ask? A footnote in small type at the bottom of the screen tells us – lung cancer patients taking *Opdivo* lived on average 3.2 months longer than the others, a very close match to the average overall increase figures provided by the *JAMA* and *BMJ* studies.

This is not to suggest that immunotherapy drugs won't have a potentially significant role to play in prolonging survival and quality of life for some cancers, particularly blood cancers with a great number of mutations like leukemia and melanoma. In April 2016, Dr. Don S. Dizon announced at the American Society of Clinical Oncology (ASCO) that doctors leading a phase 1 clinical trial had promising news. After a long-term follow-up, "newly diagnosed and previously treated patients with advanced melanoma showed that 40% of patients were alive three years after starting [PD-1 checkpoint inhibitor] pembrolizumab" (Hamid et al., 2019). The spokesperson for the trial went on to say that "new therapies that block the PD-1 are extending survival for many patients, and for some may offer the prospect of living longer than ever after a diagnosis with advanced melanoma. In a matter of a few years, these therapies have truly transformed the outlook for patients with melanoma and many other hard-to-treat cancers" (Goodman, 2016).

The excitement in the oncology world is indeed palpable as there are well over 1,000 clinical trials of checkpoint inhibitors going on worldwide at present. In their special "On Target" cancer report, *The Economist* noted that "what was a trickle, then a current, is now a torrent." It must be remembered, however, amidst the breathless talk- such as emanates from prestigious hospitals like Johns Hopkins- that, "immune therapy is a game changer. We need more research to take us the rest of the distance, but we don't think there is a single cancer that the patient's own immune system ultimately can't beat," or from Seattle's Children's Hospital that, "cancer immunotherapy is a miracle in the making," so we must look at the scientific test data more thoroughly to determine whether these claims withstand more detailed scrutiny.

Vinay Prasad, a hematologist-oncologist as the Knight Cancer Institute, Oregon Health and Science University, Portland, Oregon, in his study, "Perspective: The Precision-Oncology illusion," published in the September 2016 edition of *Nature*, provided evidence that diverged, rather starkly, from the recent exuberance over targeted, bespoke oncotherapeutic drug delivery. Prasad begins his study by noting that "precision oncology promises to pair individuals with cancer with drugs that target the specific mutations in their tumor, in the hope of producing long-lasting remission and extending their survival. The basic idea is to use genetic testing to link patients with the drugs that will work best for them, irrespective of the tissue of origin of their tumor. Enthusiasm has been fueled by reports of exceptional or super responders – individuals for whom experimental therapies seem to work spectacularly well" (Prasad, 2016). As always, unfortunately for those on the exuberant end of the spectrum, the "devil" is most definitely in the details with respect to clinical test results.

Data from some 2,600 people enrolled in a sequencing program at the MD Anderson Cancer Center in Houston, Texas, showed that just 6.4% were paired with a targeted drug for identified mutations. Similarly, the Molecular Analysis for Therapy Choice (NCI-MATCH) trial at the US National Cancer Institute enrolled 795 people who had relapsed solid tumors and lymphoma, but as of May 2016 it had only been able to pair 2% of patients with a targeted therapy. Furthermore, when patients with diverse, relapsed cancers were given drugs based on biological markers, only ~30% responded at all, and the median progression-free survival was just 5.7 months. Prasad estimated that,

> Multiplying the percentage of patients receiving targeted therapies by this response rate … that precision oncology will benefit around 1.5% of patients with relapsed and refractory solid tumours. It is on this tiny proportion of patients that the hopes for precision oncology have been built. Although many patients have undergone sequencing in the past decade (Foundation Medicine, a commercial provider of tumour profiling, has sequenced at least 18,000 patients), the number of reported cases of exceptional and super responders over that time are few. In a search of the biomedical literature with a colleague, we identified only 32 cases.
>
> *Prasad, 2016*

The results of Prasad's research, when taken as a whole, and, as he stressed, "considered objectively," underlined that "the prospects and potential of precision oncology are sobering. At best, we may expect short-lived responses in a tiny fraction of patients, with the inevitable toxicity of targeted therapies and inflated cost that this approach guarantees." Yes, precision personalized oncology is, as Prasad put it, "inspirational. What doctor or patient would not want to harness genetics to tailor a therapy to an individual?" Like motherhood and apple pie, as the saying goes, who would want to cast a shadow on this optimism? But, although we can hope things change going forward, and CAR-T therapy indeed may be the long sought after key to unlock the door for a cure for cancer, as of 2016 at least, Prasad bluntly concluded, "the proposal is neither feasible, cost-effective nor assured of future success. Yet in only one of these cases does the rhetoric so far outpace the reality that we risk fooling even ourselves" (Prasad, 2016).

This is all perhaps a trifle unsettling, but, surely, the general public can comfort itself with the knowledge that physicians possess a more sophisticated oncological understanding. They keep abreast of *JAMA's* published articles at the very least and will be forewarned and forearmed when an article such as the previously quoted one assesses and questions the overall survival, quality of life, and safety benefits associated with new cancer medicine. After all, the study concluded that,

> There is growing concern that the benefits offered by many new treatments for cancer – often discussed and promoted as "breakthroughs" – are marginal and might not be clinically meaningful to patients, despite rapidly escalating costs. Consequently, there have been calls to raise the evidence bar for market authorization of new cancer drugs.
>
> *Prasad, 2016*

In the April 2016 edition of *JAMA*, a blog entry by Sabriya Rice titled, "Uninformed docs buy into 'breakthrough hype,'" was published that discussed a recently released *JAMA* study which declared that, "physicians may be overly optimistic about the benefits of newly approved drugs, especially when the therapies are labelled as a 'breakthrough' according to a survey of nearly 700 practicing clinicians" (Rice, 2016). Researchers from the Brigham and Women's Hospital in Boston, the Dartmouth Institute for Health Policy and Clinical Practice in Lebanon, New Hampshire, and the Board of Internal Medicine in Philadelphia evaluated the impact of potentially loaded words like "breakthrough" and "promising" on clinicians. And their findings? It appeared that they had the same unfortunate impact on the 680 surveyed internists and specialists as on the average consumer. The report stated that,

> When given a hypothetical scenario in which a patient had a serious medical condition and clinicians were given two options, only 6% chose the treatment option with "early promising results." The other 94% selected the alternative drug described as 'breakthrough.' That's an unexpected finding considering that both drugs were actually breakthrough therapies.
>
> *Kesselheim et al., 2016*

The report concluded with the observation that "the findings suggest 'substantial deficits' in knowledge about FDA rapid approval programs."

1.9 NANOTECHNOLOGY: PEAK OF INFLATED EXPECTATIONS

Given this state of affairs and situated within the current atmosphere of epistemic uncertainty and doubt, nanotechnology now appears akin to a veteran technological survivor of earlier hype cycles and unbridled enthusiasm. We seek comfort in "filling in" the discontinuities in our knowledge, our blind spots, seeking cognitive ease by forming unrealistic expectations for novel solutions to medical and technological challenges. The intentions are explicable as we have seen, but we must be careful. Nanotechnology's place within the 1990s tech boom has already been noted. The *Gartner Hype Cycle Report* placed nanotechnology at that time in the "Peak of Inflated Expectations" phase of the hype cycle. This they defined as the point where technologies fail to meet expectations and quickly become unfashionable.

Cientifica Research, in their *Nanotech Opportunity Report*, asserted that nanoscience, after enduring the next phase of "Trough of Disillusionment" in the post-financial crisis of 2008, had moved well into the "Slope of Enlightenment" phase where businesses applications no longer surprise, but are seen as essential evolutionary steps in a rock-solid field of continual technological innovation (Harper, 2001).

Concurrent with the announcement of the FDA approval of CAR-T version of precision on August 30, 2017, an interdisciplinary team of ten scientists from Rice University, North Carolina State University, and Durham University announced in the August 2017 edition of *Nature* the results of their research titled, "Molecular machines open cell membranes." While this certainly did not receive the sort of "this is it" press coverage that the former FDA announcement received, we may look back on this as the beginning of a real, not overhyped, breakthrough.

But rather than beg the question, let's examine the study in question. The scientists at these three universities have developed nanomachines that drill into cancer cells and kill them in just 60 seconds. The intention, as they state in their paper, is to "show that molecular machines can drill through cellular bilayers using their molecular-scale actuation, specifically nanomechanical action. Upon physical absorption of the molecular motors onto lipid bilayers and subsequent activation of the motors using ultraviolet light, holes are drilled in the cell membranes" (García-López et al., 2017).

In one test done at Durham University, the nanomachine took anywhere between one and three minutes to breach the outer cell membrane of a prostate cancer cell and instantly destroy it. "In addition," they go on to state that, "by using molecular machines that bear short peptide addends, nanomechanical action can selectively target specific cell-surface recognition sites." By designing several different light-activated motorized molecules that can seek and destroy specific cells, such as those with prostate cancer as in the study, the scientists have, according to Dr. Robert Pal of Durham University, one of the team who developed the nanomachines, provided an "approach [which] could provide a potential step-change in non-invasive cancer treatment and greatly improve survival rates and patient welfare globally" (García-López et al., 2017). In an interview with *The Telegraph* newspaper, Science Editor Sarah Knapton, Dr. James Tour of Rice University in Houston, another member of the international team, remarked that,

> These nanomachines are so small that we could park 50,000 of them across the diameter of a human hair, yet they have the targeting and actuating components combined in that diminutive package to make molecular machines a reality for treating disease. In this study we have shown that we can drill into cells, animal cells, human cells using these nanomachines, they will attach to the surface, and then a light will be shone upon them and they will drill right into the cell.
>
> *Knapton, 2017*

When Dr. Tour stated that they are making "molecular machines a reality for treating disease," the import of this statement could hardly be overstated for the future treatment applications of not only cancer but also a variety of deadly diseases worldwide and, most importantly, *at a sustainably affordable level.*

On a personal note, Dr. Tour also described a revelation that came with this undertaking. He mentioned that,

> For many years I never had envisioned the nanomachines being used medically, I thought they were way too small, because they are much smaller than a cell, but now this work has really changed my thoughts on this and I think therapeutically this will be a whole new way to treat patients, it's going to be an excellent application for cancer treatment, not just for killing of cells but for the treatment of cells, interacting with the human body using molecular machines.
>
> *Knapton, 2017*

Dr. Tour's realization about the efficacy of nanomachines illustrates most effectively the phenomenon we have been discussing here, namely, that taking a mental "step back" to review one's earlier position, before taking the next step of changing one's entrenched belief system, is very difficult indeed. Belief perseverance can be challenging and resilient to overcome, but like a cell membrane, it is not impermeable to modification and change. Nanotechnology and nanomedicine, in particular, continue to move out of the lab to the clinic and, very soon, to human trials. As the report in *Nature* puts it in its concluding remarks, "beyond the in vitro applications demonstrated here, we expect that molecular machines could also be used in vivo, especially as their design progresses to allow two-photon, near-infrared and radio-frequency activation" (Knapton, 2017).

1.10 TARGETING CANCER

1.10.1 SPASER's against Cancer

However, what of cancers that detach from the primary tumor, which can diffuse and multiply into other regions of the body: the much-feared progression of metastases? In June of 2017, a study was published titled, "Microscopic lasers may stop tumors spreading around the body," in the journal *Nature Communications* by researchers at Georgia State University, the University of Arkansas for Medical Sciences, the University of Arkansas at Little Rock, and the Russian Academy of Science. In it, the scientists found that a nanolaser, also known as a SPASER (Figure 1.3), can serve as a hyper-bright, water-soluble probe capable of seeking out and destroying circulating tumor cells (CTCs) in the bloodstream. The SPASER, which stands for surface plasmon amplification by stimulated emission of radiation, is a nanoparticle, roughly 20 nm in diameter, consisting of a gold core with a diameter of 10 nm surrounded by a silica shell fixed with fluorescent uranine dye (used for tracing diagnostics).

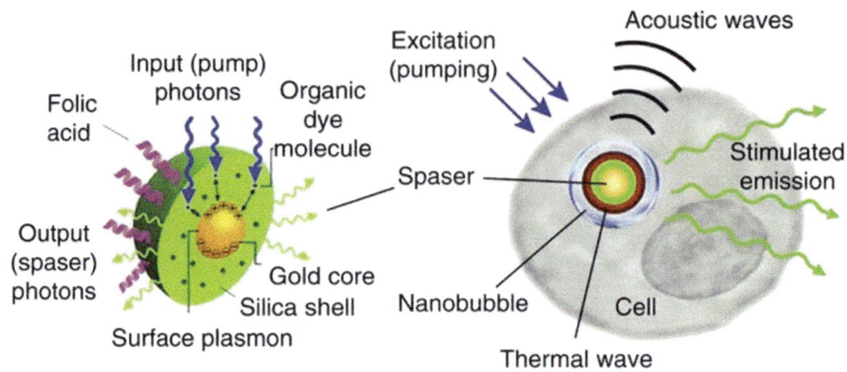

FIGURE 1.3 Composition and function of a SPASER. (Reproduced from Galanzha et al. (2017), with permission.)

Employing the functional principle of a laser, but instead of amplifying light, the system amplifies plasmons oscillating across the surface of the gold sphere. The SPASER's "guidance system" is the folic acid which covers the nanosphere. Since cancer cells, unlike most healthy cells, are covered with folate-receptor molecules, this allows for the selective molecular targeting of the CTCs in the bloodstreams. If a posse of SPASERs, either through injection or drinking a solution, is released into the body, these nanoprobes will quickly hunt down and bind to them. The SPASERs then absorb the activated laser light, heat up, and cause shock waves in the cells, which destroy the cell membrane (Galanzha et al., 2017).

Dr. Mark Stockman, director of the Center for Nano-Optics and professor of physics at Georgia State, and one of the lead researchers on this study, stated unequivocally in *ScienceDaily* that,

> There is no other method to reliably detect and destroy CTCs. This is the first. This biocompatible spaser can go after these cells and destroy them without killing or damaging healthy cells. Any other chemistry would damage and likely kill healthy cells. Our findings could play a pivotal role in providing a better, life-saving treatment option for cancer patients.
>
> *ScienceDaily, 2017*

In their September 16, 2017 edition, *The Economist* weighed in and reported that, "animal trials have been promising … researchers are now planning to test the system in people." As part of their plan to exploit this result therapeutically, the researchers plan, "to fit patients with special wrist sensors. Such a sensor will use low-level laser light to detect spaser-carrying CTCs passing through blood vessels beneath." And they stress that "even a partial slaughter would inhibit metastasis. And that would have a potent effect on treatments for all sorts of cancers" (*The Economist*, 2017).

1.10.2 Extracellular Vesicle Fingerprint Predictive Score Test

Nevertheless, what can be done at the diagnostic stage of treating potential cancer, rather than the most dangerous, full-blown metastatic stage of the disease? Would it not be preferable to have a more sensitively acute test for aggressive prostate cancer than the current Prostate Specific Antigen (PSA) test in order for men to avoid painful and often unneeded needle biopsies? As reported by Niamh Marriott in *Drug Target Review*, scientists at the University of the Alberta's Prostate Cancer Research Initiative have developed the Extracellular Vesicle Fingerprint Predictive Score (EV-FPS) test which uses machine learning to combine information from millions of cancer cell nanoparticles in the blood to recognize the unique fingerprint of aggressive prostate cancer. By evaluating a cohort of 377 Alberta men with suspected prostate cancer, researchers found that the EV-FPS test identified aggressive prostate cancer 40% more accurately than the most common PSA test the blood test – the de facto standard test today (Marriott, 2017).

Lead researcher Dr. John Lewis remarked that, "for this kind of test you want the sensitivity to be as high as possible because you don't want to miss a single cancer that should be treated. High sensitivity means that our test will miss fewer aggressive cancers. What we are hoping is that if this test says they have aggressive cancer, they go ahead and get a biopsy to confirm. If it says they don't have aggressive cancer, they can potentially skip the biopsy altogether" (Marriott, 2017). It is estimated that the successful application of EV-FPS could lead to the elimination of 600,000 unnecessary biopsies, 24,000 hospitalizations, and up to 50% of unneeded treatments for prostate cancer each year in North America alone. Lewis expects the test to cost "several hundred dollars." Compared with a biopsy that costs almost $2,000 and involves injecting around a dozen needles through the prostate, Lewis believes the blood test has the potential to provide significant savings to the health care system at the same time as preventing unwanted infections from the biopsy procedures on men who are unlikely to develop aggressive cancers.

What unites all three of these most recent announcements in nano-oncology is the fact that beyond dramatically enhancing the health care experience of individuals, each nano-based treatment modality will provide, once implemented, substantial cost savings to the health care system, as the EV-EPS test demonstrates. Let us compare this with what the previously highlighted *JAMA* study, the March 2016 "Assessment of Overall Survival, Quality of Life, and Safety Benefits Associated with New Cancer Medicines," emphasizes. Along with its major concern about the question-begging "breakthrough" language used too frequently, just as significantly, the "escalating costs" versus benefits of immunotherapeutic drug treatment is a major, and all too often overlooked, concern.

Audrey Andrews, in her 2015 article for *American Health and Drug Benefits* magazine, quotes Dr. Leonard B. Saltz, Chief of Gastrointestinal Oncology at Memorial Sloan Kettering Cancer Center, New York, as saying that he wants "the new checkpoint inhibitors to be available to his patients, but he questions how society will pay for them" (Andrews, 2015). Dr. Saltz has provided his colleagues with a rather sobering assessment of the costs for precision medicine. He estimates that the widespread use of immunotherapy drug delivery could cost just the United States $174 billion annually.

In presenting at ASCO an example of a combination immunotherapy of two checkpoint inhibitors which had encouraging progression-free survival rates compared with using one agent alone, Dr. Saltz noted that, "as a clinician I want these drugs and others like them to be available to my patients. As one who worries about how we will make them available and minimize disparities, I have a major problem, and that is that these drugs cost too much" (Andrews, 2015).

How much is "too much"? The numbers involved are staggering. "To put that into perspective, that's approximately 4,000 times the cost of gold." estimates Dr. Saltz. According to Andrews, the doctor calculated that the treatment for a typical patient receiving the immunotherapy combination that he discussed in his ASCO presentation would be $295,566. If you used only one drug, which only had a remission benefit of less than three months, the cost would be an eye-watering $158,252. Dr. Saltz projected that the total cost of the treatment to the 589,430 patients dying from metastatic cancer annually would cost society $174 billion.

Further, he emphasized, "that's $174 billion for drugs treating patients with the metastatic disease – no adjuvant therapy – for 1 year only" (Andrews, 2015). This analysis casts a very "cold eye" on the cost data and hardly presents an encouraging picture for future cost containment of immunotherapeutic cancer drug delivery. Even *The Economist's* September 2017 edition, in which the headline states unequivocally that oncological research is "Closing in on Cancer," acknowledges at the very close of their Special Report that "what such advances will not do, though, is make such treatments cheaper" (*The Economist*, 2017). Novartis's new CAR-T therapy costs $475,000.

Genome-editing treatments seem likely to be the most expensive cancer treatment the world has ever seen. And this is saying quite a lot. (*The Economist*, 2017). Vinay Prasad summed up this predicament succinctly by noting, "precision oncology is inspirational. What doctor or patient would not want to harness genetics to tailor a therapy to an individual? But travelling back in a time machine is also inspirational. Who would not want to wind back the clock to remove their cancer before it spreads?" (Prasad, 2016). Who, indeed, would not want this? But "time travelling" in immuno-oncology is an eye-wateringly expensive voyage. Even trying workarounds such as lowering dosage of checkpoint inhibitors to provide "equivalence" is, in Dr. Saltz's words, "unsustainable."

1.10.3 Chimeric Antigen Receptor T-Cell Therapy

The search is on, therefore, for an off-the-shelf CAR T-cell therapy that is not only effective but also sustainable, that is to say, a therapeutic delivery modality that is truly affordable. Nanotechnology is now being actively tested in order to provide a much-hoped-for solution. It must be remembered that Chimeric Antigen Receptor T cells are typically produced by collecting T cells from a patient, modifying them in the laboratory in order to target cancer cells, allowing them to multiply and increase in number, and then infusing them back into the patient.

As the National Cancer Institute stated in an article titled, "Nanoparticles Create Effective CAR T Cells in Living Mice," "this approach – although effective in some small clinical studies – is expensive, can take several weeks to complete, and requires special expertise and equipment. Plus, it has to be performed for each patient receiving CAR T-cell therapy" (National Cancer Institute, 2017).

To make this therapy "more accessible to more people," the NCI went on to point out that Dr. Matthias Stephan of Fred Hutchinson Cancer Research Center and his colleagues embarked on a "streamlined and inexpensive approach" for producing CAR T-cells. In order to do so, they created a nanoparticle that targets T cells and contains DNA instructions to make a CAR that identifies leukemia cells. As Dr. Terry Fry, head of Hematologic Malignancies Section of NCI's Center for Cancer Research, remarked, "the real novelty is you don't have to make CAR T-cells outside of the patient; you can deliver these nanoparticles directly to the patient. The goal is to make an off-the-shelf CAR T-cell therapy … and it's possible that this approach could accomplish that" (National Cancer Institute, 2017).

The broad-based efficacy of targeted cancer treatments such as immunotherapies and CAR T-cell therapy is very much an open question. And it is important to be aware that when one discusses the effectiveness of any new technological advance, let alone a real "breakthrough," we are discussing the issue of scalability. From the human discovery that round nodules of flint can be split and chipped to form a sharp edge millions of years ago through to the invention of the silicon microchip, scalability is key that will unlock a real "breakthrough" door, and not an impostor. The ability of conventional CAR T-cell therapies to scale up this patient-by-patient process to treat large numbers of people with cancer, as Dr. Fry bluntly puts it, is just "not practical." Although there are other hurdles to overcome with CAR T-cell therapy, scalability is, he stresses, a major "hurdle in the field" (National Cancer Institute, 2017).

Thus, Dr. Stephan and his research colleagues looked to nanoparticles to deliver DNA to specific cells, which are easier to manufacture and economical to produce. The nanoparticle that was designed by his team possesses several key elements: a gene for a CAR that binds to the CD19 molecule that is expressed in higher than normal amounts on leukemia cells, molecules that help the gene insert itself within T cell DNA, where it can be expressed efficiently, and added proteins placed on the nanoparticle's surface that selectively binds to T cells.

It is worth noting that the conventional CAR T-cell approach, as NCI reports, "involves an 'expansion' step, where CAR T-cells are grown in the lab until they number more than a billion cells. In mice treated with the nanoparticles, the researchers found the CAR T-cell population expanded rapidly, increasing 5.5-fold over 6 days." As Dr. Stephan describes it, "basically what's happening is the exact same expansion that would usually happen in a lab incubator, but it's happening inside the mice, where the tumors are" (National Cancer Institute, 2017).

Once Dr. Stephan and his research team have engaged an industry partner to assist in the synthesis of these nanoparticles on a larger scale, they are looking forward to translating their nanotechnological approach in CAR T-cell therapy in the clinical setting soon, where the safety profile will be evaluated with non-primate human models and directly in a phase-1 dose-escalation trial. Like all of the other previously mentioned nanomedical technologies, nanoparticle CAR T-cell therapy has the potential within its very design to be a far more "streamlined and inexpensive" method for the delivery of off-the-shelf cancer treatments to the maximum number of patients possible globally.

Much hard work is still to be done to make this treatment available to the public at large; however, the very discussion by medical researchers of nanotechnology delivering off-the-shelf cancer treatment means it is no longer a dream "devoutly to be wished," but a reality that is well within reach given current, rapidly developing capabilities. Nanotechnology's capacity to deliver treatment modalities is, borrowing from Dr. Saltz's vocabulary, "sustainable," and the "disparities" he refers to can be "minimized," again, not just in the United States but globally.

1.11 GLARING HEALTH CARE DISPARITIES

But what are these "disparities" that Dr. Saltz refers to, not just in the United States, but in the rest of the world? Figures compiled by the World Health Organization and the American Cancer Society paint a grim picture. Consider the following: in 2012 an estimated 14.1 million new cancer cases were reported worldwide, with this figure anticipated to rise to 23.6 million over the next 20 years. Significantly, more than 60% of the world's total cancer cases occur in Africa, Asia, Central America, and South America, while greater than 70% of the world's cancer deaths come from these same regions.

By 2030, the developing world is expected to account for 70% of newly reported cancer cases. What is more, survival rates are, to put it mildly, slight in poor countries and on the rise in wealthy ones. It is projected that by 2030, the developing world will have over nine million cancer-related deaths versus approximately 3.5 million in the developed world. Despite the fact that AIDs and other infectious diseases have received the lion's share of attention and financial support for treatment interventions in the developing world, cancer remains the biggest killer with over five million annual deaths reported. In fact, cancer kills more people worldwide than HIV/AIDS, TB, and malaria combined and was responsible for 9.6 million deaths in 2018 (National Cancer Institute, 2018; World Health Organization, 2025).

Based on this dire prognosis, Asad Moten, Daniel Schafer, and Mauro Ferrari, in an article published in the University of Edinburgh's *Journal of Global Health* titled, "Redefining global health priorities: improving cancer care in developing settings," state plainly that,

> Disproportionate access to prevention and treatment, in addition to an ageing and growing population and decreasing mortality rates resulting from infectious diseases, will cause new cancer cases and cancer mortality to continue to rise in coming years. The global health community must take decisive action to bridge the cancer divide between wealthy and poor nations.
>
> *Moten et al., 2014*

And what kind of action to bridge this huge disparity do the authors recommend? Besides the expected short-term solutions such as "increased access to drugs for treatment and palliation," and increasing "the number of community health professionals and facilities to provide prevention and early detection," they also stress the need for the "development and implementation of innovative health care delivery options to support rapid scale-up, while integrating access to patients in resource-poor regions."

Nanomedical therapeutics can, the report concludes, provide the principal choice for the recommended "innovative health care delivery options to support rapid-scale up." Moten and his research colleagues furnish specific reasons to support their endorsement of nanomedicine as a foundational element in "bridging the cancer divide" between the two solitudes: wealthy and poor countries. They note that,

> Conventional medicine is largely inaccessible to patients in the developing world because of its dependence on technology, laboratory facilities, and/or highly trained medical professionals. Conversely, nanomedicine, based predominantly on prevention and early management of disease, bypasses the need for scarcely available, expensive laboratory equipment and medical expertise. Additionally, nanotherapeutics are fabricated in a highly-specific manner at a nanoscale level using bottom-up molecular manufacturing methods, which makes these nanomedicine drugs easier to manufacture, distribute, and subsequently administer to patients, all of which translate into improved health outcomes in resource-limited settings.
>
> *Moten et al., 2014*

It is encouraging that realistic potential candidates to bridge the cancer divide, such as the scalable nanotherapeutic delivery mechanism Dr. Stephan and his research team have created with targeted nanoparticles to enhance and make more effective CAR T-cell immunotherapy, are on the horizon.

If we drill down a bit further and look at specific continents, Africa, for instance, we discover a glaring example of cancer health care disparity. In its report "Closing in on cancer," *The Economist*, besides trumpeting the recent, cutting-edge developments in the four pillars of cancer treatment – surgery, chemotherapy, radiation, and more recently, immunotherapy, doesn't entirely ignore this issue; namely, the rather dire fact that, "much of the world has yet to see progress on cancer. It's not just that the resources needed are not there. Nor is the hope." Dr. Meg O'Brien, director of global cancer treatment at the American Cancer Society (ACS), states candidly that, "nobody in Africa knows anyone who survives cancer. It is considered 100% fatal and is associated with so much suffering" (*The Economist*, 2017).

The report concludes that big pharma companies unsurprisingly neglect the market, "because of the volumes – low volumes which mean that the procuring agencies pay higher prices than they should for old, off-patent drugs." Support for this claim can be found in the IMS Institute for Healthcare Informatics study, "Global Oncology Trend Report: A Review of 2015 and Outlook to 2020," in which they report that of the 49 oncology New Active Substances analyzed that were initially launched between 2010 and 2014, "fewer than half were available to patients by the end of 2015 in all but six countries: the U.S., Germany, U.K., Italy, France and Canada." The study concluded rather somberly that, "targeted immunotherapies are available in most developed countries, but none of the emerging markets outside of the European Union has yet registered these treatments" (IMS Institute for Healthcare Informatics, 2016).

Also unsurprising is the news that the cancer mortality statistics are stark. The number of patients with cancer in low- and middle-income countries is rising so rapidly that BIO Ventures for Global Health (BVGH), a non-profit organization based in Seattle, Washington, states in their report "Africa's Emerging Cancer Crisis: A Call to Action" that 60% more Africans die from cancer than succumb to malaria. Furthermore, they add, over 20% of African countries have no access to cancer treatment at all, and most cancers, when they are diagnosed, are late-stage. To put some numbers on this latter point, the five-year survival rate for women with breast cancer in Gambia is 12% and in Uganda 46%, compared with 90% in America (Dent et al., 2017).

Dr. O'Brien sums up this calamitous situation by stating that, "right now we have more patients waiting for treatment than we have treatments." She asserts emphatically that, "one of the first things we need to do is get drugs on the ground" (*The Economist*, 2017). Without any meaningful change to this sort of status quo, Moten and his co-authors forecast, "a global health-care disaster in the making." Likewise, the BVGH authors conclude their alarming report by declaring forcefully that, "Africa's emerging cancer crisis is a developing tragedy that can no longer be ignored. African leaders and the global community must mobilize now to develop sustainable solutions. Cooperation, capacity and innovative business models are critical to providing access to cancer care for patients in Africa" (Dent et al., 2017).

1.12 NANOMEDICINE AND GLOBAL HEALTH CARE EQUIVALENCY

Even if we assume that health, governmental, and business organizations can somehow cooperate and innovate to "get the drugs on the ground," what might follow? Looking at the issue longer term, and from a broader perspective, Goldin and Kutarna assert that, with respect to costly gene therapy,

> It may enable us to add lots of new drugs to our kit but it also teases us with the possibility that we might one day throw out our whole pharmaceutical bag altogether. Why deliver little chemical warriors into our bloodstream, with all the unruly side-effects, when we can alter the body's behavior at a cellular – maybe even molecular – level so it returns itself to health? This new medical model would have titanic economic consequences.
>
> *Goldin and Kutarna, 2016*

In discussing throwing out "our whole pharmaceutical bag altogether," these academics are pointing directly to the immense economic consequences of disruptive technologies like nanomedicine and nanotherapeutic health delivery, which will provide a new medical model.

There is little doubt that such nanomedical disease treatment modalities will dislocate the conventional, well-established pharmacotherapeutic paradigm. On this matter, the economist Adam Smith's astute observation that the long wave of economic progress advances irrespective and almost in spite of the shorter-term focus of corporations and governments (systemically based as they are on quarter-to-quarter financial results and two/four-year election cycles) should remind us that irresistible, long-term forces are at play for both nanomedicine and Global Health Care Equivalency (GHCE). These will likely come to fruition despite the inevitable shorter-term operative challenges and obstacles. As we have already seen, the signal that contains the DNA of a true paradigm shift is always stronger than the noise that surrounds the disruption when entrenched ways of thinking are in the uncomfortable throes of being overturned.

But irresistible does not imply inevitable. Again, it must be stressed that scalability is an essential component of any truly breakthrough technology, whether we are talking of the medieval moldboard plow, penicillin, or the Internet. However, other factors are just as vital: cost-effectiveness of transmission and delivery, convenient access, and a broad-based knowledge base for future development. Moving forward, nanomedicine, in concert with every innovative technology in history, must be able to meet all of these criteria despite all of the short-term developmental challenges it faces.

One of the biggest challenges that nanomedicine must confront – whether we are talking about the specifics of lipid-based nanocarriers, viral particles, polymeric nanocarriers, inorganic nanoconstructs, or drug conjugates – is its inherent multidisciplinary nature. The interplay between various academic researchers, clinicians, industry, and regulatory bodies will require an effective communication effort that fuses the strengths of the disparate disciplines of chemistry, physics, biology, engineering, and medicine into a secure knowledge base from which to draw upon. In this way, "cognitive gaps" are not allowed to be "filled" in and fester into ignorant intransigence based on false stereotypes and confirmation bias.

For inasmuch as the diverse nature of nanomedicine's necessarily collaborative structure is a challenge to keep it on the "front-foot" to propel its developmental momentum, it is also one can argue its greatest strength. We would do well to remember that *the* truly great breakthrough in modern genetics, breaking the genetic code, i.e., the discovery of the precise chemical structure of genes and enzymes, the molecular mechanisms of protein synthesis, the mechanisms of protein synthesis, and of the mechanisms of gene replication and mutation, was a multidisciplinary effort.

A critical element in the breaking of the genetic code, as Capra and Luisi remind us, was that "physicists moved into biology. Max Delbruck, Francis Crick, Maurice Wilkins, and several other protagonists had backgrounds in physics before they joined the biochemists and geneticists in their study of heredity" (Capra and Luisi, 2016). The epochal result was due to the essential fact that, "these scientists brought with them a new perspective and new methods that thoroughly transformed genetic research" (Capra and Luisi, 2016). Nanomedicine, and eventually, GHCE itself, like the rapid advance of molecular biology in the 1950s, currently, and will in the years to come, benefit from the enriched terrain of its participants' multidisciplinary collaboration.

Recent trends in nanomedicine support such optimism. According to the *BCC Research* report titled "The Maturing Nanotechnology Market: Products and Applications," the global market for nanotechnology "should reach $90.5 billion by 2021 from $39.2 billion in 2016 at a compound annual growth rate (CAGR) of 18.2%, from 2016 to 2021." In a second report dealing more specifically with nanomedicine titled "Nanoparticles in Biotechnology, Drug Development and Drug Delivery," the global market for nanoparticles in the life sciences was estimated at over $29.6 billion for 2014. This market is forecast to grow to more than $79.8 billion by 2019, resulting in a healthy compound annual growth rate (CAGR) of 22%, with the biggest increase likely to be seen in the area of drug delivery systems (McWilliams, 2016).

In commenting on these figures, Tarun Pandotra, in his own report, "Nanomedicine: Delivering on its promise?" observes that, "nanotechnology has come a long way in the last five years and now nanomedicine … is taking its place in the fight against unmet diseases. With the promise of medical applications such as targeted cancer treatments, microscopic sensors and even devices that mimic human organs in the future, enthusiasm for the field is backed by serious research and development funding from government and industry" (Pandotra, 2017).

Titles of reports on nanotechnology using words like "maturing" as in the *BCC Research* report above reflect an understanding that, as Margaret Harris, editor of *Physics World*, remarked in a recent article, "nanotechnology is growing up, and we are starting to expect grown-up things from it" (Harris, 2017). And that is what observers like Harris are now getting in nanotechnology – "grown-up things." The field has come a long way and while the road ahead will have obstacles to be negotiated before a fully robust scaling-up process achieves release-velocity, the initial technological building blocks are already in place to make the dream of GHCE achievable. There are tangible reasons for excitement. This enthusiasm should, however, be tempered with the awareness of exogenous factors in play, such as proposed budget cuts to science in the world's most prosperous countries.

For instance, proposed 2018 budget cuts to science in the United States where the government aims to slash funding to the National Institutes of Health by $6 billion (representing about one-fifth of its budget) are indeed worrisome (Achenbach, 2017). Though this may be offset to a certain extent by recent legislation put forward by the spending panel in the U.S. House of Representatives (Kaiser, 2019). In Canada, the government's 2017 federal budget failed to provide any new money for the three federal science funding agencies, while a national report titled *Canada's Fundamental Science Review*, on the state of federal funding for fundamental research in Canada, declared that the country's scientific enterprise is in serious decline (Advisory Panel for the Review of Federal Support for Fundamental Science, 2017).

The pathway to envisaged GHCE is being cleared within nanomedicine by advances on myriad fronts, both within the laboratory and between various academic, governmental, and industry participants. However, no scientific field of inquiry and discovery operates in a cultural vacuum. A 2015 Pew Research Center survey of citizens and a representative sample of scientists connected to the American Association for the Advancement of Science (AAAS) revealed that 75% of AAAS scientists stated that too little science, technology, engineering, and math (STEM) education for grades K-12 was a major factor in the public's limited knowledge about science (Funk et al., 2015).

An overwhelming majority of scientists see the public's limited scientific knowledge as a problem for science. And limited knowledge, scientific or otherwise, produces the sort of cognitive gaps where social complexity overwhelms cognitive capacity. These knowledge deficits must be addressed early on, from grades K-12, to provide the nutrient-rich intellectual subsoil that engenders socially supported scientific advancement in general, and in particular, innovation in nanomedicine.

This problem, fortunately, has not gone unnoticed by some leaders in the scientific community. An example of an organization taking the lead in addressing scientific knowledge deficits, *Science World* in Vancouver, British Columbia, launched a program in the spring of 2017 titled *Symbiosis*, which they referred to as a first-of its-kind learning ecosystem for Canada. In their press release, the organization noted that,

> High-tech innovations are rapidly transforming 21st century societies and the Canadian marketplace is increasingly dominated by novel, knowledge-based jobs requiring high levels of literacy in science, technology, engineering and math (STEM). Failing to prepare the next generation to be STEM literate threatens the health of our youth, the economy and the places we live. STEM literacy needs to be integrated into the broader context of what it means to be a 21st century citizen.... Driven by a diverse network of cross-sector partners, Symbiosis will become a vibrant model for scaling the kinds of learning and careers needed in a knowledge-based economy.
>
> *Science World British Columbia, 2017*

The aim of *Symbiosis*, as Scott Sampson, President of *Science World*, notes, is "taking the providers of STEM learning and networking them together." It is worth noting here that *Symbiosis* has been selected by STEM Learning Ecosystems, a US-based organization, to join a "growing movement," and it will be the first member of this initiative outside the United States.

Likewise, at the City University of New York Advanced Science Research Center (ASRC) in Manhattan, New York, nanoscience students are studying in a program based upon "systems thinking," which, as Rein et al. explained in *Nature Nanotechnology*,

> Involves the translation of concepts and ideas across disciplinary barriers, and also across length scales, from nano to macro to global; it involves a consideration of how fundamental insights connect to applications, and how various theoretical and experimental parts connect and can be integrated into a functional whole. Following biologists and chemists, nanoscientists have recently started to appreciate the value of a systems approach.
>
> *Ulijn and Riedo, 2016*

The Advanced Research Center is structured around five thematic areas: nanoscience, photonics, structural biology, neuroscience, and environmental science. In an intriguing concept, each of these areas is distributed across different floors of the purpose-built ASRC building. It begins with nanoscience on the ground floor, with each discipline – in the order of the preceding list – located on subsequent floors above, encompassing 15 orders of magnitude in length scale in total (Figure 1.4). The intention is to provide students a better appreciation "of scale and connectivity between disciplines" and it is a new didactic testing ground for improved research methodologies in applied nanoscience. The authors provide an example of how this approach operates at ASRC:

> While nanoscience graduate students on the first floor of the ASRC study the properties of water in nanoscale confinement and aqueous phase molecular assemblies, environmental science colleagues on the fifth-floor study global water management challenges. Students from across these floors are encouraged to interact in order to develop new ways to tackle global water issues; our aim, in other words, is to integrate researchers who view a single topic (in this case, aqueous systems).... Together, these students are developing new nanomaterials for water purification, as well as nanotech-based sensing systems that enable data collection in large-scale field applications.
>
> *Ulijn and Riedo, 2016*

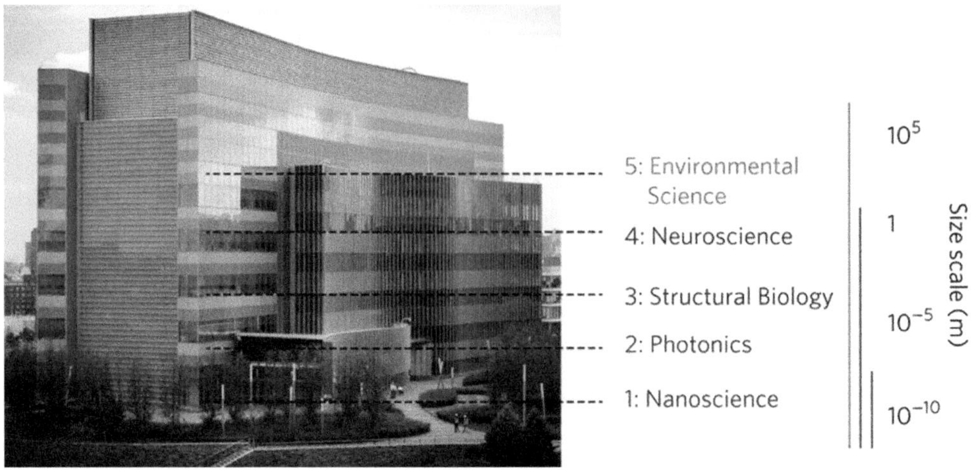

FIGURE 1.4 City University of New York (CUNY) Advanced Science Research Center (ASRC), Manhattan, New York integrates 15 orders of magnitude scale. (Reproduced from Ulijn and Riedo (2016), with permission.)

This sort of integrated research culture exposes students to the role that nanoscience can play across disciplines. It is in such a unique cross-disciplinary setting that systems thinking in the sciences can nurture true collaboration and cooperation to increase the velocity of disruptive change. With growing movements at the grassroots level like *Symbiosis* and innovative educational programs such as the City University of New York's ASRC, the future for turning the dream of GHCE into a concrete reality appears bright indeed.

1.13 CONCLUSION

We began this chapter by observing that we live in a turbulent age of disruptive innovation set against the forces of reaction, an age of cooperative partnership versus competitive self-interest. We also took a brief retrograde glance at scientific history as it affected the difficulty, even for some of the greatest minds mankind has ever produced, of changing belief systems when confronted with evidence of error. It can be said, with some degree of confidence, I believe, that while paradigm shifts are difficult, "mind shifts" are even harder.

If we glance back to the Renaissance one last time, we find in the age that produced the great Copernican paradigm shift, "something else happened," as Goldin and Kutarna underline in their book. Indeed, this "something else" is what they correctly referred to as "*collective* genius. Every person possesses a unique fragment of capability; collective genius happens when society nurtures and connects those diverse fragments. Diverse minds bent to a problem can spark original ideas and contributions that accelerate individual breakthroughs or take those breakthroughs forward" (Goldin and Kutarna, 2016). Efforts must be doubled and redoubled as we move forward in this age of disruption to increase the public's scientific knowledge and, as was noted, this cannot begin soon nor early enough in our children's schooling. For GHCE to become a reality sooner rather than later, society must increasingly nurture and connect the "diverse fragments" that Goldin and Kutarna allude to.

Renowned scientist K. Eric Drexler, who was instrumental in establishing nanotechnology as a legitimate discipline of scientific and technical research, echoes the above comments in his recent book, *Radical Abundance: How a Revolution in Nanotechnology Will Change Civilization*. What Drexler says about the challenges facing future adoption of molecular manufacturing (MM) at the conclusion of his book equally applies to nanomedicine and GHCE:

> The current state-of-the-art is more than enough to support a drive for next-generation molecular systems on the road to atomically precise manufacturing. Indeed, I am persuaded that advances in recent years now provide a platform that can support rapid progress. The greatest challenge today is to put the pieces together – not only components and computational tools, but also engineering concepts and the research teams that can bring them into physical reality.
>
> *Drexler, 2013*

"*The greatest challenge today is to put the pieces together.*" The human journey has been forever one in which the forces of innovation and collaboration have overcome the forces of dismissal and skepticism. Nanomedicine today has matured, "grown-up," and currently possesses the "collective genius" in order "to put the pieces together" on the road to the GHCE. The book you hold in your hands is a testament to that journey and the contributors who follow will give forceful evidence of how these pieces might be put together for the benefit of all mankind.

REFERENCES

Achenbach J. "Trump's budget calls for seismic disruption in medical and science research." *The Washington Post*, March 16, 2017. https://www.washingtonpost.com/national/health-science/trumps-budget-would-slash-scientific-and-medical-research/2017/03/15/d3261f98-0998-11e7-a15f-a58d4a988474_story.html (accessed March 28, 2025).

Advisory Panel for the Review of Federal Support for Fundamental Science. "INVESTING IN CANADA'S FUTURE: Strengthening the Foundations of Canadian Research." 2017. http://www.sciencereview.ca/eic/site/059.nsf/vwapj/ExecSummary_April2017-EN.pdf/$file/ExecSummary_April2017-EN.pdf (accessed March 28, 2025).

Andrews A. Treating with checkpoint inhibitors—Figure $1 million per patient. *American Health & Drug Benefits*. 2015;8(Special Issue):9.

Born M. *Max Karl Ernst Ludwig Planck. 1858–1947. Obituary Notices of Fellows of the Royal Society*. 1947;6(17):161–188.

Brown BR. *Planck: Driven by Vision, Broken by War*. Oxford University Press, 2015.

Capra F, Luisi PL. *The Systems View of Life: A Unifying Vision*. Cambridge University Press, 2016, p. 25.

Crowe K. "Many new cancer drugs don't save lives." CBC Health News, October 07, 2017. https://www.cbc.ca/news/health/second-opinion171009-1.4344943 (accessed August 24, 2019).

Davis C, Naci H, Gurpinar E, Poplavska E, Pinto A, Aggarwal A. Availability of evidence of benefits on overall survival and quality of life of cancer drugs approved by European Medicines Agency: Retrospective cohort study of drug approvals 2009–13. *BMJ*. 2017;359:j4530.

Dent J, Manner CK, Milner D, Mutebi M, Ng'ang'a A, Olopade OI, Rebbeck TR, Stefan DC. "Africa's Emerging Cancer Crisis: A Call to Action." *BVGH-BIO Ventures for Global Health*, 2017. https://bvgh.org/wp-content/uploads/2017/07/Africas-Emerging-Cancer-Crisis-A-Call-to-Action.pdf (accessed March 28, 2025).

Drexler K. *Radical Abundance: How a Revolution in Nanotechnology Will Change Civilization*. Gildan Media LLC, 2013.

Durant W, Durant A. *The Age of Faith. Vol. IV*. Simon & Schuster. 1950.

EvaluatePharma. "World Preview 2017, Outlook to 2022." Evaluate Ltd., 2017. https://www.skepticalraptor.com/blog/wp-content/uploads/2017/11/EvaluatePharma-Pharma-Preview-2017.pdf (accessed March 28, 2025).

Feero WG. Introducing "Genomics and Precision Health". *JAMA*. 2017 May 9;317(18):1842–1843.

Funk C, Rainie L, Page D. "Public and Scientists' Views on Science and Society." *Pew Research Center*, January 29, 2015. https://www.pewresearch.org/science/2015/01/29/public-and-scientists-views-on-science-and-society/ (accessed March 28, 2025).

Galanzha EI, Weingold R, Nedosekin DA, Sarimollaoglu M, Nolan J, Harrington W, Kuchyanov AS, Parkhomenko RG, Watanabe F, Nima Z, Biris AS, Plekhanov AI, Stockman MI, Zharov VP. Spaser as a biological probe. *Nature Communications*. 2017;8:15528.

García-López V, Chen F, Nilewski LG, Duret G, Aliyan A, Kolomeisky AB, Robinson JT, Wang G, Pal R, Tour JM. Molecular machines open cell membranes. *Nature*. 2017;548(7669):567–572.

Goldin I, Kutarna K. *Age of Discovery: Navigating the Risks and Rewards of Our New Renaissance*. St. Martin's Press, 2016.

Goodman A. "Pembrolizumab Survival Benefit Proves Durable in Patients with Advanced Melanoma." *The ASCO Post*, May 25, 2016. https://ascopost.com/issues/may-25-2016/pembrolizumab-survival-benefit-proves-durable-in-patients-with-advanced-melanoma/ (accessed March 28, 2025).

Greenwood J. "FDA's Historic Gene Therapy Approvals Could Be Beginning of the End of Blood Cancer." *BIOtech Now*, October 23, 2017. https://www.vbiognostics.com/fdas-historic-gene-therapy-approvals-could-be-beginning-of-the-end-of-blood-cancer/ (accessed March 28, 2025).

Hamid O, Robert C, Daud A, Hodi FS, Hwu WJ, Kefford R, et al. Five-year survival outcomes for patients with advanced melanoma treated with pembrolizumab in KEYNOTE-001. *Ann Oncol.* 2019 Apr 1; 30(4):582–588.

Harper T. "The Nanotechnology Opportunity Report™." *Cientifica Research*, 2001. https://www.nanotech-now.com/nanotechnology-opportunity-report.htm (accessed March 28, 2025).

Harris M. "When nanotechnology grows up." *Physics World*, May 2017. https://physicsworld.com/ (accessed March 28, 2025).

IMS Institute for Healthcare Informatics. "Global Oncology Trend Report: A Review of 2015 and Outlook to 2020." June 2016. https://www.iqvia.com/-/media/iqvia/pdfs/institute-reports/global-oncology-trend-report-2016.pdf (accessed March 28, 2025).

Kahneman D. *Thinking, Fast and Slow.* Farrar, Straus and Giroux; 2011. p. 114.

Kaiser J. House panel proposes $2 billion increase for NIH. *Science.* 2019. https://www.sciencemag.org/news/2019/04/house-panel-proposes-2-billion-increase-nih (accessed March 28, 2025).

Kesselheim AS, Woloshin S, Eddings W, Franklin JM, Ross KM, Schwartz LM. Physicians' knowledge about FDA approval standards and perceptions of the "Breakthrough therapy" designation. *JAMA.* 2016;315(14):1516–1518.

Knapton S. "Nanomachines that drill into cancer cells killing them in just 60 seconds developed by scientists." *The Telegraph.* 31 August 2017. https://www.telegraph.co.uk/science/2017/08/30/nanomachines-drill-cancer-cells-killing-just-60-seconds-developed/ (accessed August 07, 2025).

Koestler A. *The Act of Creation.* Last Century Media, 2014.

Marriott N. "New blood test uses nanotechnology to predict prostate cancer." *Drug Target Review*, June 14, 2017. www.nanoappsmedical.com/new-blood-test-uses-nanotechnology-to-predict-prostate-cancer/ (accessed March 28, 2025).

Maturana HR, Varela FJ. *The Tree of Knowledge: The Biological Roots of Human Understanding.* Shambhala, 2008.

McWilliams A. "The Maturing Nanotechnology Market: Products and Applications." *BCC Research: Market Research Reports & Industry Analysis*, November 2016. https://www.bccresearch.com/market-research/nanotechnology/nanotechnology-market-products-applications-report.html (accessed March 28, 2025).

Moten A, Schafer D, Farmer P, Kim J, Ferrari M. Redefining global health priorities: Improving cancer care in developing settings. *Journal of Global Health.* 2014;4(1):010304.

Nanotechnology Products Database. 2019. https://product.statnano.com/ (accessed March 28, 2025).

National Cancer Institute. "Nanoparticles Create Effective CAR T Cells in Living Mice." May 5, 2017. https://www.cancer.gov/news-events/cancer-currents-blog/2017/nanoparticles-engineer-immune-cells (accessed March 28, 2025).

National Cancer Institute. "Cancer Statistics." April 27, 2018. https://www.cancer.gov/about-cancer/understanding/statistics (accessed March 28, 2025).

Nye B. (2017, October 04). Bill Nye says climate change deniers need to 'respect facts' [Interview, Transcript]. In *The Current.* Toronto, Ontario: CBC. https://www.cbc.ca/radio/thecurrent/the-current-for-october-4-2017-1.4320118/bill-nye-says-climate-change-deniers-need-to-respect-facts-1.4320277 (accessed March 28, 2025).

Pandotra T. "Nanomedicine: delivering on its promise?" *Pharmaphorum*, October 20 2017. https://pharmaphorum.com/views-and-analysis/nanomedicine-delivering-promise/ (accessed March 28, 2025).

Planck MK. Letter to: Wood RW. 1931 Oct 7. In: Hermann A. *Frühgeschichte der Quantentheorie. Mosbach*: Physik Verlag; 1969.

Planck MK. *Scientific Autobiography and Other Papers.* Philosophical Library, 1949.

Prasad V. Perspective: The precision-oncology illusion. *Nature.* 2016;537(7619):S63.

Rice S. "Uninformed docs buy into 'breakthrough' hype." *Vital Signs Blog*, Modern Healthcare, April 12, 2016. https://www.modernhealthcare.com/article/20160412/BLOG/160419978/blog-uninformed-docs-buy-into-breakthrough-hype (accessed March 28, 2025).

Rovelli C. *Seven Brief Lessons on Physics.* Penguin Books Limited, 2015.

Salas-Vega S, Iliopoulos O, Mossialos E. Assessment of overall survival, quality of life, and safety benefits associated with new cancer medicines. *JAMA Oncology.* 2017;3(3):382–390.

Sargent T. *The Dance of Molecules: How Nanotechnology Is Changing Our Lives.* Thunders Mouth Press, 2006.

Schumpeter. "Who's afraid of disruption?" *The Economist.* 2017 Sep 30. https://www.economist.com/business/2017/09/30/whos-afraid-of-disruption (accessed March 28, 2025).

Schwab K. "The Fourth Industrial Revolution: what it means, how to respond." *World Economic Forum*, January 14, 2016. https://www.weforum.org/agenda/2016/01/the-fourth-industrial-revolution-what-it-means-and-how-to-respond/ (accessed March 28, 2025).

Schwartz P. "Nanotechnology: This Could Change Everything." *Red Herring.* 2001 Jun;(99):Special Issue.

Science World British Columbia. "Science World Introduces Symbiosis: A First-of Its-Kind Learning Ecosystem for Canada." May 25, 2017. https://www.bccolleges.ca/news/posts/science-world-introduces-symbiosis-a-first-of-its-kind-learning-ecosystem-for-canada (accessed March 28, 2025).

ScienceDaily. "Nanolaser can detect, kill circulating tumor cells to prevent cancer metastases, study finds." August 21, 2017. https://www.sciencedaily.com/releases/2017/08/170821102714.htm (accessed March 28, 2025).

Szabo L. "Widespread Hype Gives False Hope to Many Cancer Patients." *Kaiser Health News*, April 27, 2017. https://khn.org/news/widespread-hype-gives-false-hope-to-many-cancer-patients/ (accessed March 28, 2025).

The Economist. "The Third Industrial Revolution." 2012. https://www.economist.com/leaders/2012/04/21/the-third-industrial-revolution (accessed March 28, 2025).

The Economist. "Technology Quarterly: Closing in on cancer: New therapies, new priorities." *The Economist.* 2017;3-12. https://www.economist.com/leaders/2017/09/16/closing-in-on-cancer (accessed March 28, 2025).

The Max Planck Center-EPFL Center for Molecular Nanoscience and Technology. https://www.mpg-epfl.mpg.de/1007405/Focus (accessed March 28, 2025).

Turok N. *The Universe Within: From Quantum to Cosmos.* Allen & Unwin, 2013.

Ulijn RV, Riedo E. "Learning to 'Think Systems'." *Nat Nanotechnol.* 2016 Sep 7;11(9):824.

World Health Organization. "Cancer." February 3, 2025. https://www.who.int/news-room/fact-sheets/detail/cancer (accessed March 28, 2025).

2 Healthcare's Current Landscape, the Future, and Mars

Raphael Rakowski

CONTENTS

DOI: 10.1201/9781003680093-2

2.1 CURRENT HEALTHCARE CONTEXT – A FOCUS ON THE ISSUES WITH THE HEALTHCARE DELIVERY SYSTEM

When thinking holistically in envisaging a just and sustainable future for our planet and its growing number of inhabitants, an efficient and robust healthcare strategy is a core consideration. Ironically, as we contemplate the confluence of conflicting macro forces, including population growth, climate change, life-extension technologies, the recent rapid surge in AI advances via GPT, corrosive lifestyles, etc., we find ourselves challenged by the collision of good and bad news.

The good news is that there will be a lot more people to experience life on Earth and they will likely live longer. The bad news is that the percentage of individuals with the burden of chronic disease will explode, while their expectations to live for an extended time with that burden will grow as well. Put simply, our planet with its air, water, and food resources has already been strained by population growth and will be stretched even further by a growing segment of the populace that struggle with chronic health-related issues.

Likely the best place to begin in providing a vision for the future of healthcare is to share the status of where we find healthcare *delivery* today (the "it" of healthcare) and the implications of our current reality for potential "future states." I wish to set the specific focus on healthcare *delivery* as opposed to sexy (current and prospective) healthcare topics such as genomics/CRISPR (Liu et al., 2023), 3D tissue printing (Nie et al., 2023), molecular manufacturing (Freitas, 2009), immunotherapy drugs (Shi and Lammers, 2019), and a host of other compelling technologies (Patra et al., 2018). This is because it is in the delivery of healthcare that we all confront most of the non-sexy, financial, regulatory, operational, and human issues that constrain the idyllic promise of our future healthcare, which advanced technologies may provide us.

The summarized missive contained within this chapter cannot do justice to the vast pools of information that comprise the subject of healthcare delivery; however, the ten selected key drivers are worthy of note. Many areas might have been included but were not, including regulatory barriers, infectious diseases, point-of-care diagnostics technology, genomics, and the role of poverty in healthcare. In my own sense of how best to prioritize, I biased my selection on those areas that have a pervasive and vexing nature and avoid simple solutions.

2.2 PART ONE (A) – HEALTHCARE IS PERVASIVE

As a subject matter, healthcare *delivery* is likely one of the most pervasive domains that our society and economy contend with, as it touches every single person from the beginning to the end of their lives. In nearly every large and small community across the United States, there are hospitals, physician offices, ambulances, pharmacies, walk-in clinics, surgical centers, laboratories, and a host of other healthcare associated delivery assets.

Healthcare is ubiquitous because we, along with our friends and relatives, regularly experience births, deaths, and myriad health issues that intersect with healthcare delivery. We hear about medical challenges and compare notes with others about these experiences. We hear about healthcare in the news and read about it in newspapers and online. Cancer frequently touches our lives, as does the birth of a child who we are in orbit with. We have kids who break bones in sporting events and other kids who have close calls with peanut allergies. We have parents with heart disease and diabetes who are frequent fliers in hospitals. Finally, we watch TV programs and movies that focus on the delivery of healthcare and extol the virtues of the heroic clinicians who care for patients and regularly save lives.

Since healthcare expenses represent 18% of our nation's GDP and are an extremely hot button issue with voters, politicians talk about healthcare and healthcare reform at every opportunity. It is one of the only national policy subjects that has never left the top of the political discourse agenda. Stated most simply, we are surrounded by healthcare delivery via the experience of just being alive. Even if we are healthy and have not directly interacted with the healthcare delivery system as a patient, as we see our paychecks ravaged by healthcare-related expenses; this too is a pervasive reminder of the overwhelming influence of healthcare delivery in our day-to-day lives.

2.3 PART ONE (B) – HEALTHCARE IS COMPLEX

Healthcare delivery is a complex labyrinth of doctors, insurance companies, technologies, hospitals, nursing homes, medical procedures (Physician Insurers Association of America, 2009), medications, diseases, research efforts, regulators, public policy, patients, nutrition, lifestyles, and the profound impacts that families and related social determinants have on the efficacy of interventional healthcare outcomes.

It is safe to posit that there are precious few individuals on our humble planet who can intelligently speak to each of the myriad *components* of healthcare delivery that constitute the *whole*. It appears that one of the most curious aspects of the healthcare delivery system we rely on for our medical care is that *almost no one* can explain how the entire system operates in a manner that average people, legislators, and the media can grasp.

Over the last 50 years, the growing complexity of healthcare delivery and how it is paid for ("reimbursement") begged for *compartmentalization*, and that in fact is precisely what happened. Each component of the healthcare delivery puzzle has evolved its own knowledge base, experts, research, strategies, payment systems, cultures, and nomenclature.

Listed below is but a very small sample of the many facets of the healthcare delivery system as a whole:

Clinicians who medically care for us: e.g., geriatricians, pediatricians, surgeons, *primary* care physicians, nurses, nurse practitioners, therapists, physical therapists, medical technicians, etc.

Sites where we obtain our medical care: e.g., hospitals, skilled nursing facilities, physician clinics, urgent care centers, ambulatory surgery centers, walk-in clinics, radiology centers, cancer care centers, reference laboratories, pharmacies, ambulances, etc.

Payers who foot the bill for our medical care: e.g., private insurance companies, Medicare, Medicaid, employers, providers of medical care who serve as insurance companies by taking financial risk within new accountable care organizations, patients without insurance who pay out of pocket for their own care, etc.

Providers of domestic and community clinical services: e.g., home health aides, home nursing, paramedics, disease management companies, etc.

Medical issues that require intervention from the healthcare delivery system**:** e.g., bacterial and viral infections, genetic diseases, cancer, broken bones, heart attacks, stroke, lacerations, burns, loss of function, cognitive impairment, etc.

Providers of therapeutic goods**:** e.g., pharmaceutical companies, medical device companies, durable medical equipment companies, oxygen suppliers, etc.

Providers of diagnostic tools**:** e.g., medical diagnostic companies, laboratories, point of care diagnostic devices, etc.

Providers of surgical tools**:** e.g., operating room equipment, surgical robots, etc.

Technologies that guide and oversee medical care and its payment**:** e.g., electronic health records, etc.

Research organizations**:** e.g., The National Institutes for Health, universities, pharmaceutical companies, medical device companies, etc.

The partial list of healthcare delivery stakeholders listed above provides a small glimpse into the source of complexity that drives healthcare delivery compartmentalization. This compartmentalization manifests itself as knowledge silos (thinking, decisions, and actions), which have led to a healthcare "system" that is far from being one. What we experience when we engage with healthcare delivery is a mind-numbing series of *disconnected* activities, handoffs, anonymous experiences, incentives, and outcomes. This silo-driven complexity has led to medical advice and care that is highly variable, as are the outcomes that emanate from them.

2.4 HIGHLIGHTING COMPLEXITY – CARE VARIABILITY

To bring this critical point to life, a favorite pastime of mine was to take a patient with a known malady to multiple healthcare providers. When I did, the identical signs and symptoms were presented, and yet, to my surprise, each of those providers did not come up with the same diagnosis (Figure 2.1). This exercise was followed up by taking another patient with an agreed-upon diagnosis, signs, and symptoms to multiple clinicians, each of whom did not prescribe the same treatment regimen. I had mistakenly assumed that there would be a "best practice" treatment approach once the patient's circumstances, needs, and diagnosis were clear. I was wrong.

Everyone can agree that having the most *robust and relevant information and experience* should be the strategy for generating a patient's diagnosis and treatment regimen. This is certainly not the case in the practice of medicine today. The level of variability in patient diagnoses

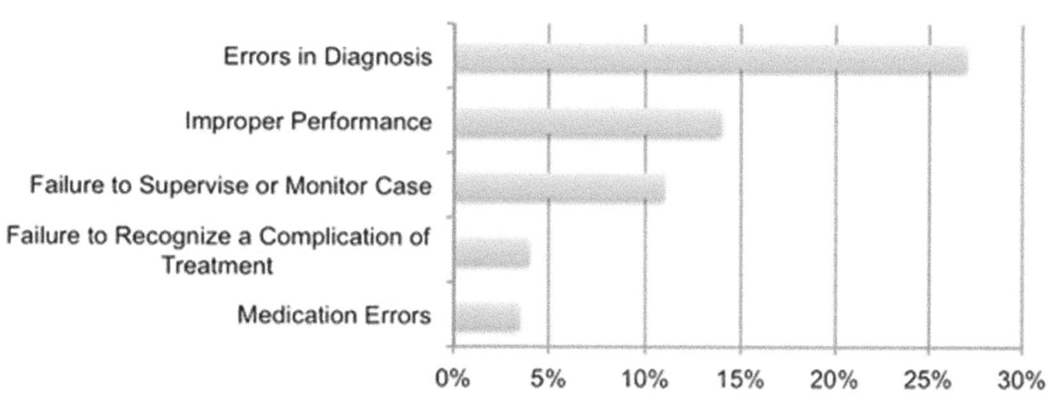

FIGURE 2.1 Top alleged medical errors named in claims where the patient expired. (Reproduced from Graber (2017), with permission.)

is stunning. Once a potentially incorrect diagnosis is established, the subsequent treatment path is fraught with irregularities, and poor outcomes are indeed possible, which are funded by precious financial healthcare resources.

2.5 HIGHLIGHTING COMPLEXITY – THE HUMAN BODY

One of the many challenges that the healthcare system faces is derived from the immense physiological complexity of the human body, and its nested myriad of macro-, micro-, and more recently, nano-systems that need to be comprehensively understood by the clinicians, who are empowered to diagnose and treat patients when medical issues arise. Compounding this physiological complexity are the multitude of viral and bacterial diseases that intrude on this physiology, often doing so in unique/non-intuitive ways. Further, add to this base level of understanding the signs and symptoms of the debilitating influences of chronic diseases that fuel lifelong patient health challenges with diabetes, heart disease, asthma, and COPD. If this were not enough, add the impacts of environmental factors, including allergens, toxins (and even peanuts), anthropogenically generated ambient atmospheric and aqueous contaminants, microplastics, nanoparticles, forever chemicals, etc. on human health. Finally, clinicians need to elucidate the health impacts of a host of caustic patient behaviors and social influencers that encompass alcohol/drug abuse, smoking, poor diet, sedentary activity levels, and social isolation.

Consequently, it is easy to understand that this level of root-cause complexity is no simple matter, and why healthcare variability is so pervasive. With so many factors at play, even the most adept human mind, combined with years of hands-on experience, can generate an inaccurate medical diagnosis. It is also no wonder that the popular TV show, "House," romanticized the Sherlock Holmes-like skills of an inspired physician, who had to navigate extraordinary levels of complexity to arrive at a diagnosis when faced with mysterious symptoms.

2.6 FUTURE STATE – HOW WE WILL CONTEND WITH COMPLEXITY AND ARTIFICIAL INTELLIGENCE?

How does this complexity of today serve to author and drive solutions for the future of healthcare tomorrow? Artificial intelligence (AI) in conjunction with quantum computing (QC) will likely be key ingredients of a future state that will simultaneously transform both patients and the clinicians who care for them. AI strategies and technologies will collect and analyze large-scale evidence of complex variables and patterns to reveal the most likely diagnoses and suggest evidence-based treatments. This, of course, is good news for patients and the positive outcomes that they seek.

Conversely, sitting on the flip side of this positive AI coin is a very valid concern that physicians will likely have to struggle with. If, in the future, a well (data) fed AI program can more reliably diagnose and define treatment for patients than an experienced physician, it is possible that the future role of many physicians will begin to resemble that of a pharmacist. Over the many years of their profession, the roles of pharmacists have evolved; moving from patient diagnosis and drug formulation to executing the orders of other clinicians (i.e., filling prescriptions written by doctors). Similarly, AI may usher in an age where the stepping down of physician critical thinking and decision-making becomes commonplace, and new applications in healthcare emerge (Table 2.1). Kue et al. (2023) state that, "A.I.-based effective intervention can reduce medical errors by 30–40% and solve 20% of unmet clinical demand."

In this light, physicians who sacrifice a great deal to secure their positions in the medical care community will not easily shift to a world where "an algorithm" tells them what to do with a patient in their care. This will likely be a fertile battlefield that finds the *old way* (individual physicians

TABLE 2.1
Ten AI Applications That Could Change Healthcare

Application	Potential Annual Revenue by 2026 (in Billions)	Key Drivers for Adoption
Robot-Assisted Surgery	$40 B	Technological advances in robotic solutions for more types of surgery
Virtual Nursing Assistants	$20 B	Increasing pressure caused by medical labor shortage
Administrative Workflow Assistance	$18 B	Easier integration with existing technology infrastructure
Fraud Detection	$17 B	Need to address increasingly complex service and payment fraud attempts
Dosage Error Reduction	$16 B	Prevalence of medical errors, which leads to tangible penalties
Connected Machines	$14 B	Proliferation of connected machines/devices
Clinical Trial Participant Identifier	$13 B	Patent cliff; plethora of data; outcomes-driven approach
Preliminary Diagnosis	$5 B	Interoperability/data architecture to enhance accuracy
Automated Image Diagnosis	$3 B	Storage capacity; greater trust in AI technology
Cybersecurity	$2 B	Increase in breaches; pressure to protect health data
Total	**~$150 B**	

Source: Derived from Accenture, "ARTIFICIAL INTELLIGENCE: Healthcare's New Nervous System." https://www.ehidc.org/sites/default/files/resources/files/Accenture-Health-Artificial-Intelligence.pdf

making decisions based on their judgment and experience) at odds with the *new way* (AI algorithms rapidly and specifically defining both diagnoses and treatment regimens). In many respects, this battle has already begun with the advent of new electronic health records that "recommend" preferred treatment protocols for patients.

2.6.1 The Subject vs. Object of the Future

Predicting the future is often fraught with risk (e.g., errors or miscalculations), as the ecosystem of forces that shapes the future is fluid. Anyone who ventures into that ecosystem and who is capable and committed to shaping that future (i.e., they are the "subject" vs. the "object" of that future) can (left to their own devices) overwhelm the forces at play. Simply stated, the "actors" that drive the future are often more influential than the forces at play, prior to the future being realized. Steve Jobs is a superb example of what happened to personal computing due to his role and intervention in the computing ecosystem.

2.6.2 Avoiding the AI Hype

AI is nothing more than an algorithm that mathematically predicts patterns based on those that already exist in a dataset. It is not "artificial intelligence." It is not intelligence. It is math. Most patterns have an inherent margin of error. Years ago (1980–1986), I worked in the robotics field when everyone predicted that robots would take over all factories and humans would no longer be required in manufacturing. Akin to AI, there were pockets of logical-use robots (e.g., painting, welding, solar panels, and microchips) that required high-volume repetitive precision, where they were effective. However, in most other applications, robots could not compete with humans, which remains the case today. Stated most simply, the future will include AI but will not be dominated by it, unless and until an actor emerges with an overwhelmingly unique and influential solution that facilitates the creation of an AI-dominated future.

2.6.3 Potential Applications of AI in Unique Clinical Use Cases

2.6.3.1 Pharmacists

The traditional role of the pharmacist required a great deal of training in the practice of "compounding" drugs, in cases when physicians prescribed ingredients and combinations thereof in non-standardized ways. Once pharmaceutical companies created these more standardized products, the role of compounding all but disappeared. The actual role of the pharmacist is to "catch" profound errors and risks to patients, who are increasingly taking 5–20 drugs a day. In this context, every doctor, specialist, or (algorithm) who/(that) prescribes a drug is not aware of the integrated effects of the potential drug combinations, which in some cases can be deadly. There may well come a time when algorithms gain expertise in "watching over" all of the medications of patients and the integrated effects of their combinations, which can "intervene" and address any issues that may arise. However, until then, human "pharmacists" remain a critical element in healthcare delivery (even though patients do not recognize the role that they actually play behind the scenes, as they witness only a mechanical process in play when picking up their prescription).

2.6.3.2 Diagnosis and Treatment

For the most common/repetitive/well-understood clinical issues (e.g., ear infections, pneumonia, rashes, etc.), the roles of AI in diagnoses and guiding treatments are likely to be significant and extensive. For other clinical issues that are more specialized and complex, human interventions will continue to be meaningful.

2.6.3.3 Surgery

Surgery will continue to require surgeons (humans assessing and intervening); however, over time, many surgical interventions will be justified/approved by AI, due to the frequent overuse of surgery and the non-standardized ways in which it is performed. This is particularly the case for interventional cardiology and spinal surgeries.

2.6.3.4 Cancer

The diagnosis and treatment of cancer will likely be dramatically influenced by AI, as cancer treatment is already the most protocol-driven disease intervention. AI is likely to play an exceptionally significant role in improving the local community treatment of cancer, to meet the healthcare pathways that are sought in large cancer centers (e.g., Sloan Kettering, MGH, Mayo, etc.). In my view, this AI development will serve to "democratize" cancer care and create a positive revolution for patients. That said, my prediction is contingent on how large a continued role cancer-related surgery will play in the future. Immunotherapy and other biologic cancer interventions will lend themselves to an AI-influenced future. New and earlier stage cancer diagnoses (Stage I and II) will also align well with an AI-driven future.

2.6.3.5 Mental Health/Addiction

Mental health and addiction are true frontier challenges for healthcare today. A great deal of (missing) science needs to be discovered/developed to address this immense and (metastatically) growing crisis. This yet-to-be-discovered science will need to understand and reconcile two intersecting forces: (i) the organic nature/operation of the brain and (ii) the impact of psychologically damaging patient experiences that together influence and drive a host of mental health issues (e.g., depression, PTSD, addiction, and ADD).

I predict that promising clinical research work currently being conducted with mushroom-based hallucinogens (e.g., psilocybin and ketamine) will find its way into wide-scale treatment over the next seven years. Unlike all previous mental health interventions, the mechanism of action in these compounds appears to create patients' "recontextualization" of their underlying traumas, which often drive and feed their mental health issues. Once more, proven science, in terms of diagnosis and treatment, is in place; AI can and should play a role.

2.6.3.6 Evolution/Revolution

My final point involves a synthesis of all the above. Back in the day when Ross Perot and EDS began to "automate" large systems (e.g., passport control, the post office, the IRS, airline reservations, etc.), there was a quiet recognition that the historically "manual" ways of doing things were fraught with errors and inefficiencies. As these automated "systems" came online, they delivered on the promise of significantly improved efficiencies and productivity. However, many of us began to "feel" a nagging sense of dehumanization creep into our collective consciousness. After decades of this automation, healthcare began to adopt electronic systems. Unlike the other use cases, healthcare's understanding and strategies for the use of this platform were laden with "Frankenstein-like" abominations of tools and systems that burdened the physician-patient relationship (despite the well-meaning intention to improve this relationship).

The staggering cost of this so-called electronic medical record automation has now reached over $10 billion per year, and yet we continue to suffer from poor healthcare outcomes. Sad artifacts of this automation are burned-out clinicians and patients feeling like they are inanimate objects moving down an assembly line conveyor belt. I share this view, as all the predictions about the positive impacts of this automation were off in meaningful ways. Thus, my caution is to appropriately temper prognostications regarding AI in healthcare.

2.7 INCREASED TRANSPARENCY – ANOTHER RESPONSE TO COMPLEXITY

Another front worth highlighting that will likely manifest in response to the current burden of healthcare complexity is the increased use of strategies and tools that provide patients (and their families) with enhanced information and the simplification of their medical needs, to clarify roles and accountability for their own healthcare expenses.

To enable this improved and simplified access to patient medical data, greatly enhanced *interoperability* between electronic medical records (e.g., enabling ubiquitous access to patient medical data across all providers and sites of care) will be required. For patient access to be pervasive and reliable, with multiple providers of medical care documenting patient medical data on different technology platforms, this seamless integration of patient medical records from different sites, platforms, and providers will be essential.

There will also be a host of new consumer-like apps to support patients and their families as they assume larger roles in their own healthcare. These apps will enable: (i) increased access to healthcare providers, (ii) improved convenience with the healthcare delivery experience, (iii) an enhanced understanding of medically related data, and (iv) healthcare shoppers/navigators to assist with the selection of their medical providers based on their quality and value ratings. This process of increased transparency has already commenced, and large technology players including Apple, Google, and Amazon will certainly play roles in this space.

As relates to transparency, patients and their caregivers have straightforward needs when it comes to their medical care, which may be summarized as follows:

- *Healthcare status* – Am I OK?
- *Maintenance* – What should I be doing to stay healthy?
- *Diagnosis* – What is wrong with me?
- *Treatment* – How will I get better?
- *Provider competency* – How qualified is the doctor or hospital caring for me?
- *Economics* – What will it cost me to get the care I need?

In this simple light, healthcare data will likely be translated to a more consumer-friendly vocabulary, as will the supportive platforms that facilitate it. That said, while the need and logic for the provision of this type of information is obvious, there are non-trivial technology and regulatory barriers that remain in the path of integrating, simplifying, and communicating healthcare data and patient medical records.

2.8 PART TWO – DAUNTING ECONOMICS

In 2023, the United States generated $4.8 trillion of the ~>$10 trillion in global healthcare expenditures at ~48% of global spending, comprising 17.6% of its GDP (Statista, 2025), while the rest of the world consumes 10.8% of its GDP on healthcare (WHO, 2022). Despite spending more on healthcare than any other industrialized nation, the United States ranks sixth in healthcare quality and outcomes (Roy and Girvan, 2021). This disparity between spending and outcomes is a blaring signal of systemic deficits in the way healthcare delivery is configured, paid for, and used in the United States.

With no reliable strategy in place to improve the costs or outcomes of our healthcare delivery system, by 2030 current trends in healthcare spending are projected to comprise 19.6% of the U.S. economy (down slightly from 19.7% in 2020), according to the Office of the Actuary at the Centers for Medicare & Medicaid Services (CMS). Healthcare spending is projected to reach $6.8 trillion by 2030, up from $3.9 trillion today (Table 2.2) (CMS, 2022).

As the percentage of U.S. GDP continues to climb, government sponsored healthcare programs (i.e., *Medicare*, for seniors, and *Medicaid*, for low-income Americans) will soon collide with matters of national security. With global tensions always threatening peace and stability, military spending will always rise to the top of strategic decision-making. As government sponsored healthcare spending rises, the rationing of non-military resources is an inevitable outcome. As other programs (e.g., education) are at an elevated risk of being rationed, there is the very real potential for even more acute knowledge and income disparities between individuals, which will further exacerbate polarization and tension.

This tension is well underway and is being intensified by the growing burden of college student debt that has potential long-term life altering consequences. One cannot decouple the increased costs of education from the unbridled growth in healthcare expenditures. Put simply, **with 18.3% of the U.S. GDP being consumed by healthcare spending, we are at the precipice of a strategic tipping-point with our national priorities**. This will require immediate and meaningful public debate, biased by facts and immune to the forces of political rhetoric.

Until now, as healthcare costs have continued to escalate above the rate of inflation, this debate has been muted by government borrowing to fund federal budget deficits. Clearly, one might posit that we are paying for our healthcare delivery inefficiencies with money that we are borrowing from the Chinese (who we fund with the money we send them in exchange for the low-cost goods they produce in their factories). In the absence of a complete transformation of the $3.9 trillion U.S. healthcare delivery system, the rationing of healthcare or other nationally funded programs appears inevitable.

How important are education, research, defense, the National Parks System, and social security? How important is the provision of $40,000 in hospital care for a patient during their last two weeks of life? Are donations to the arts more important than medications for the poor? Do we really need to go to Mars? These are critical questions that need to be addressed. Regardless of the answers, the debate is made more civil if we harness the innovation engine of American ingenuity and technology. As our highest priority, we need to redesign our healthcare delivery system using a new anchoring strategy. This strategy needs to result in a strategically acceptable proportion of

TABLE 2.2
Healthcare Spending in the United States Is Projected to Reach $6.8 Trillion by 2030, up from $3.9 Trillion Today (Derived from CMS, 2022)

YEAR	2023	2024	2025	2026	2027	2028	2029	2030
SPENDING ($T)	$3.9	$4.1	$4.4	$4.7	$5.0	$5.3	$5.6	$6.8

our nation's GDP being used for healthcare. Given the global nature of healthcare writ large, and similar issues being faced around the world, this new strategy can and will create the potential for massive new U.S. exports.

2.9 FUTURE STATE – FORMIDABLE HEALTHCARE EXPENDITURES

If we were to adopt a *glass half-full* orientation for our future, we would see the pressures of healthcare spending becoming so acute that they trigger a moonshot-like focus on holistically transforming the cost of healthcare. This, to bring about greater access, improved quality, and reduced disparity with no meaningful compromise to strategic spending for defense and education.

Many elements of such a transformation are described in this chapter. In this glass half-full view of the world, we would employ our innovative skill sets, fueled by technology, to dramatically reduce the cost of healthcare. Conversely, if we take a *glass half-empty* view of the future, the continued escalation of medical costs may result in the significant rationing of healthcare in the United States. This might involve a dramatic increase in the numbers of patients seeking lower cost alternatives (e.g., medical tourism in Mexico, India, and other developing countries), or relocation (depending on one's situation) to countries with more affordable healthcare.

2.10 PART THREE – THE GROWING COST AND QUALITY BURDEN OF U.S. HOSPITAL CARE

Perhaps the single largest opportunity for shifting the trajectory to the future of healthcare is centered on where the most costly healthcare delivery is domiciled. In 2022, U.S. hospitals spent $1.2 trillion in caring for patients at ~6,093 sites across the country (AHA Hospital Statistics, 2022). An immense complex of land, buildings, equipment, staff, and fixed costs accounts for more than 60% of the price tag of a hospital stay (Martin et al., 2012) (Table 2.3). Said another way, the fixed costs of U.S. hospitals represent roughly twice the federal education budget and ~75% of the federal defense budget.

TABLE 2.3
A Complex of Land, Buildings, Equipment, Staff, and Fixed Costs Accounts for More than 60% of the Price Tag of a Hospital Stay in the United States

Hospital Expenses	Percentage
Government Administration	1%
Home Healthcare	3%
Retail and Other Products	3%
Nursing Home Care	5%
Net Cost of Health Insurance	6%
Investments	6%
Other Professional Services	7%
Other Health, Residential, and Personal Care	8%
Retail, Rx Drugs	10%
Physician/Clinical Services	20%
Hospital Care	31%
Total	**100%**

Source: Derived from Martin et al. (2012).

U.S. hospital spending represents roughly one-third of healthcare expenditures and continues to grow, despite evidence of excess hospital capacities. A recent *NEJM* paper (Wiler et al., 2017) reflects on a thoughtful challenge to the continued reliance on large centralized medical care hubs for non-routine medical care. This high fixed-cost approach to caring for acutely ill patients may indeed have reached a tipping point in its evolution. Perhaps the greatest single statistic to highlight in this chapter relates to the safety of hospitals, which has been summarized by the number of individuals who expire due to preventable hospital errors every hour, every day (50 deaths per hour = ~34,000 deaths per month) (Accelerated Nursing.net., 2025).

Preventable deaths due to medical errors are a national epidemic and represent the second largest cause of death in the United States. At this level of service/product failure, if hospitals were hotels, they would be shuttered, and if they were car companies, they would likely be unable to retain their customers. There are a host of reasons for the issues that hospitals experience with their care model. Some are systemic, while others are the result of human error. In my view, one of the most significant issues that drives this level of adverse events is the transition to the *industrialization* of medical care delivered in hospitals. As cost pressures increased, many hospitals began to embrace strategies that are employed in manufacturing environments (e.g., Toyota's Lean Manufacturing System) to improve efficacy and throughput.

While the logic behind this approach makes some sense, one of the unintended consequences involves challenges in accountability by individual clinicians. Unlike a worker who installs a transmission on a Buick assembly line, who does not need to know about the other workers interacting with the rest of the car, a nurse on third shift working with a patient needs to be highly integrated with the entire care-giving team. Every medical team member should recognize that their actions can significantly impact the decisions and actions of other clinicians and, ultimately, the patient's outcome. This dynamic is explored in greater depth in Section 2.22.

2.11 FUTURE STATE – HOSPITALS AND ADVANCED MEDICAL CARE

While the centralized delivery of medical care was once considered the ultimate expression of efficiency, today's hospital economics are driving new care models away from large brick-and-mortar models. Akin to what Amazon has done with the decentralization of retail shopping, healthcare in the future will see a significant reduction in the number of hospitals, with those remaining focused on complex medical care.

There is a nascent movement underway to begin shifting advanced medical care away from hospitals to new, lower-cost sites of care, which encompass patient homes, skilled nursing facilities, and group living accommodations. As this transition evolves and gains momentum, a growing breadth of technology will enable virtual care models that centralize clinicians (via telemedicine), while decentralizing the community services provided to patients through new, stringently managed clinical supply and service chain networks that can be rapidly dispatched to wherever patients require immediate care. This will be explored in greater depth in Section 2.18.

2.12 PART FOUR – ALIGNING INCENTIVES

There is widespread agreement that the key stakeholders in the U.S. healthcare delivery system are not ideally aligned. Those who pay for medical care (i.e., employers, insurers, and patients who want to pay as little for healthcare as possible) are often at cross-purposes with those who deliver it (i.e., physicians and hospitals who need to maximize their revenues and incomes to ensure financial sustainability). Over the last 30 years, multiple strategies have been deployed to better align the incentives between stakeholders; however, experience with these approaches has been mixed at best. A classic example that highlights this challenge may be best illustrated with the introduction of the hospital payment incentive innovation referred to as "DRGs" (diagnoses-related groups).

A DRG is a hospital reimbursement system initiated by health insurers to reduce the growing burden of hospitalization expenses through the creation of incentives for hospitals to reduce their costs. The underlying goal of DRGs was to align the hospital business model with the needs of employers and insurers. The DRG model was a simple one and conceptually elegant: create a fixed price for reimbursing hospitals for each diagnosis category they treat, regardless of the actual costs they incur. This idea was intended to incentivize hospitals to reduce their costs to achieve a profit margin on this fixed price paid to them.

As the DRG incentive scheme kicked in, hospitals quickly recognized they could no longer bill insurers, patients, and employers for medical care based on *how long* patients stayed in the hospital (this is called "per-diem" billing). Instead, since the movement to DRGs forced them to put their costs under a microscope, they realized their largest single cost was *fixed* (i.e., their brick-and-mortar buildings and administrative expenses, which equated to +60% of the costs of a hospital stay). Once this fixed cost burden was evident, it became imperative for hospitals to *discharge* patients from the hospital as quickly as possible to generate a profit margin.

Not surprisingly, as the DRG incentive gathered steam, the time patients stayed in the hospital reliably declined, where the current average length of hospital stay is ~4.7 days (Definitive Healthcare, LLC, 2024) (down from 8.7 days in 1990) (CDC, 2017). This would be fine, of course, if 4.7 days were the appropriate amount of time required for patients to recover from an acute illness; however, it is not. Therefore, as DRGs gained in popularity, so did the practice of discharging patients to skilled nursing facilities to "complete" their care elsewhere.

As DRG usage amplified, the growing costs for care in these skilled nursing facilities were borne by patients, insurers, and employers. Another artifact of the trend for decreasing hospital length of stay and the increased utilization of skilled nursing facilities was the need to readmit patients to the hospital soon after they were discharged. Depending on the diagnoses, short-term hospitalization rates in the United States range from 10% (e.g., for cellulitis) to 20% (e.g., for pneumonia and congestive heart failure).

Given the complexity of the healthcare system and its many stakeholders, the intention of DRGs was well-founded. However, the unintended consequences were not, which included higher net costs in contrast to the approach that preceded it. Many refer to this outcome of increased net costs from the practice of saving expenses in one area, without recognizing the impact it will have in another (siloed) area as the "squeezing the balloon" effect.

Another classic strategy for incentives alignment with good intentions but unsavory or unintended consequences had its focus on the *deductibles and co-pays* that patients must pay out-of-pocket for their own care. The intent of this extra financial burden on patients was to improve accountability for their own health and high costs for the utilization of medical care. Simply put, insurers forced patients to put "more financial skin in the game." In many cases, this had the intended consequence of reducing high-cost (avoidable) medical care utilization, albeit in other cases (as with the DRGs) some surprises surfaced.

One of these relates to the *medication co-pay*. To save healthcare expenses for insurers, many patients found themselves having to pay more in co-pays for their chronic medications. As certain patients were unable to keep up with these co-pays each month, they began to skip their medications out of financial necessity. Many of these patients ended up in the hospital due to the exacerbation of their chronic disease triggered by the disruption of their medication. The math on this illustrated example is simple. The patient saved $50 on their medication co-pay that month but had to pay a $500 hospital co-pay. The insurer saved $300 on the medication for the month as well but had to pay the hospital $12,000 for the patient's hospital stay.

2.13 FUTURE STATE – POSITIONING INDUCEMENTS

There are multiple movements underway to innovate the current incentive systems.

2.13.1 Value-Based Payments

One strategy for accomplishing this goal is to shift to models where medical providers are paid for *quality and outcomes* (aka *value-based payments*), rather than simply paying them for the service they provide (*aka fee for service payments*). To illustrate this simple idea, let us imagine calling a plumber to your home because you have a leaky pipe. Under today's healthcare delivery incentive structure system, the plumber arrives at your home, works on your pipe, and then leaves, despite the pipe continuing to leak. The plumber then sends a $400 bill to your insurance company for the two hours he spent working on your pipe. The insurance company then sends the plumber a check for $400 for his time and effort, and then sends you a $100 bill for your co-pay portion of the bill.

In a value-based payment approach, the plumber gets paid $350 for fixing your pipe or gets paid nothing if your pipe continues to leak. You and the insurance company save money and the successful plumbers have customers and revenue to rely on, while plumbers with poor outcomes either improve their skills or get forced out of business.

While still in the early stages of development, the intent of the value-based payments approach is to reward providers of medical care with a reimbursement approach that reinforces lower costs and higher quality. Most healthcare providers today recognize that value-based payments will gain momentum and are tooling up to prepare for this inevitable approach to medical care payments.

2.13.2 Accountable Care Organizations

Another strategy that is gaining momentum to better align incentives between stakeholders is referred to as accountable care organizations (ACOs). If hospitals and physician groups are confident that they can better manage their patients' cost of medical care, they create and participate in a structure from which they get paid a fixed amount from insurers each month for each patient they care for. This ACO structure is intended to provide the funding that those hospitals and physician groups require to cover all the costs of a patient's care, while leaving enough money to generate a profit for those providers. From the perspective of the insurer, there is now a clearly aligned incentive for those providers to not waste or overuse healthcare resources. If the ACO is structured correctly, providers who better manage patient healthcare are financially rewarded, while payers save money and patients enjoy better quality. ACOs have had mixed results; however, the trend to explore them continues.

2.13.3 Concierge Medicine

For those Americans who have the financial capacity to pay for medical care how, when, and where they want it, there are growing healthcare provider movements to provide concierge-like healthcare services that are paid for by patients out of their own pockets. These services include primary care physicians who make themselves available 24/7 and pediatricians who will make house calls on demand. There is even a service for the privileged few that provides a fully stocked medical emergency room and surgical suite in a member's home, which can be activated on-demand by a member requiring urgent medical care.

While still a relatively small portion of the healthcare delivery landscape, this growing movement signals the potential for a multi-tiered system based on personal economic capacities. In the short term, this concierge trend will likely continue, as cost and service pressures on medical providers force them to seek higher sources of revenues and profits with less administrative complexity.

2.14 PART FIVE – LIFESTYLE ISSUES AND INSURANCE POOLING

Americans experience significant and growing challenges with their health due to non-medical issues that drive the need for medical care. Poor diet, smoking, alcohol/drug abuse, social isolation, and limited education are but a few of the lifestyle and social determinants of increased healthcare utilization. This is noteworthy as a significant portion of our healthcare delivery system is consumed by Americans who, through their own behavior, require a high degree of costly medical interventions.

Admittedly, this is an area of potential ethical complexity at the heart of an ethical conundrum. This is reflected in the voices of some Americans who do not struggle with the lifestyles and social issues of others who drive increases in healthcare costs.

> "Why should I, a healthy and responsible tax-paying, happily-married, employed American who goes to the gym, eats a sensible diet, and doesn't smoke, be forced to pay for the healthcare costs of an obese, unemployed/uneducated smoker who abuses alcohol?"

This is a perfectly valid question, the heart of which involves *pooling*. For insurance to operate as a sustainable business model, most of the insured members must have a low need for insurance reimbursement, while insurance payments for those who do require them are paid for by the low utilizers of the insurance product.

Life insurance is the simplest example to help illustrate the nature of pooling. Young and middle-aged workers make a reasonable monthly payment for many years to provide a future death benefit payout for their beneficiaries. As these insured individuals begin to pass away, significant payments must be made by the insurance company to the beneficiaries. The money for these large payments is derived from the continuous instream of new younger insured members. Imagine if 95% of a pool of insured members were 90 years old and no new younger members bought insurance. It would take less than three years for the insurance company to become insolvent and cease in its capacity to make death payment benefits.

The creation of a mixed pool of (younger) and healthier Americans with (older) and unhealthier Americans enables the healthcare funding needs of the more-costly population. This pooling function is an essential element of how to insure and pay for the healthcare needs of those who require medical care. While this pooling approach is a well-developed bedrock of our insurance system, how should we view younger people with corrosive lifestyle issues, who author their own needs for high-cost medical care?

2.14.1 Obesity

Obesity is one of the greatest drivers of preventable chronic diseases and healthcare costs in the United States, with a current estimate for these costs amounting to ~$173 billion per year (Ward et al., 2021). Another approximation of the annual medical costs for obese individuals (which can be ~$1,429 higher than those of normal weight) totals $147 billion. Obese adults spend 42% more on direct healthcare, with costs for severely or morbidly obese adults that are 81% higher than healthy adults. Further, obesity is associated with job absenteeism and productivity losses, at a cost that ranged from $13.4 to $26.8 billion in 2016 and with lower productivity while at work, costing employers from $271 to $542 per obese worker per year (Cawley et al., 2021).

Ubiquitous access to low-cost, high sugar/high fat foods combined with sedentary lifestyles has ushered in a national epidemic of obesity in both adults and children (Stobbe, 2021). Currently, in the United States, 41.9% of adults (2017–2020) and 19.7% of youth (2017–2020) are obese, and their risks of becoming diabetic are well known (Warren et al. 2022). There is also a 75–78% incidence of hypertension for individuals with obesity (Shariq and McKenzie, 2020). As a conscious, progressive, and responsible society, how do we wrestle with the accountability for obesity? Who is to blame? Who should pay for the medical costs that reliably arrive after years of Happy Meals? The patient or McDonald's?

2.14.2 Alcohol Abuse

The currently annual healthcare expenditures for alcohol-related issues in the United States amount to $7.6 billion (Peterson et al., 2021), while the total annual cost of alcohol problems is $249 billion per year compared to $193 billion for illicit drug use (U.S. Department of Health and Human Services, 2023). Approximately 25% to 40% of all patients in U.S. general hospital beds (not in maternity or intensive care) are being treated for alcohol-related complications (Center on Addiction and Substance Abuse, 1994).

Again, as a sympathetic society, how should we think about alcoholism? Is it a disease? If so, does not everyone with this disease deserve sympathy and access to medical care? How should we view the vendors of alcohol? Are they complicit in the disease? Should we tax them to pay for the healthcare issues they contribute to as is done for tobacco companies?

2.14.3 Smoking

The total economic cost of smoking in the United States is estimated at more than $864.5 billion a year (Nargis et al., 2022). This includes nearly $240 billion in direct medical care for adults and more than $372 billion in lost productivity due to premature death and exposure to secondhand smoke (Max et al., 2012; Shrestha et al., 2022; Xu et al., 2015). To offset some of this cost, state governments collect $27.2 billion each year from tobacco taxes and legal settlements. This tax revenue represents less than 10% of the annual societal cost of smoking (Campaign for Tobacco-Free Kids, 2018).

Cigarette smoking causes a wide variety of preventable diseases. Its prevalence has declined substantially since the first U.S. Surgeon General's report (from 43% in 1965 to 18% today); however, it remains a leading cause of preventable death in the United States, where it is responsible for more than 480,000 deaths per year, including nearly 42,000 deaths from secondhand smoke exposure (CDC STACKS, 2014). Nicotine, like alcohol, is a known source of profound addiction, and as such, cigarette smokers, like alcohol abusers, are on a reliable path for elevated healthcare issues. The costs associated with these issues are a burden for non-smokers and non-drinkers, who pay elevated healthcare costs for these addictions by others. Is this fair? Issues surrounding the medical care burdens of lifestyle-related behaviors are challenging on multiple ethical, cost, and policy fronts. For the time being, we continue to increase our national debt to fund the healthcare costs associated with these issues.

2.15 FUTURE STATE – HOW ADDICTION-DRIVEN LIFESTYLE ISSUES WILL BE ADDRESSED

In the future, AI and advanced gene sequencing will likely play significant roles in the identification of specific gene expressions for people who are predisposed to addictive behavior. Given the magnitude of the problem and the (financial) value of potential solutions, pharmaceutical interventions will be a likely strategic R&D imperative (Jiang et al., 2023). In the short term, insurance companies will increasingly move to models to better "risk-adjust" health insurance premiums to reflect lifestyle issues. As mentioned earlier, there may be a role for hallucinogens in the treatment of addiction as well.

2.16 PART SIX – END OF LIFE CARE

One inevitable fact of life is that it will end one day. Death, dying, and making peace with one's life are not really healthcare issues. Although, to date, there is no "cure" for death, intense efforts on a global scale are dedicated to the significant extension of human longevity (Guo et al, 2022; Johnson et al., 2022; Vijg and Austad, 2023; Yang et al., 2023). Death and dying are currently a social and spiritual issue that spills over into the healthcare domain. Another driver of healthcare costs is the use of the healthcare system to care for patients who are near or at the end of life.

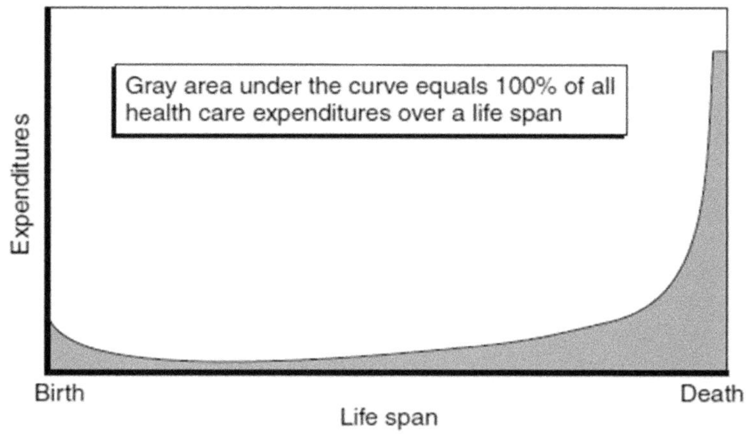

FIGURE 2.2 Majority of American healthcare expenditures are concentrated near the end of life. (Reproduced from Lynn and Adamson, Living Well at the End of Life: Adapting Health Care to Serious Chronic Illness in Old Age. Santa Monica, CA: RAND Corporation, 2003. https://www.rand.org/pubs/white_papers/WP137.html, with permission.)

While high-quality cost data for this issue is often a source of debate, most researchers agree that for those Americans who have an unlikely path to recovery from (terminal) illness, hospitals, surgery, and other interventions remain commonplace.

In the United States, 13% of all healthcare spending is consumed in the last year of a person's life (Figure 2.2). A great deal of this spending is focused on heroic attempts to extend life for brief periods of time using hospitals and life-extending technologies. The question of what level of medical care is appropriate at the end of life resides in the domain of medical ethics, which is certainly worthy of vibrant debate (Cohen et al., 2023; Di Risio and Thompson, 2023; Ho et al., 2023). That said, humanity is deeply enamored with advanced technologies and the exciting prospects that these innovations might bring.

As a society that represents a polyglot of cultural beliefs associated with the responsibilities for death and the sanctity of life, we continue to rely on expensive healthcare resources to continue physical life, as we consistently resist coming to grips with our own mortality. In this light, assisting families with advanced care planning and, when appropriate, transitioning to hospice and palliative care is a national healthcare resource preservation priority. At the time of this writing, hospice utilization growth has accelerated with only 21.6% of Medicare beneficiaries utilizing hospice in 2000 growing to 51.6% in 2019 (Teno et al., 2022).

2.17 FUTURE STATE – END OF LIFE CARE

Healthcare is for the living. In the future, end of life care will be transformed as hospice strategies will graduate, allowing patients and families to explore the end of their lives divorced from the connection of death with healthcare. While the role of technology in this domain is not clear-cut, AI will help patients to better understand and grapple with their remaining time on Earth. We can imagine that for some, it will be best that this information stays confidential. For others, knowing how much time is left might improve the quality of that remaining time.

2.18 PART SEVEN – HEALTHCARE ACCESS

Imagine a world where you cannot obtain an affordable cup of coffee where and when you want one? Now imagine a world where Amazon, iTunes, Netflix, and Uber do not respond to your needs, on demand. That world was one that we safely left behind and replaced with on-demand access to our *wants and desires*. What about on-demand access to the most important of our *needs ... healthcare*?

The only reliable on-demand site for medical care, regardless of one's level of illness, time of day, or day of week, is the *hospital*. As described above, hospitals are the highest-cost sites of care and are built for 24/7 access. They have the scale to staff for all manner of medical issues, tests, procedures, and emergencies, which is, of course, the good news. The not so good news is that if your child has an ear infection in the middle of the night and he/she needs an antibiotic prescription, a trip to the hospital will cost you and your insurer dearly.

It is intuitively obvious that when patients require medical care, *delayed access* to that care is an issue that has a host of consequences, not the least of which is inconvenience. While health insurers work to dissuade patients from using hospitals (due to their cost structure), they have not provided a reliable on-demand 24/7 alternative. Perhaps one of the greatest opportunities and challenges facing the sustainability of healthcare delivery is the reliable provision of patient access to appropriate medical care that aligns with their time-sensitive needs. Ideally, medical care should be *brought to when and where* patients need it; i.e., their homes, the workplace, on airplanes, and in hotels. In the absence of this, patients will seek medical care from hospitals, as they know hospitals are open for business whenever they require them.

A recent example of how access to medical care challenges patients, hospitals, and insurers is best illustrated in a recent high-profile story that made the national news: "An ER visit, a $12,000 bill – and a health insurer that wouldn't pay. A new insurance policy expects patients to diagnose themselves" (Kliff, 2018a).

> Brittany Cloyd was doubled over in pain when she arrived at Frankfort Regional Medical Center's emergency room on July 21, 2017. "They got me a wheelchair and wheeled me back to a room immediately," said Cloyd, 27, who lives in Kentucky. Cloyd came in after a night of worsening fever and an increasing pain on the right side of her stomach. She called her mother, a former nurse, who thought it sounded like appendicitis and told Cloyd to go to the hospital immediately. The doctors in the emergency room did multiple tests including a CT scan and ultrasound. They determined that Cloyd had ovarian cysts, not appendicitis. They gave her pain medications that helped her feel better, and an order to follow up with a gynecologist. A few weeks later, Cloyd received something else: a $12,596 hospital bill her insurance denied – leaving her on the hook for all of it.
>
> Health insurance coverage guides and bills belonging to Brittany Cloyd. "We have a mortgage, we have bills, we have student loans," says Cloyd, who works for the Kentucky government and has a 7-year-old daughter. "There is absolutely no way I could pay a $12,000 bill. I don't even have $1,000 sitting around." Cloyd has her health insurance coverage through her husband's job. His company uses Anthem, one of the country's largest health insurance plans. In recent years, Anthem has begun denying coverage for emergency room visits that it deems "inappropriate" because they aren't, in the insurance plan's view, true emergencies.
>
> The problem: These denials are made after patients visit the ER, sometimes based on the diagnosis after seeing a doctor, not on the symptoms that sent them, as in Cloyd's case.
>
> *Kliff (2018b)*

2.19 FUTURE STATE – PATIENT ACCESS TO MEDICAL CARE

What does the future state look like toward addressing this patient access challenge? Patients will have 24/7 access to medical care; however, the nature of this care will change. In the future, rather than patients seeking out doctors in brick-and-mortar clinics, urgent care centers, or hospitals, medical care delivery will transition to a *more virtual and personalized model.* This virtual model will be designed with centralized *medical command centers* staffed by physicians and nurses, overseeing new mobile on-demand medical resources, brought to patients, wherever and whenever they need them.

2.19.1 Medical Command Centers

To make 24/7 healthcare access a reality, in the future, physicians will be centralized in medical command centers and deliver their portion of medical care *telemedically.* They will obtain medical information from patients using remote monitoring and examination technologies that create a virtual presence. Patients will be able to access this type of care from their home, at work, at kiosks, or retail clinics in drug stores in their community.

This centralized clinician-staffing approach overcomes any local inefficiencies that challenge the financial sustainability of a 24/7 community-based brick-and-mortar medical clinic. The truth is, the reason that urgent care centers and other lower acuity medical treatment sites are not open 24/7 today is simple: there are simply not enough local patients to cost-justify the expense.

2.19.2 Mobile On-Demand Medical Resources

Once a centralized examination has been conducted by a physician, additional diagnostic tools might be required (e.g., X-rays and blood tests). These services will become mobile and be brought to patients. Once a diagnosis is established, physicians will requisition treatments that are brought to patients, which may include nursing services, IV medications, physical therapy, and behavioral health support. The providers of these services will be part of a specialized mobile network that can be reliably dispatched to patients' homes in less than one hour. This model of care already exists and is being deployed across the United States with clinical organizations including the Mayo Clinic and Kaiser Permanente to help shape the dialogue of the future state of healthcare access challenges.

2.20 PART EIGHT – MEDICAL TECHNOLOGIES – A MIXED BLESSING

As diagnostic, pharmaceutical, medical device, and surgical technologies continue to innovate to combat disease and the deleterious impacts of aging, the R&D costs for these innovations are quite significant and growing. For example, a new drug typically takes ten years to bring to market and at a cost of $1 billion. A significant portion of these costs relates to FDA and other regulatory approvals. To justify the extraordinary investment required to bring innovation to market, pharma, device, and diagnostic companies must recoup their investments through the profits generated by sales of their technologies. The cost of an individual drug, device, procedure, or test is often greater than the annual salaries of the patients who utilize these technologies.

Here is a small sample of commonly used technologies that are facets of the delivery of medical care today:

MRI scan - for diagnostics that are not accurate enough via X-Ray – $400 to $12,000 per test (Woodard, 2022).

Opdivo – a chemotherapy drug for lung cancer and melanoma – $12,500 per month (Incollingo, 2015).

Hospitalization for elderly patients with leukemia – annualized cancer (acute myeloid leukemia) attributable costs for EOL cancer phase $239,000 (with an average survival rate of two months after hospitalization) (Mariotto et al., 2020).

Hip replacement for an 80-year-old man – $39,300 (Mallord, 2020).

Heart valve replacement – $164,000 (CostHelper, Inc., 2023).

There is no doubt that breakthrough diagnostic and therapeutic tools have had incredibly positive impacts on people's lives; however, the expenses associated with these technologies create a large and growing healthcare expense burden. At what point does the cost of a given technology and the value of a life find themselves at odds with each other?

To date, once a new technology is approved by regulators, private insurers must decide whether they will reimburse for these innovations. If they select to adopt the payment, then their medical costs may go up substantially, which then forces them to raise insurance rates for employers and their employees. Thus, the introduction of new diagnostics and therapeutics is one of the cost drivers of U.S. healthcare.

If, on the other hand, insurers decide not to adopt the payment for these new technologies, employers may seek other insurers that provide coverage for those technologies. This, of course, would result in insurers losing customers, with the concurrent loss of revenues and profits. Given these dynamics, commercial insurers will opt to include new technologies as part of their medical coverage but do so recognizing that they will pass this increased cost to their members.

For patients who are older than 65 and covered by Medicare, the dynamic of new (costly) technology reimbursement is likely with commercial insurers, but with a different twist. As Medicare decides to adopt payment for these technologies, the payment for this decision comes from two sources: the increased burden on taxpayers and the increased debt the country takes on to fund budget losses that are heavily influenced by growing healthcare costs.

Either way, while one of the most provocative sources of innovation, U.S. medical technologies that increase the quality of life and longevity, have financial sustainability consequences that have not yet been fully grappled with as a society.

2.21 FUTURE STATE – MEDICAL TECHNOLOGIES

The future state of medical technologies will attempt to utilize a more rational approach as a facet of the following context. AI-driven algorithms that attempt to capture macro-societal cost-benefit analyses will become a decision support tool that drives the (reimbursement) adoption of future medical technologies. One of the potential outcomes of this approach may have a chilling effect on the elderly, while biasing younger people who still pay taxes and make meaningful contributions to the economy. There will also be a diminutive, albeit high-value segment of the global population who possess the financial resources to pay for new technologies out of their own pockets. This will be the real market for innovation (regardless of how insurers decide to adopt or reject these technologies).

2.22 PART NINE – THE FALSE PROPHET OF MANUFACTURING MODELS VS. A HEALING COVENANT

Over the last 50 years, as the number of hospitals, patients, and spending have grown, so has the need to contain costs. As the hospital sector sought cost-reduction models to learn from, they turned to the Toyota lean manufacturing system approach. While highly effective for making cars, lean manufacturing strategies did not envision human beings as the product moving down an assembly line.

The manufacturing process metaphor that drove hospital efficiency created an unexpected artifact. As the idea of efficiency took hold, patients were "handed off" from clinician to clinician during "workflows" designed to conserve labor and time. As patients had the experience of being treated like a transmission in a Toyota Camry, their capacity to engage in their own care became significantly challenged. Older patients recall a doctor fully in charge of their care, who attended to them until they were well. This time-tested patient-physician relationship was disrupted by the emergence of this new factory-like mindset in healthcare.

Beyond the challenges for patients, physicians, and nurses committed to and trained in the healing arts, they found themselves experiencing intensified job stress and reduced satisfaction levels. Much of this stress stemmed from the reduced connection to patients and the satisfaction gained from staying with a patient until they were healed. This was always an element of the *healing covenant* that comprised a sacred bond between clinician and patient. The new lean manufacturing paradigm coupled with reduced clinician compensation has created a crisis in clinician burnout, which is but only one of the myriad challenges that face the advent of a sustainable future healthcare system.

2.23 FUTURE STATE – A RETURN TO THE HEALING COVENANT

In the future, clinicians will employ telemedicine and related technologies to stay connected with patients throughout their course of care, and in the process return to the satisfaction that they sought when they committed their lives to patient care. In this future, care models will involve dedicated teams that remain with patients until their medical issues are resolved. In the future, *handoffs* will be viewed as patient experience failures vs. the way healthcare is delivered today.

2.24 PART TEN – HEALTHCARE DELIVERY JOBS

Due to the large and continuously growing demand for healthcare services in the United States, the healthcare industry is a vibrant source of employment and a large part of the U.S. economy. Employment in the healthcare industry has been expanding for decades. Not only has the number of workers employed in healthcare increased, but the percentage of the nation's private-sector workforce thus employed has increased as well. It was 2017 that marked the first year when healthcare surpassed manufacturing and retail; "the most significant job engines of the 20th century to become the largest source of jobs in the U.S. In 2000, there were seven million more workers in manufacturing than in health care. At the beginning of the Great Recession, there were 2.4 million more workers in retail than healthcare. In 2017, health care surpassed both" (Thompson, 2018) (Figure 2.3).

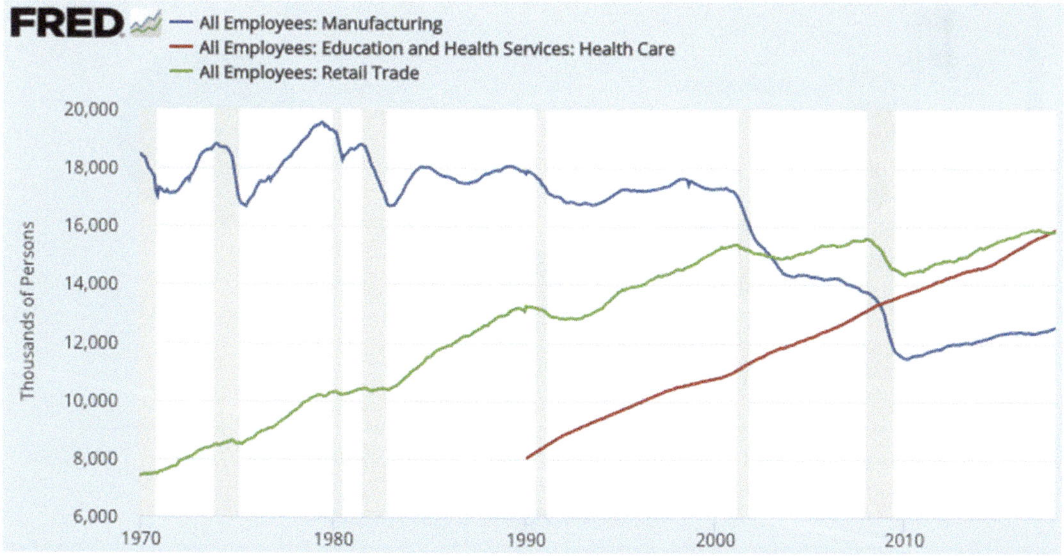

FIGURE 2.3 Rapid growth in health sector jobs consists of non-clinical positions. (Reproduced from Federal Reserve Economic Data, 2018. Public domain.)

2.25 FUTURE STATE – HEALTHCARE JOBS

There are multiple drivers behind the healthcare employment growth boom, which include an aging population and accelerating technological advances. While the most destabilizing forces for labor over the last generation have been globalization and automated, to date, healthcare has been substantially resistant to both due to its quintessentially localized nature.

Recently, the growth in healthcare employment stems more from administrative than physician jobs. The number of non-doctor workers in the health industry has exploded over the last two decades (Dawson, 2017). Most of these jobs are not clinical roles, like registered nurses. They are primarily administrative and management jobs, including receptionists and office clerks, and it is not always clear how these workers tangibly improve health outcomes for patients (Kocher, 2013). The two fastest-growing occupations, personal-care aides (who perform non-medical duties for older Americans, such as bathing) and home-health aides (who assist the elderly with medical care), are projected to account for one in every ten new jobs. Further, the entire healthcare sector is projected to account for a third of all new employment (Figures 2.4 and 2.5).

Today, healthcare employees work in hospitals and medical clinics; thus, patients are forced to come to them. In the future, clinicians will make face-to-face house calls and virtually round on patients using telemedicine. Healthcare jobs will focus on telemedically supported remote care, giving

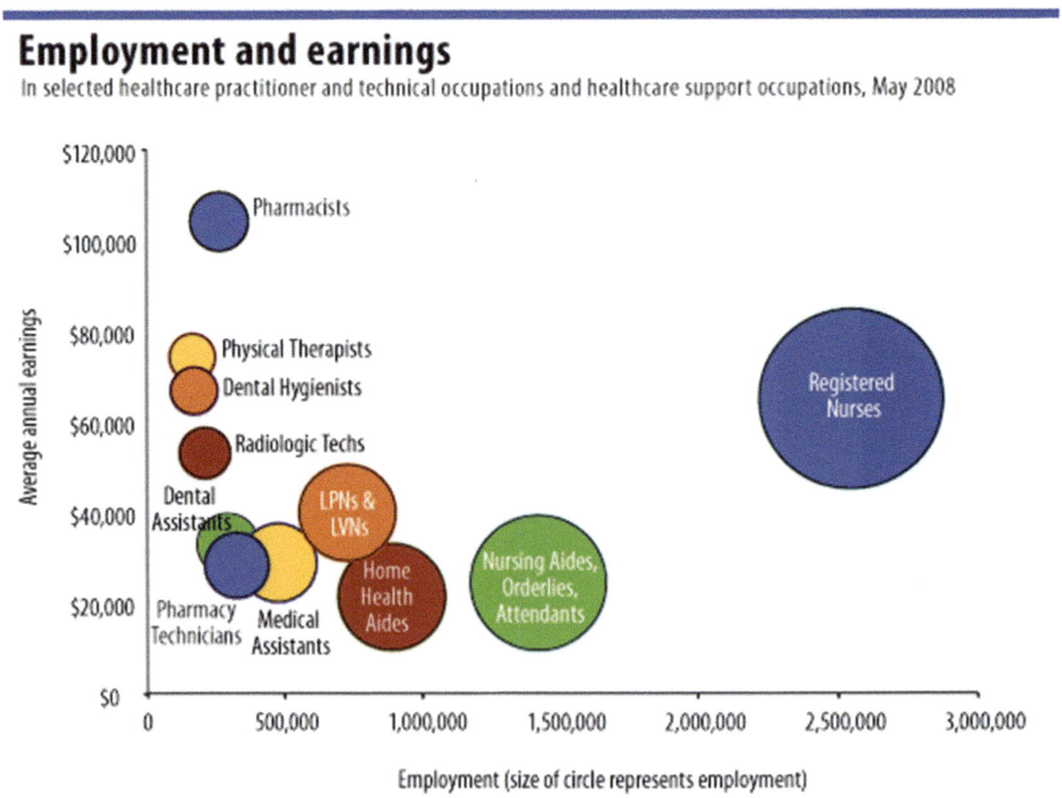

FIGURE 2.4 Employment and earnings in healthcare occupations. (Reproduced from U.S. Bureau of Labor Statistics, 2008. Public domain.)

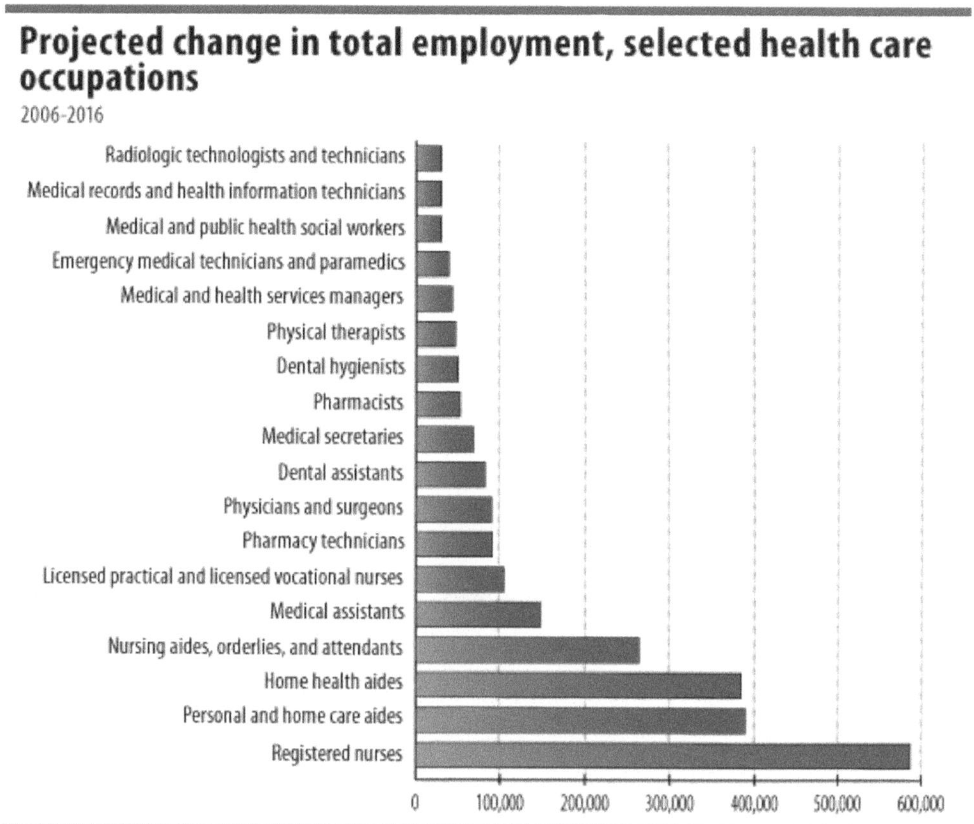

FIGURE 2.5 Estimated changes in specific healthcare occupations. (Reproduced from U.S. Bureau of Labor Statistics, 2008. Public domain.)

technology-enabled diagnoses and care models in the home and at work. Medical services traditionally domiciled in hospitals (e.g., nursing procedures, X-rays, laboratory testing, physical therapy, etc.) will become mobile, and these mobile-delivered services will create an entire pool of new jobs. Hospice care workers will also be in great demand, as will a new breed of technology-enabled non-licensed home healthcare aides. An adjunctive aspect to this type of patient support might be the use of humanoid robots (e.g., Beomni AI, Beyond Imagination) (Showen, 2021). Having said this, as cited earlier, my experience with robotic infiltration in the manufacturing economy gives me pause.

As mentioned beforehand, AI will likely play a growing role in the delivery of medical care. It is unclear whether the growing use of AI will take on the limited role of decision-support for physicians, or take a cost-saving trajectory, where nurse practitioners and other lower licensed clinicians armed with AI serve as substitutes for high-priced physicians. Regardless, we can anticipate a tsunami of growth in analytics that will drive more evidence-based medical care delivery.

2.26 CLOSING THOUGHTS – ABOUT MARS

Mars is a faraway planet that will one day likely be inhabited by humans. In a future that will require human health management; most of the ideas shared in this chapter will become integral elements of medical care delivery. There will be no large hospitals on Mars, and technologies (e.g., AI, nanotechnology, and nanomedicine) will be maximized. Individuals will have the capacity to quantify and understand exactly what is wrong with them and what care will be required. As AI platforms are robustly populated, healthcare cost curves will begin to resemble the trajectory of computer processing economics. There will be a premium placed on rapid diagnoses, transparency, consistency, and remote care delivery. The tethering of medical care experts/know-how to remote nodes where care is required will be a mature capability.

Mars, anyone?

REFERENCES

Accelerated Nursing.net. How Safe Are You in Hospitals? 2025. https://www.accelerated-nursing.net/hospital-safety/ (accessed June 01, 2025).

AHA Hospital Statistics. 2025. https://www.aha.org/statistics/fast-facts-us-hospitals

Campaign for Tobacco-Free Kids. *Broken Promises to Our Children: The 1998 State Tobacco Settlement 20 Years Later*. Washington: Campaign for Tobacco Free Kids, 2018.

Cawley J, Biener A, Meyerhoefer C, Ding Y, Zvenyach T, Smolarz BG et al. Job absenteeism costs of obesity in the United States: National and state-level estimates. *Journal of Occupational and Environmental Medicine*. 2021;63(7):565–573.

CDC STACKS. The Health consequences of smoking—50 years of progress: a report of the Surgeon General. 2014. https://stacks.cdc.gov/view/cdc/21569 (accessed September 04, 2025)

Center on Addiction and Substance Abuse, Columbia University. *The Cost of Substance Abuse to America's Health Care System, Report 1: Medicaid Hospital Costs*. 1994.

Centers for Disease Control and Prevention (CDC). 2017. https://www.cdc.gov/nchs/data/hus/2017/082.pdf (accessed August 10, 2025).

CMS Office of the Actuary Releases 2021-2030 Projections of National Health Expenditures. March 28, 2022. https://www.cms.gov/newsroom/press-releases/cms-office-actuary-releases-2021-2030-projections-national-health-expenditures (accessed February 09, 2025).

Cohen MG, Althouse AD, Arnold RM, White D, Chu E, Rosenzweig M et al. Primary palliative care improves uptake of advance care planning among patients with advanced cancers. *Journal of National Comprehensive Cancer Network*. 2023;21(4):383–390.

CostHelper, Inc. How Much Does Heart Valve Replacement Cost? 2023. https://health.costhelper.com/valve-replacement.html (accessed April 16, 2023).

Dawson G. Why There Are No Bipartisan Solutions to Exorbitant Healthcare Costs in the USA. Real Psychiatry blogspot. August 7, 2017. https://real-psychiatry.blogspot.com/2017/08/why-there-are-no-bipartisan-solutions.html (accessed April 16, 2023).

Definitive Healthcare, LLC. Hospital average length of stay by state. Oct 23rd, 2024. https://www.definitivehc.com/resources/healthcare-insights/average-length-of-stay-by-state (accessed August 10, 2025).

Di Risio M, Thompson A. Current practices in managing end-of-life existential suffering. *Current Opinion in Supportive and Palliative Care*. 2023 Jun 1;17(2):119–124.

Freitas RA Jr. Meeting the challenge of building diamondoid medical nanorobots. *International Journal of Robotics Research*. 2009;28:548–557.

Guo J, Huang X, Dou L, Yan M, Shen T, Tang W et al. Aging and aging-related diseases: From molecular mechanisms to interventions and treatments. *Signal Transduction Target Therapy*. 2022;7(1):391.

Ho CY, Lim NA, Rahman NDA, Chiam M, Zhou JX, Phua GLG et al. Physician-patient boundaries in palliative care. *BMC Palliative Care*. 2023;22(1):41.

Incollingo BF. Considering cost: What's an immunotherapy worth? *CURE Immunotherapy Special Issue* 2015;1(1). https://www.curetoday.com/view/considering-cost-whats-an-immunotherapy-worth (accessed August 11, 2025).

Jiang X, Yan C, Zhang H, Chen L, Jiang R, Zheng K et al. Oral probiotic expressing human ethanol dehydrogenase attenuates damage caused by acute alcohol consumption in mice. *Microbiol Spectrum*. 2023;11:e0429422.

Johnson AA, English BW, Shokhirev MN, Sinclair DA, Cuellar TL. Human age reversal: Fact or fiction? *Aging Cell*. 2022;21(8):e13664.

Kliff S. "An ER visit, a $12,000 bill – and a health insurer that wouldn't pay. A new insurance policy expects patients to diagnose themselves." *Vox*. 2018a. https://www.vox.com/policy-and-politics/2018/1/29/16906558/anthem-emergency-room-coverage-denials-inappropriate (accessed August 10, 2025).

Kliff S. "Hospitals keep ER fees secret. We uncovered them." *Vox*. 2018b https://www.vox.com/2018/2/27/16936638/er-bills-emergency-room-hospital-fees-health-care-costs (accessed Nov. 07, 2021)

Kocher R. The Downside of Health Care Job Growth. *Harvard Business Review*, September 23, 2013. https://hbr.org/2013/09/the-downside-of-health-care-job-growth

Kue C, Williams T, Perimal-Lewis L, McCauley V. A systematic review of digital health and AI in healthcare: Improve patient safety, lower medical errors, and enhance the quality of care via AI. Flinders Digital Health Research Centre, College of Science and Engineering, Flinders University. 2023. https://www.researchgate.net/publication/381460236_A_Systematic_Review_of_Digital_Health_and_AI_in_Healthcare_Improve_patient_safety_lower_medical_errors_and_enhance_the_quality_of_care_via_AI (accessed June 01, 2025).

Liu X, Li G, Liu Y, Zhou F, Huang X, Li K. Advances in CRISPR/Cas gene therapy for inborn errors of immunity. *Frontiers in Immunology*. 2023;14:1111777.

Mallord K. Hip Replacement Recovery Time for Elderly. *Senior Living Help*, 2020. https://www.seniorliving-help.co/hip-replacement-recovery-time-for-elderly/ (accessed April 16, 2023).

Mariotto AB, Enewold L, Zhao J, Zeruto CA, Yabroff KR. Medical care costs associated with cancer survivorship in the United States. *Cancer Epidemiology, Biomarkers & Prevention*. 2020;29(7): 1304–1312.

Martin AB, Lassman D, Washington B, Catlin A, National Health Expenditure Accounts Team. Growth in US health spending remained slow in 2010; health share of gross domestic product was unchanged from 2009. *Health Affairs (Millwood)*. 2012;31(1):208–219.

Max W, Sung HY, Shi Y. Deaths from secondhand smoke exposure in the United States: Economic implications. *American Journal of Public Health*. 2012;102(11):2173–2180.

Nargis N, Hussain AKMG, Asare S, Xue Z, Majmundar A, Bandi P et al. Economic loss attributable to cigarette smoking in the USA: An economic modelling study. *Lancet Public Health*. 2022;7(10):e83 4–e843.

Nie M, Nagata S, Oda H, Takeuchi S. Millimeter-thick 3D tissues constructed by densely cellularized core-shell microfluidic bioprinting. *Biofabrication*. 2023 Apr 25;15(3).

Patra JK, Das G, Fraceto LF, Campos EVR, Rodriguez-Torres MDP, Acosta-Torres LS et al. Nano based drug delivery systems: Recent developments and future prospects. *Journal of Nanobiotechnology*. 2018;16(1):71.

Peterson C, Li M, Xu L, Mikosz CA, Luo F. Assessment of annual cost of substance use disorder in US hospitals. *JAMA Network*. 2021;4(3):e210242.

Physician Insurers Association of America. Data sharing project: Medical professional liability claims data, 1985–2009. Rockville (MD): *PIAA*; 2009:123.

Roy A, Girvan G. Key Findings from the 2021 FREOPP World Index of Healthcare Innovation. June 25, 2021 https://freopp.org/whitepapers/key-findings-from-the-2021-freopp-world-index-of-healthcare-innovation/ (accessed, February 09, 2025).

Shariq OA, McKenzie TJ. Obesity-related hypertension: A review of pathophysiology, management, and the role of metabolic surgery. *Gland Surgery*. 2020;9(1):80–93.

Shi Y, Lammers T. Combining nanomedicine and immunotherapy. *Accounts of Chemical Research*. 2019;52(6):1543–1554.

Showen M. "Your Doctor Is in:" Physicians Test Humanoid Robot for Eldercare. *Business Wire*, December 06, 2021. https://www.businesswire.com/news/home/20211206005120/en/%E2%80%9CYour-Doctor-is-in%E2%80%9D-Physicians-Test-Humanoid-Robot-for-Eldercare

Shrestha SS, Ghimire R, Wang X, Trivers KF, Homa DM, Armour BS. Cost of Cigarette smoking–attributable productivity losses, U.S., 2018. *The American Journal of Preventive Medicine*. 2022 Oct;63(4):478–485.

Statista. U.S. national health expenditure 1960-2023. 2025. https://www.statista.com/statistics/184688/us-national-health-expenditure-since-1960/ (accessed August 07, 2025).

Stobbe M., "Study: Childhood obesity in U.S. accelerated during pandemic." AP News, September 16, 2021. https://apnews.com/article/coronavirus-pandemic-childhood-obesity-ef3d426b5580b72f76eb1207be1af24b (accessed February 20, 2023).

Teno JM, Ankuda CK, Keohane L, Meyers D, Bunker J, Mitchell S et al. Pathway to hospice: How has place of care before hospice changed with the growth of hospice in the United States? *Journal of Palliative Medicine*. 2022;25(11):1661–1667.

Thompson S. Health Care Just Became the U.S.'s Largest Employer. *The Atlantic*, January 9, 2018. https://www.theatlantic.com/business/archive/2018/01/health-care-america-jobs/550079/

U.S. Department of Health and Human Services, Addiction and Substance Misuse Reports and Publications, 2023. https://www.hhs.gov/surgeongeneral/reports-and-publications/addiction-and-substance-misuse/index.html (accessed August 08, 2025).

Vijg J, Austad SN. Biological restraints on indefinite survival. *Cold Spring Harbor Perspectives in Medicine.* 2023;13(4):a041200.

Ward ZJ, Bleich SN, Long MW, Gortmaker SL. Association of body mass index with health care expenditures in the United States by age and sex. *PLoS One.* 2021;16(3):e0247307.

Warren M, Beck S, West M. The State of Obesity: Better Policies for a Healthier America, 2022. *Trust for America's Health (TFAH).* https://www.tfah.org/wp-content/uploads/2022/09/2022ObesityReport_FINAL3923.pdf (accessed August 11, 2025).

WHO. Global spending on health: Rising to the pandemic's challenges. 2022. https://apps.who.int/iris/bitstream/handle/10665/365133/9789240064911-eng.pdf (accessed February 19, 2023).

Wiler JL, Harish NJ, Zane RD. Do Hospitals Still Make Sense? The Case for Decentralization of Health Care. *NEJM Catalyst.* December 20, 2017.

Woodard D., How Much Does an MRI Cost? GoodRx Health, July 7, 2022. https://www.goodrx.com/health-topic/diagnostics/how-much-does-an-mri-cost (accessed April 16, 2023).

WHO World Health Organization Global Health Expenditure database. 2022. https://apps.who.int/nha/database (accessed February 19, 2023).

Xu X, Bishop EE, Kennedy SM, Simpson SA, Pechacek TF. Annual healthcare spending attributable to cigarette smoking: An update. *American Journal of Preventive Medicine.* 2015;48(3):326–333.

Yang JH, Hayano M, Griffin PT, Amorim JA, Bonkowski MS, Apostolides JK et al. Loss of epigenetic information as a cause of mammalian aging. *Cell.* 2023;186(2):305–326.e27.

3 What Is Nanomedicine?

Simon R. Corrie, Jeffrey V. Rosenfeld, and Julia Walker

CONTENTS

3.1 INTRODUCTION

Nanomedicine involves the application of nanoscale materials for the prevention, diagnosis, treatment, and monitoring of disease. These nanomaterials currently include an extensive range of nanoparticles; however, in the future they will likely comprise autonomous nanorobots that are deployable in vivo for imaging, biosensing, diagnostics, precision drug delivery, and nanosurgical procedures on individual cells, organelles, and even DNA. The conceptual design, synthesis, and evaluation of such nanomaterials also requires interdisciplinary collaboration beyond nanoscience itself and typically encompasses elements of the chemical, pharmaceutical, biomedical, and engineering sciences. Nanoscale materials possess attributes that behave much differently than their bulk material counterparts (e.g., gold, silver), even though they share identical chemical compositions. These include far higher surface area to volume ratios and significant differences in their biochemical, magnetic, optical, and electronic properties (Albanese et al., 2012; Chaudhuri and Paria, 2012; Cimalla et al., 2013; Dreaden et al. 2012; Elsabahy and Wooley, 2012; Kamaly et al., 2012; Tang et al., 2012).

Nanomedicine is a rapidly evolving discipline with >50 FDA-approved products, many more in clinical trials, and a vast array of materials being designed and tested in pre-clinical studies worldwide. This chapter provides a relatively brief overview of this rapidly evolving field, with a focus on outlining the breadth (e.g., drug delivery, imaging, sensing) rather than the depth (e.g., detailed chemistry) of activities, which have already made significant impacts in medical practice.

Interactions between nanomaterials and their environment are dependent on the interplay between the properties of both the nanoparticles and their surroundings. Morphology (size/shape/density) and surface chemistry (e.g., electrostatics, van der Waals forces, degree of hydrophilicity/hydrophobicity) are key factors that determine how the nanoparticles will behave, in terms of the formation of a protein "corona", cellular endocytosis, biodistribution patterns, and biodegradation/

DOI: 10.1201/9781003680093-3

clearance mechanisms (for a detailed review, see Nel et al., 2009). Size is a critical factor in the control of human physiological clearance processes, with small nanoparticles (<10 nm) generally cleared via the kidneys and larger nanoparticles (>10 nm) being cleared through the liver and the reticuloendothelial system (Choi et al., 2007; Fox et al., 2009; Rolfe et al., 2014; Sadauskas et al., 2007).

The first issue that is encountered when a nanoparticle comes into contact with the bloodstream following injection is the adsorption of proteins onto its surface. The key problem here involves the denaturation of the protein, which leads to a signal cascade mechanism that results in immune recognition and removal or nanoparticle aggregation. It is practically impossible to completely avoid this non-specific protein adsorption (Tenzer et al., 2011); however, hydrophilic surface coatings can either reduce the total protein content adsorbed or alter the composition of the resulting corona. Current research efforts remain focused on the development of universal design criteria that can lead to predictable nanomedical behaviours upon exposure of various nanomaterials to biological milieu; however, it is clear that size, morphology, and surface chemistry can affect the immediate fate of nanomaterials, from the moment of injection into the bloodstream (Figure 3.1).

3.2 NANOMEDICAL FDA APPROVAL PROCESS

Currently, the approval process for therapeutic nanomedicines in humans is regulated in the United States by the US Food and Drug Administration (FDA) and is essentially the same as that for any other regulated drug, device, or biologic (for a detailed review, see Eifler and Thaxton, 2011). There are similar organizations in other countries with comparable roles, along with online databases (e.g., clinicaltrials.gov), which include the details of clinical trials that are increasingly becoming mandatory worldwide. Under FDA regulations, following the invention of a new nanomaterial, pre-clinical testing involving animal studies is undertaken to generate safety, efficacy, and toxicity profiles and to identify appropriate dose ranges that might be applicable for clinical trials involving humans.

For nanomedicines, it is increasingly becoming apparent that an elucidation of the fundamental chemical and physical properties of the materials involved, along with demonstrated scalability and reproducibility, is extremely important to establish at the early stage. Inherently, these issues are still at the research stage; hence, interactions between academics, corporations, health experts, and regulators are crucial to arrive at reasonable and comprehensive guidelines. Subsequent to animal trials, the efficacy, safety, and physicochemical properties of the nanomaterials are combined to produce an Investigational New Drug (IND) application for FDA consideration. If approved, human trials can be initiated to determine their safety and efficacy as a nanomedicine. The trials pipeline is demarcated as Phase I (dosing, toxicity, excretion in healthy subjects), Phase II (safety and efficacy in a larger group of patients with the target illness), and Phase III (multi-center, randomized, placebo-controlled) trials. If a nanomedicine successfully passes these hurdles with minimal toxicity and improved efficacy over existing medicines, a New Drug Application (NDA) can be filed with the FDA to request final approval for sending the product to market. Further "post-marketing" (Phase IV) studies may be requested, based on specific outcomes of the clinical trials.

In the case of nanomedicines developed for diagnostic applications, the regulatory processes follow a similar pathway, yet there are deviations based on whether the product is designed for in vitro diagnostics (IVD) or operates as an implantable device. In either case, the product is classified as Class I (low risk) (e.g., elastic bandages), Class II (medium risk – typically not required to sustain life) (e.g., condoms, pregnancy test kits), or Class III (unacceptably high risk and/or required to sustain life) (e.g., implantable pacemakers). For non-exempt devices (i.e., many Class I devices are subject only to "general controls"), either Premarket Notification (510(k)) or Premarket Approval (Investigational Device Exemption (IDE) application) needs to be undertaken to market the product.

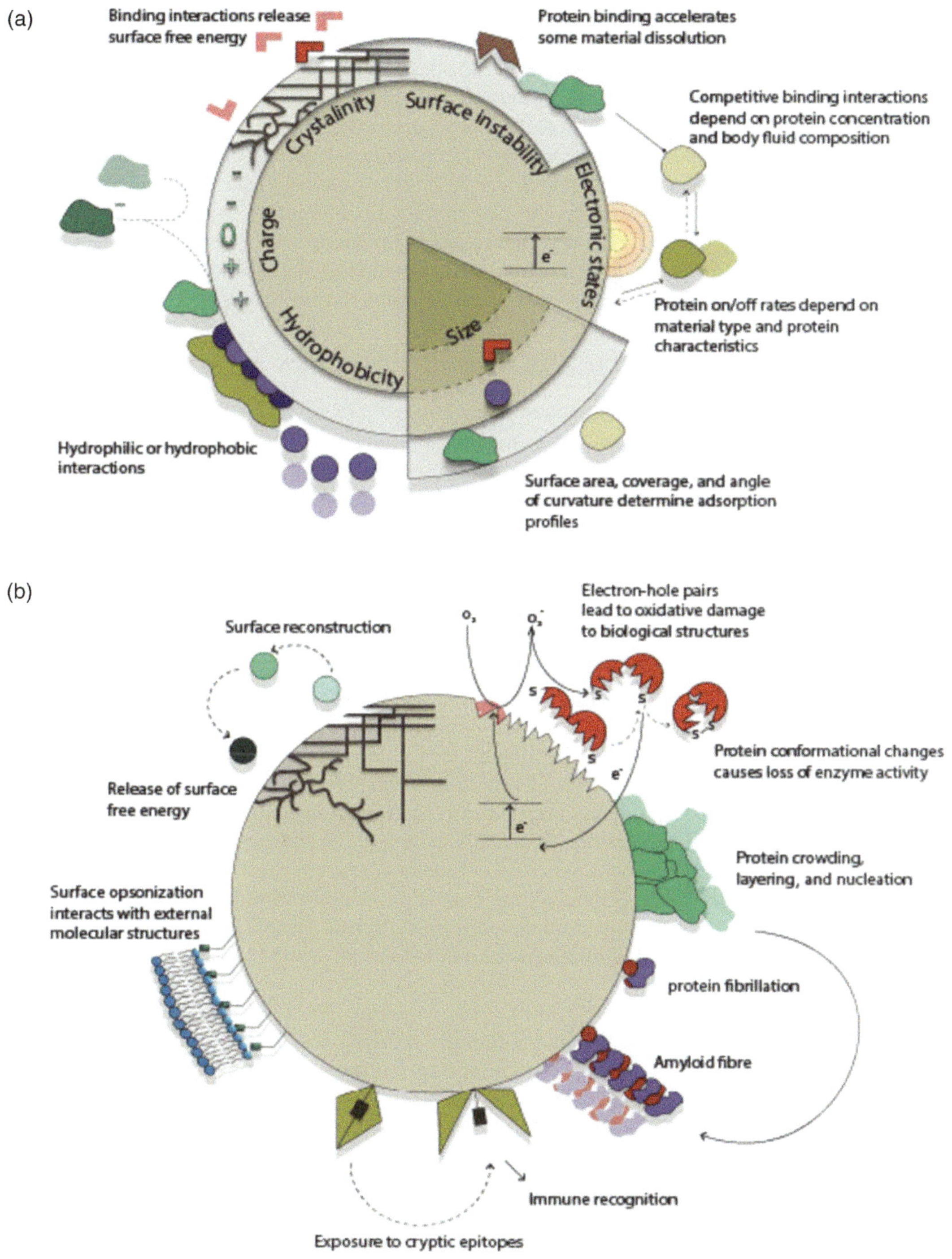

FIGURE 3.1 Effects of nanoparticle composition on biological interactions. (A) Nanoparticle showing the different physical attributes which can affect nanoparticle interactions with a biological environment: charge (or electrostatics), size, crystallinity, surface instability, and electronic states on the surface of the nanoparticle. (B) Demonstrates how physical properties may impact the formation of a protein "corona" through protein layering on the surface. This can result in the formation of amyloid fibers. Biodistribution patterns are affected by surface opsonization by ligands present on the surfaces of nanoparticles. (Image created by Julia Walker.)

The goal of the 510(k) pathway is to demonstrate "substantial equivalence" to an existing FDA-approved product, in which case clinical trials may not be required given an agreed upon minimal risk. For any Class III device, clinical trials need to be designed and approved by both an Institutional Review Board (IRB) and the FDA, in the form of an IDE application. Interestingly, IVD devices are often Class II or Class III, as the risks associated with diagnostic errors may be considerably high; for example, in the case of false-negative tests for pre-cancerous lesions. Furthermore, there may be additional requirements depending on where the IVD test would be performed (e.g., in a hospital lab, the physician's office, or at home). In the case of implantable sensors, for Class II or III device applications, the FDA is particularly interested in issues involving sensor breakdown/migration in vivo, the clinical utility for a specific indication, and biocompatibility aspects related to the materials and the telemetry/communication process required to obtain results.

3.3 POLYMERIC NANOMEDICINES AS DRUG DELIVERY AGENTS

The most common reasons to develop nanomedicine-based therapies rather than traditional small-molecule drugs include improved circulation time in the blood and/or improved targeting efficiency for a specific organ or tumor, either of which can lead to reduced toxicity and/or side-effects. Frequently, the nanomedicine consists of an active drug compound that is integrated into a nanoparticle to achieve these effects; however, there are also a few examples of nanoparticles that are therapeutic agents in and of themselves.

For example, Copaxone® (glatiramer acetate) was initially approved in 1996 as an immunomodulator for the treatment of multiple sclerosis. It is composed of a random copolymer of L-glutamic acid, L-alanine, L-tyrosine, and L-lysine, where the mechanism of action does not involve any other agents in combination. The following section describes a range of polymeric constructs designed for therapeutic applications, as these are the most popular materials to date across the spectrum of approved, in-trial, and research-based activities in the field.

The most basic class of polymeric nanomedicines involves simple singular polymer chains, either directly as the therapeutic or bound to a drug to alter its biodistribution, efficacy, and/or toxicity profile. For clinicians and patients, this can result in reduced drug side-effects, the use of higher doses for enhanced therapeutic effects, and/or reduced frequency of drug administration. A key challenge that has been encountered in some trials and approved products is the generation of immune responses to the polymer-drug conjugate, which can reduce therapeutic efficacy over time. The most common polymer employed is polyethylene glycol (PEG), which is a neutral, hydrophilic polymer that enhances circulation time and delays clearance via the liver and reticuloendothelial system.

Examples of FDA-approved PEGylated drugs include Neulasta® (PEGylated granulocyte colony stimulating factor; for chemotherapy-induced neutropenia) (Benbrook, 2015), PLEGRIDY® (PEGylated IFN beta-1a; for relapsing multiple sclerosis) (Hu et al., 2012), and ADYNOVATE (PEGylated anti-hemophilic factor VIII; for the prevention/treatment of bleeding episodes in patients with hemophilia A) (Ing et al., 2016). In all cases, PEGylation was found to significantly prolong the circulation time of the original drug. Hydrophilic polymers other than PEG have also been successfully employed; for example, poly(glutamic acid) has been conjugated to paclitaxel and is currently being investigated as an alternative to paclitaxel in the context of non-small cell lung cancer (NSCLC) (Paz-Ares et al., 2008). This illustrates that while PEGylated polymers have been most extensively utilized in nanomedicine, there are other available polymers that might facilitate the emergence of new properties for nanomedicine applications.

Beyond polymer-drug conjugates, more complex polymeric systems have been designed which allow an element of control over the delivery of drugs in vivo. Indeed, polymerization strategies provide avenues for the synthesis of an almost unimaginable array of nanomaterials, based on the unprecedented levels of control that can now be achieved over the composition, 3D architectures, and purity of the desired products. These can range from simple polymer chains, to block copolymers that aggregate in a controlled manner, to complex multifunctional polymeric nanoparticles. By varying the degree of hydrophobicity of a polymeric nanoparticle, both the encapsulation process and controlled release of a therapeutic agent can be designed.

A good example of this is Eligard®, which is a testosterone inhibitor used for prostate cancer treatment, which comprises the drug leuprolide encapsulated within a polylactide-co-glycolic acid (PLGA) nanoparticle (Berges, 2005). PLGA is a biodegradable material that temporally decomposes into non-toxic monomers, which is composition dependent. Another interesting case study is that of CRLX101, which is a DNA topoisomerase I inhibitor used in cancer treatment that consists of the drug camptothecin encapsulated within a cyclodextrin-PEG copolymer nanoparticle (Gritli et al., 2016).

In its free form, camptothecin is quite insoluble, toxic, and has poor in vivo stability; however, in nanoparticle form it is ~1000 fold more soluble, with impressive anti-tumor effects over a broad range of animal studies (Conley et al., 2015; Lin et al., 2016; Pham et al., 2016; Svenson et al., 2011). Further to these promising results in animal studies, human trial results have shown encouraging tolerance of the nanomedicine. However, enhanced efficacy in comparison to controls, either in isolation or in combination with other drugs, has not yet been demonstrated (Chao et al., 2017; Clark et al., 2016; Gritli et al., 2016; Tian et al., 2017; Voss et al., 2017). Polymeric liposomes likely represent the most evaluated nanomedicines to date, with many examples of FDA-approved formulations (Figure 3.2).

These nanoparticles, along with the closely related micelles, are hierarchical in structure; hence, they represent a more complex class of nanomedicines in contrast to the polymer-drug conjugates or single-polymer nanoparticles outlined above; however, they are also relatively simple to synthesize. Micelles consists of self-assembling polymeric amphiphiles that are utilized for the controlled delivery of hydrophobic drugs, where the polymer selection largely defines the final nanoparticle size and shape. The internal core is a hydrophobic cavity that can be used to encapsulate poorly water-soluble drugs, while the external surface is polar and water-soluble.

Currently, Estradiol (Estrasorb™) is the only FDA-approved micelle which is used as a topical treatment for the moderate to severe symptoms of menopause. Transdermal delivery avoids first pass metabolism and gastrointestinal side-effects, which can destabilize serum levels for

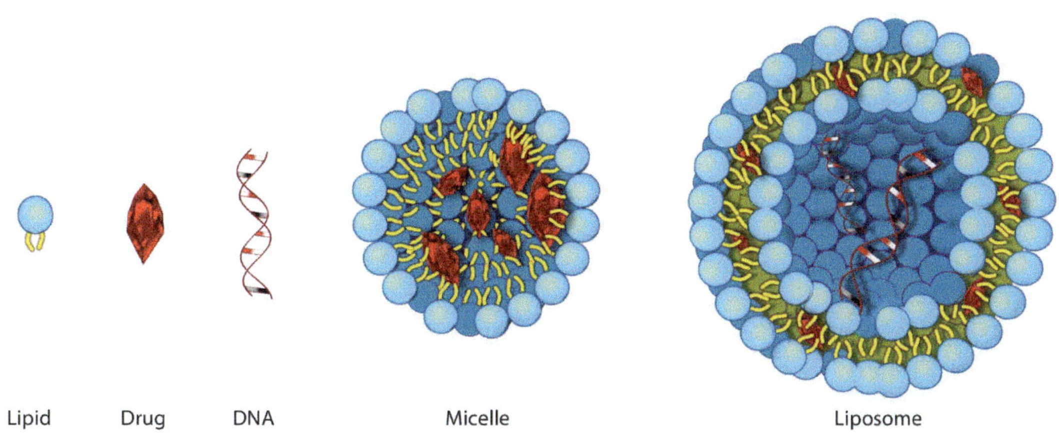

FIGURE 3.2 Polymeric micelles and liposomes. These nanoparticles are hierarchical in structure, from the phospholipid to form a micelle or a liposome. The final architecture is designed for the type of drug to be delivered. Hydrophobic drugs may be encapsulated within the hydrophobic cavity of the micelle. Liposomes can encapsulate both hydrophobic and hydrophilic drugs in the bilayer or core, respectively. (Image created by Julia Walker.)

1–2 weeks (Taylor and Gercel-Taylor, 2008). In contrast, liposomes consist of a phospholipid bilayer that envelops an aqueous core. Since their initial emergence as the first FDA-approved nanomedicine (e.g., liposomal formulations of doxorubicin, amphotericin B), there have been a range of improvements for the synthesis of this class of nanoparticles, which enable the delivery of both hydrophobic and hydrophilic compounds. These are arguably the most easily synthesized class of nanoparticles that can be used to create new potential combinations to improve therapeutic delivery, and there are many clinical trials underway across several disease indications (Allen and Cullis, 2013).

Protein-based nanoparticles were originally designed to replicate the properties of naturally circulating blood proteins, to facilitate the dissolution and transport processes of drugs across biological barriers. Abraxane® is the best example to date, which consists of albumin-bound paclitaxel that eliminated the requirement of a toxic solvent used to administer the free drug (cremophor) (Green et al., 2006). Initially approved by the FDA in 2005 for the treatment of breast cancer following the unsuccessful combination treatment of relapse, Abraxane® has also shown enhanced pharmacokinetics and tumor inhibition when compared to the Cremophor-based therapy, due to enhanced endothelial binding and transcytosis of the nanoparticle (Desai et al., 2006). Several other albumin-bound nanoparticles (NABs) have been entered into clinical trials, which seek to improve the therapeutic efficacy of other drugs, suggesting that this may be a nanomedical platform technology. Current examples include nab-docetaxel, nab-rapamycin, and the nab-heat shock protein inhibitor.

3.4 PHOTOTHERMAL AND MAGNETOTHERMAL TREATMENTS

Photothermal treatments are designed to generate highly localized hyperthermia at a specific site to eradicate diseased cells while minimizing effects on healthy cells. Photothermal agents (PTAs) are molecules that efficiently transform light energy into heat, where metallic nanoparticles are uniquely suited to this process. These materials need to show high light absorbance in the near-infrared range (NIR is the biological imaging window in which absorption/scattering by water, proteins, and biological tissues is minimal), high photothermal conversion efficiency, high photostability, tunable size/shape range, good biocompatibility, and low toxicity.

To date, the most employed materials are metallic nanoparticles that include gold and iron oxide derivatives; however, carbon nanomaterials (e.g., carbon nanotubes, graphene) and polymers (e.g., dye-containing micelles, conjugated polymers, etc.) are also of great interest. While metallic nanoparticles possess high photothermal conversion efficiency and stability, their biocompatibility and toxicity profiles are less promising. However, clinical trials have been undertaken that involve gold/silica nanoparticles, which show promising treatment responses in animal models for cancer (Day et al., 2011) and atherosclerosis (Kharlamov et al., 2015).

The coating of metal nanoparticles with silica is a common approach to reduce toxicity while improving biodistribution profiles. A key limitation with photothermal treatments is that light can be attenuated or blocked by certain biological structures. For example, treatments applied to the brain (or deep tissues) require the invasive use of fiber optic sources as the skull dampens the penetration of NIR light.

In contrast to photothermal treatments, the conversion of externally applied electromagnetic energy into heat provides unique advantages in terms of tissue depth and spatiotemporal control. As there is no attenuation of the electromagnetic energy by biological tissues, internal structures can be targeted at sites where light cannot reach. Furthermore, based on the design of the magnetically active nanomaterials, the spatial control of the magnetothermal effect is at the nanoscale. Studies have been undertaken that focused on combination therapies, including chemotherapeutic agents that are more effective under heating effects, localized magnetically triggered drug delivery, and the remote targeting of cells to induce gene expression or other metabolic activities (Noh et al., 2017). To date, these investigations have proceeded with animal models; however, clinical trials will likely ensue in the near future.

3.5 NANOMATERIALS FOR IMAGING AND BIOSENSING

Biomedical imaging is a cornerstone of clinical diagnostics, which enables clinicians and patients to "see" evidence of a disease process in vivo, in the context of the surrounding anatomy and normal physiological processes. Moving beyond X-ray technologies (which enable the detailed observation of highly dense structures – primarily bones), techniques based on magnetic resonance imaging (MRI), positron emission tomography (PET), ultrasound (US), and near-infrared (NIR) fluorescence imaging have opened avenues to extremely high-resolution imaging (e.g., spatial, temporal, or molecular resolution), which allow abnormalities to be identified significantly earlier than they would otherwise be.

Each modality has its advantages and disadvantages. NIR fluorescence imaging offers the highest resolution and sensitivity (defined here as the highest degree of contrast enhancement for minimal nanomaterial concentration); however, NIR light cannot penetrate very far into biological tissues; thus, it is presently limited to the imaging of subcutaneous tissues or for image-guided surgery. US imaging offers the high temporal resolution of dynamic processes (e.g., blood flow, guiding the insertion of catheters) and can be applied to deep tissues; however, it does not have particularly high spatial resolution. Nevertheless, a key advantage of US imaging is that it is widely available and easy to use in hospitals and small medical clinics – much more so than PET and MR imaging systems.

PET/MRI modalities offer the high spatiotemporal resolution of tissues, with PET offering sufficient sensitivity to detect very small tumors in situ. A limitation of PET imaging is that radioisotopes must be administered to facilitate the imaging process; however, this is now standard practice in radiation oncology. MRI can be performed without the administration of any such agents to obtain detailed anatomical images; however, contrast agents are required to extract molecular level data. The role of nanomaterials across the spectrum of imaging modalities is to provide "contrast enhancement", or in other words, to make the differences between normal/abnormal tissues significantly more obvious.

In the area of MRI, extensive work has been undertaken to develop imaging and drug-delivery approaches that rely on iron oxide nanoparticles or superparamagnetic iron oxide nanoparticles (SPION). This class of nanoparticles can be precisely controlled and induced to perform highly precise hyperthermia for the in vivo thermal ablation of tissues and cells (Boehm and Chen, 2009); there have been several clinical trials undertaken and products approved. The aim of using these materials in MRI is to provide contrast enhancement in metabolically active tissues, which may highlight and delineate abnormal structures such as tumors.

Despite EU authorization of several iron oxide nanoparticles, only a few have received FDA approval to date (e.g., Feraheme® for the treatment of iron deficiency anemia, and Feridex® and GastroMARK™ as MRI contrast agents); two of which were later withdrawn from the market. Issues that arose ranged from the lack of clinical uptake to the small yet significant frequency of adverse reactions including lumbar pain and hypotension. However, it is anticipated that because of the trials and tribulations of the original group of iron oxide-based MRI contrast agents, future clinical approvals will be relatively straightforward to achieve, as they are now relatively well-understood.

Radioisotopes for PET imaging (and related methods) have long been employed for cancer diagnostics and imaging; however, nanoparticles can have a dramatic effect by increasing signal:noise background, while altering the biodistribution profiles of key compounds. The key advantage of radioisotope nuclear imaging is the sensitivity gained in comparison to other methods, as only small amounts of contrast agent are required to generate a sufficient signal. Through the incorporation of radioisotopes into nanomaterials, significant payloads can be delivered to sites of interest (either through passive or active targeting), to further reduce the dose required.

Simultaneously, since only minimal quantities of imaging molecules are required to generate a signal, there is ample space within high surface area-to-volume ratio nanomaterials to incorporate other functional molecules for additional therapeutic and diagnostic applications. Traditionally, radioisotopes are immobilized into/onto nanomaterials via chelation chemistry; however, a range of "chelation-free" approaches are being developed, along with the direct bombardment of nanomaterials with radioactive atomic precursors. In the development of these ultrasensitive imaging approaches, the issue of non-specific nanoparticle adsorption is important, as it can lead to false-positive clinical results.

Consequently, even for scientifically valid imaging studies, there must be a demonstrable benefit for the patient (e.g., actionable result leading to improved long-term survival) to obtain clinical approval for use. Bradbury and Wiesner completed the first-in-human clinical trials using radiolabeled nanomaterials (Phillips et al., 2014).They employed organosilica nanoparticles that were designed as particle tracers, but also exhibited some degree of tumor uptake in metastatic lesions (although optimizing the dose for tumor imaging was not a design criterion for this study).

The nanoparticles consisted of an ultra-bright core particle coated with PEG and cRGDY-I^{124} integrin-binding peptides. The Phase 1 study, which investigated the pharmacokinetics, metabolic profiles, blood/urine tests, and serial PET scans, found that the nanoparticles were well-tolerated with no adverse toxic effects. Given that these nanoparticles could be decorated with all manner of targeting ligands, fluorescent dyes, and drug compounds, this study paved the way for future trials and products using this promising nanoparticle scaffold (Phillips et al., 2014).

Ultrasound is a common and popular imaging technique due to its low cost, ability to non-invasively image deep tissues, and the wide availability of portable devices. To observe contrasts in US images, there must be acoustic impedance at material interfaces; thus, contrast agents have been developed to exploit this phenomenon. The combination of US and nanomaterials is not new, as commercially available gas-filled "nanobubbles" have been employed for many years as contrast enhancers for vascular imaging. For short-term imaging (minutes), this approach works very well, as the nanobubbles naturally collapse and the gases are released via normal respiratory processes.

However, new advances are being sought for longer-term contrast agents, as well as drug delivery agents that can be "triggered" to release drugs in response to the application of US. An array of silica-based nanomaterials has emerged to address these challenges, ranging from hollow shells to multi-layered solid spheres, all of which show longer US lifetimes (Ma et al., 2015). There are also examples of silica nanoparticles that can be filled with drugs and designed to burst in response to US exposure, enabling the spatiotemporal control of drug release. Hall and colleagues fabricated a range of US-responsive materials; initially creating organosilica shells with mechanical properties suited to US-induced bursting, and secondly through the design of a nanomaterial that increased its US contrast in response to lipase-induced degradation (Lin et al., 2009; Tsao and Hall, 2016).

Fluorescence imaging is a further example of a relatively common technique in vascular imaging, which is being revolutionized through the integration of bright and stable nanoparticles. While there are several FDA-approved molecular fluorophores for fluoroscopy applications, nanoparticles loaded with NIR dyes can be targeted to specific sites, in the same way that nanomedicines can be targeted via the control of size, shape, surface chemistry, etc. A key limitation of fluorescence imaging is the relatively shallow penetration depth of light into tissue, which limits applications to skin imaging, the intraoperative imaging of underlying tissues, or the use of invasive fiber optic probes.

However, the Phase 1 clinical study of Bradbury et al. (Phillips et al., 2014), which involved C-dots, was also in fact an example of a fluorescence imaging contrast agent being translated for clinical use. The nanomaterial was not only modified with a radiolabeled peptide but also contained a bright, near-infrared dye (Cy5). In pre-clinical studies, this group also demonstrated the use of C-dots for sentinel lymph node (SLN) mapping in breast cancer models, which paved the way to biopsy-free SLN procedures in humans (Benezra et al., 2011).

Fluorescence-guided surgery is also currently being utilized to assist surgeons with distinguishing between normal and diseased tissues (e.g., identification of tumor margins), to remove all damaged tissue while minimizing trauma to healthy tissue, blood vessels, nerves, etc. For example, 5-aminolevulinic acid (Gliolan®) is a small molecule that is orally administered and then converted into protoporphyrin IX, which preferentially accumulates in glioma cells compared with normal tissues. Nevertheless, this fluorophore suffers from issues that include photobleaching and poor penetration depth as it emits outside of the NIR window. However, new studies have shown improvements in the technique through the use of nanoparticle contrast agents involving enhanced fluorescence in the NIR region and also in combination with other imaging modalities (Kircher et al., 2012; Ni et al., 2014). The development of fluorescent nanoparticles with NIR fluorescence emission is an active area of research toward the development of novel imaging agents with high contrast properties.

3.6 BIOSENSORS

While the application of nanomedicine does not often encompass biosensors, these devices play increasingly important roles in clinical practice. The key factor that dictates the utility of a biosensor is whether it can provide useful data that guides treatment decisions within a reasonable time, typically within an hour. Furthermore, the continuous monitoring of health-related biomarkers is becoming more popular, which is leading to the invention of wearable and in vivo biosensor techniques. Aside from their clinical utility, these devices may also give patients more control over their medical data, as it is directly generated by the patient following sample collection, rather than in a hospital or commercial laboratory.

Nanoparticles and nanomaterials are playing increasing roles in next generation IVD, particularly due to their ability to improve assay detection limits (down to single-molecule sensitivity) combined with dramatic increases in multiplexing potential. The SIMOA technology by Quanterix (Cohen and Walt, 2017) is perhaps the most advanced example, in which an array of nano-wells etched into a fiber optic cable is filled with individual nanoparticles for analysis. The nanoparticle size is selected such that each well will only accommodate a single nanoparticle. The capture/detection reactions are performed on the nanoparticles in such a way as to statistically limit the number of events per nanoparticle to the single-molecule level. Thus, one can simply count the number of "positive" wells based on secondary labels (typically fluorescence microscopy) to determine the number of captured molecules in the experiment. Such techniques are potentially calibration-free in the same way that gravimetric (mass counting) or coulometric (electron counting) approaches do not require calibration.

3.7 WEARABLE AND IN VIVO DEVICES

Another emerging area in biosensing is the generation of technologies that allow for the continuous monitoring of health-related biomarkers from on or within the body (Figure 3.3). Logic dictates that continuous monitoring is inherently an in situ technique; hence, the transducing surface and related chemistry must be in constant contact with the analyte solution. In the wearables space, several consumer health technologies have been developed (e.g., for sport and fitness-related applications), while clinical applications such as for diabetes are emerging.

The Wang Laboratory at the University of California, San Diego (UCSD) has developed a range of tattoos, mouthguards, and contact lenses that monitor an assortment of small molecules including glucose and lactate. An example of a continuous sweat "tattoo" biosensor was developed by Jia et al. (2013) for the measurement of exercise-induced lactate (Figure 3.3). The device consists of a screen-printed electrode on a flexible substrate, with lactate oxidase immobilized onto the working electrode with multi-walled carbon nanotubes acting as the transducer surface, and tetrathiafulvalene (TTF) added to enhance the low voltage electrocatalytic conversion of lactate.

In vitro testing, with the sensor attached to both rigid and flexible substrates, revealed that the amperometric response was stable to repeated mechanical bending and was unaffected by the presence of physiological concentrations of other metabolites (e.g., creatinine, ascorbic acid, glucose, and uric acid), with a linear response rate for lactate of 1–20 mM (covering the lactate reference range) (Burtis et al., 2006). Epidermal testing was also performed over a ~30 min period of exercise,

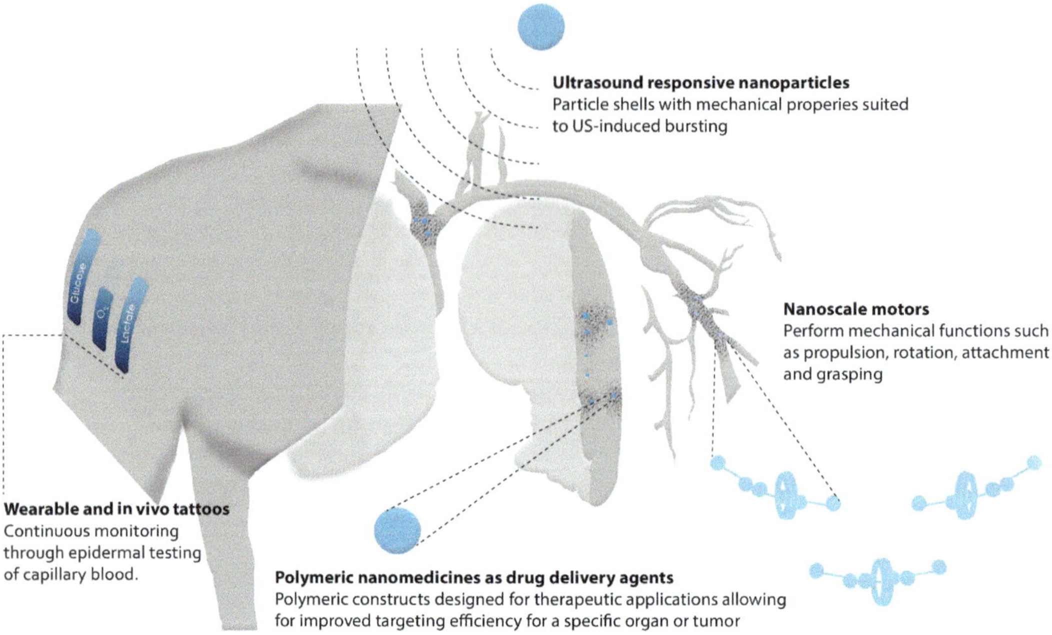

FIGURE 3.3 Nanomaterials for imaging, biosensing, drug delivery, theranostics, and nanosurgery. Ultrasound responsive nanoparticles have opened avenues for high-resolution imaging (spatial, temporal, or molecular resolution). Biomimetic nanoscale motors show potential for medical applications such as the transport of molecules, selective biosensing, and nanosurgery. Polymeric nanomedicines as drug delivery agents can be used for retention in specific organs. This allows for longer in vivo circulation times compared to small-molecule drugs. Wearable and in vivo tattoos for glucose, lactate, pH, and other small drug candidates and analytes enable the real-time continuous monitoring of analyte concentrations. (Image created by Julia Walker.)

with excellent agreement to laboratory testing. The in situ colorimetric analysis of pH and metal ions has also been demonstrated for sweat (Yu and Sun, 2020).

Such devices might be extremely useful for a range of non-invasive applications, particularly if advances are made that facilitate sensitive protein detection. However, this approach is limited as sufficient electrolyte fluid (sweat) must be in contact with the sensor for the generation of the amperometric signal; thus, it currently relies on active sweating. While wearable sensors are minimally invasive and convenient for users, the key challenge is that blood is generally considered the most valuable biological sample, and skin is an effective barrier that does not readily allow for mass transport. An alternative is to inject "nanosensors" into the body for the continuous monitoring of health biomarkers in vivo.

Clark and co-workers have developed a micellar nanoparticle platform to generate optical nanosensors that can detect each of the analytes in a standard hospital blood panel (Awqatty et al., 2014). They have extended this platform for the monitoring of drug levels as well (e.g., lithium) (Cash et al., 2015), which opens the door to new applications for therapeutic drug monitoring. Plaxco's group coated electrode surfaces with short DNA molecules that bind specific antibiotics and chemotherapy drugs (aptamers), which have demonstrated the real-time monitoring of these drugs in animal models over many hours (Arroyo-Currás et al., 2017; Ferguson et al., 2013).

3.8 BLUE SKY NANOMEDICINE

There are unlimited opportunities for the application of nanostructures in biomedicine (Figure 3.3).

3.9 NANOMATERIALS-BASED THERANOSTICS

In conjunction with the emergence of nanomaterials in both therapeutic and diagnostic (imaging and sensing) areas, there are also concerted efforts toward the development of "theranostic" nanomaterials that can be used for both imaging and therapy in vivo. Although there are a number of exciting examples of products in pre-clinical trials, none are yet in clinical trials (Mura and Couvreur, 2012). Examples of theranostic agent development can be seen across the spectrum of material categories highlighted previously. In most cases, the primary goal of imaging techniques is to monitor the distribution of nanomedicines within the body, such that at an optimal post-injection time (e.g., when the tumor:background ratio is highest), a triggered and controlled drug delivery process can be initiated to achieve an optimal result.

This requires the nanomaterials to be multifunctional, combining the incorporation of an imaging agent (e.g., a "contrast agent" for MR, PET, US, or NIR-fluorescence imaging), a therapeutic agent, and a controlled release chemistry. Further, these nanomaterials must meet stringent safety and biocompatibility requirements. These nanomaterials are also likely to require tissue-specific targeting ligands to be integrated within their structures (e.g., antibodies, peptides, etc.). Although researchers have developed numerous "active targeting" approaches over the last decade, the degree of success to date remains debatable (Wilhelm et al., 2016).

3.10 NANOMACHINES AND NANOSURGERY

Nanomachines are nanoscale motors that perform mechanical functions such as propulsion, rotation, attachment, and grasping (tweezer effect). Their designs have often been inspired by the myriad micro- and nanoscale devices observed in nature. There will be myriad potential medical applications for nanomachines in the future, including transport molecules, drug delivery, biosensing nucleic acids and proteins, cell sorting, micromanipulation, and nanosurgery (Wang, 2009). Future nanosurgeries may extend to the clearance of atheromatous plaques from arteries, single cell biopsy, and tissue repair or reconstruction. Nanomachines might be self-propelled or externally activated and propelled by ultrasonic, electrical, or magnetic energy. Some of the challenges toward the development of future nanomachines may include the achievement of specific trajectories in vivo, prolonged activities within the human body, and wireless bidirectional communications. The ultimate challenge will be to construct complex nanomachines that integrate a power module, an engine, a controller ("brain"), and sensor and effector components. An extreme example of minimally invasive surgery might be manifest via the deployment of swarms (perhaps in the billions) of precisely controlled nanorobots.

3.11 NANOMEDICAL PAIN MANAGEMENT

There is a great demand for more effective acute and chronic pain management. Nanomedicine might provide potential solutions in the form of monitoring systems using nanosensors for the point-of-care detection of pain biomarkers or drug levels. Precise and sustained drug delivery and gene therapies using dendrimer particles are being investigated. These advances are aimed at personalized care at the cellular level (Chakravarthy et al., 2018).

3.12 BRAIN–COMPUTER INTERFACE

Various classes of brain–computer interfaces (BCIs) are rapidly being developed. Electrodes are placed in contact with neurons in the brain (or axons in the peripheral nerves) and used for recording and stimulation. Electrode arrays are connected to a computer interface for the processing of the recorded data or control of stimulation. BCIs have been developed for the control of robotic and prosthetic arms in paralyzed patients, and more recently, to reactivate the paralyzed arm itself. BCIs are also being developed for the restoration of vision and memory (Rosenfeld and Wong, 2017).

Microelectrodes have the capacity to directly stimulate thousands of adjacent neurons, whereas nanoelectrodes will have the potential to increase the selectivity of stimulation to single neurons or several adjacent neurons. Thousands, millions, or billions of nanobots interfacing directly with neurons could potentially provide exponentially more detailed recording and stimulation than is currently available. This would open the possibility of connecting the human brain to the "cloud" for more direct and complex intercommunications (Martins et al., 2019). Wireless communications between nanorobots and a more superficial intermediary receptor interface would likely be required. There are clearly myriad significant engineering challenges to resolve; however, researchers are already working on these problems.

Flexible electronics offers a means of conforming electrodes to non-planar surfaces, and recently, a microporous mesh has been syringe injected into a rodent brain. The mesh unfolded within the brain matter or ventricle and provides a way of delivering flexible electronics. Populating the mesh with nanoelectrodes would allow multiplexed neural recording in vivo and enable the potential for wireless interfaces (Liu et al., 2015).

Preliminary research in mice has indicated that magnetoelectric nanoparticles may couple intrinsic neural activity–induced electric fields with external magnetic fields. These nanoparticles cross the blood–brain barrier (BBB) through the application of a magnetic gradient. This technology may allow for the control of intrinsic fields deep within the brain. This opens the possibility of treatments for Parkinson's disease and a host of other neurodegenerative disorders (Guduru et al., 2015).

REFERENCES

Albanese A, Tang PS, Chan WCW. The effect of nanoparticle size, shape, and surface chemistry on biological systems. *Annual Review of Biomedical Engineering.* 2012;14:1–16.

Allen TM, Cullis PR. Liposomal drug delivery systems: From concept to clinical applications. *Advanced Drug Delivery Reviews.* 2013;65(1):36–48.

Arroyo-Currás N, Somerson J, Vieir PA, Ploense KL, Kippin TE, Plaxco KW. Real-time measurement of small molecules directly in awake, ambulatory animals. *Proceedings of the National Academy of Sciences of the United States of America.* 2017;114(4):645–650.

Awqatty B, Samaddar S, Cash KJ, Clark HA, Dubach JM. Fluorescent sensors for the basic metabolic panel enable measurement with a smart phone device over the physiological range. *Analyst.* 2014;139(20):5230–5238.

Benbrook DM. Biotechnology and biopharmaceuticals: Transforming proteins and genes into drugs, 2nd edition. *Clin Infect Dis.* 2015;60(2):331–332.

Benezra M, Penate-Medina O, Zanzonico PB, Schaer D, Ow H, Burns A et al. Multimodal silica nanoparticles are effective cancer-targeted probes in a model of human melanoma. *Journal of Clinical Investigation.* 2011;121(7):2768–2780.

Berges R. Eligard®: Pharmacokinetics, effect on testosterone and PSA levels and tolerability. *European Urology Supplements.* 2005;4(5):20–25.

Boehm FJ, Chen A. Medical applications of hyperthermia based on magnetic nanoparticles. *Recent Patents on Biomedical Engineering.* Bentham Science Publishers. 2009;2(2):110–120(11).

Burtis CA, Ashwood ER, Burns DE. *Tietz Textbook of Clinical Chemistry and Molecular Diagnostics.* 4th ed. St Louis, Missouri, USA: Elsevier Saunders, 2006.

Cash KJ, Li C, Xia J, Wang LV, Clark HA. Optical drug monitoring: Photoacoustic imaging of nanosensors to monitor therapeutic lithium in vivo. *ACS Nano.* 2015;9(2):1692–1698.

Chakravarthy KV, Boehm FJ, Christo PJ. Nanotechnology: A promising new paradigm for the control of pain. *Pain Medicine.* 2018;19(2):232–243.

Chao J, Lin J, Frankel P, Clark AJ, Wiley DT, Garmey E, et al. Pilot trial of CRLX101 in patients with advanced, chemotherapy-refractory gastroesophageal cancer. *J Gastrointest Oncol.* 2017 Dec;8(6):962–969.

Chaudhuri RG, Paria S. Core/shell nanoparticles: Classes, properties, synthesis mechanisms, characterization, and applications. *Chemical Reviews.* 2012;112(4):2373–2433.

Choi HS, Liu W, Misra P, Tanaka E, Zimmer JP, Ipe BI et al. Renal clearance of quantum dots. *Nature Biotechnology.* 2007;25(10):1165–1170.

Cimalla P, Werner T, Gaertner M, Mueller C, Walther J, Wittig D, et al. Magnetomotive imaging of iron oxide nanoparticles as cellular contrast agents for optical coherence tomography. *Optical Coherence Tomography and Coherence Techniques VI*, SPIE Proceedings Optical Society of America, Bouma B and Leitgeb R, eds., 2013, Vol. 8802, paper 88020M.

Clark AJ, Wiley DT, Zuckerman JE, Webster P, Chao J, Lin J et al. CRLX101 nanoparticles localize in human tumors and not in adjacent, nonneoplastic tissue after intravenous dosing. *Proceedings of the National Academy of Sciences of the United States of America*. 2016;113(14):3850–3854.

Cohen L, Walt DR. Single-molecule arrays for protein and nucleic acid analysis. *Annual Review of Analytical Chemistry*. 2017;10:345–363.

Conley SJ, Baker TL, Burnett JP, Theisen R, Lazarus D, Peters CG et al. CRLX101, an investigational camptothecin-containing nanoparticle-drug conjugate, targets cancer stem cells and impedes resistance to antiangiogenic therapy in mouse models of breast cancer. *Breast Cancer Research and Treatment*. 2015;150(3):559–567.

Day ES, Thompson PA, Zhang LN, Lewinski NA, Ahmed N, Drezek RA et al. Nanoshell-mediated photothermal therapy improves survival in a murine glioma model. *Journal of Neuro-Oncology*. 2011;104(1):55–63.

Desai N, Trieu V, Yao ZW, Louie L, Ci S, Yang A et al. Increased antitumor activity, intratumor paclitaxel concentrations, and endothelial cell transport of Cremophor-free, albumin-bound paclitaxel, ABI-007, compared with Cremophor-based paclitaxel. *Clinical Cancer Research*. 2006;12(4):1317–1324.

Dreaden EC, Alkilany AM, Huang X, Murphy CJ, El-Sayed MA. The golden age: Gold nanoparticles for biomedicine. *Chemical Society Reviews*. 2012;41(7):2740–2779.

Eifler AC, Thaxton CS. Nanoparticle therapeutics: FDA approval, clinical trials, regulatory pathways, and case study. *Methods Mol Biol*. 2011;726:325–338.

Elsabahy M, Wooley KL. Design of polymeric nanoparticles for biomedical delivery applications. *Chemical Society Reviews*. 2012;41(7):2545–2561.

Ferguson BS, Hoggarth DA, Maliniak D, Ploense K, White RJ, Woodward N et al. Real-time, aptamer-based tracking of circulating therapeutic agents in living animals. *Science Translational Medicine*. 2013;5(213):213ra165.

Fox ME, Szoka FC, Frechet JMJ. Soluble polymer carriers for the treatment of cancer: The importance of molecular architecture. *Accounts of Chemical Research*. 2009;42(8):1141–1151.

Green MR, Manikhas GM, Orlov S, Afanasyev B, Makhson AM, Bhar P, Hawkins MJ. Abraxane, a novel Cremophor-free, albumin-bound particle form of paclitaxel for the treatment of advanced non-small-cell lung cancer. *Ann Oncol*. 2006 Aug;17(8):1263–1268.

Gritli I, Garmey E, Eliasof S, Tellez A, Davis ME, Yun Y. Polymeric nanoparticles and cancer: lessons learnt from CRLX101. In: Braddock M, editor. *Nanomedicines: Design, Delivery and Detection*. Cambridge: Royal Society of Chemistry; 2016. p. 199–232.

Guduru R, Liang P, Hong J, Rodzinski A, Hadjikhani A, Horstmyer J et al. Magnetoelectric 'spin' on stimulating the brain. *Nanomedicine*. 2015;10(13):2051–2061.

Hu X, Miller L, Richman S, Hitchman S, Glick G, Liu SF et al. A novel PEGylated interferon beta-1a for multiple sclerosis: Safety, pharmacology, and biology. *Journal of Clinical Pharmacology*. 2012;52(6):798–808.

Ing M, Gupta N, Teyssandier M, Maillère B, Pallardy M, Delignat S, Lacroix-Desmazes S; ABIRISK consortium. Immunogenicity of long-lasting recombinant factor VIII products. *Cell Immunol*. 2016 Mar;301:40–48.

Jia W, Bandodkar AJ, Valdés-Ramírez G, Windmiller JR, Yang Z, Ramírez J et al. Electrochemical tattoo biosensors for real-time noninvasive lactate monitoring in human perspiration. *Analytical Chemistry*. 2013;85(14):6553–6560.

Kamaly N, Xiao Z, Valencia PM, Radovic-Moreno AF, Farokhzad OC. Targeted polymeric therapeutic nanoparticles: Design, development and clinical translation. *Chemical Society Reviews*. 2012;41(7):2971–3010.

Kharlamov AN, Tyurnina AE, Veselova VS, Kovtun OP, Shur VY, Gabinsky JL. Silica-gold nanoparticles for atheroprotective management of plaques: Results of the NANOM-FIM trial. *Nanoscale*. 2015;7(17):8003–8015.

Kircher MF, de la Zerda A, Jokerst JV, Zavaleta CL, Kempen PJ, Mittra E et al. A brain tumor molecular imaging strategy using a new triple-modality MRI-photoacoustic-Raman nanoparticle. *Nature Medicine*. 2012;18(5):829–834.

Lin CJ, Lin YL, Luh F, Yen Y, Chen RM. Preclinical effects of CRLX101, an investigational camptothecin-containing nanoparticle drug conjugate, on treating glioblastoma multiforme via apoptosis and antiangiogenesis. *Oncotarget*. 2016;7(27):42408–42421.

Lin PL, Eckersley RJ, Hall EAH. Ultrabubble: A laminated ultrasound contrast agent with narrow size range. *Advanced Materials*. 2009;21(38–39):3949–3952.

Liu J, Fu TM, Cheng ZG, Hong GS, Zhou T, Jin LH et al. Syringe-injectable electronics. *Nature Nanotechnology*. 2015;10(7):629–636.

Ma M, Chen HR, Shi JL. Construction of smart inorganic nanoparticle-based ultrasound contrast agents and their biomedical applications. *Science Bulletin*. 2015;60(13):1170–1183.

Martins NRB, Angelica A, Chakravarthy K, Svidinenko Y, Boehm FJ, Opris I, Lebedev MA, Swan M, Garan SA, Rosenfeld JV, Hogg T, Freitas RA Jr. Human brain/cloud interface. *Frontiers in Neuroscience*. 2019;13:112.

Mura S, Couvreur P. Nanotheranostics for personalized medicine. *Advanced Drug Delivery Reviews*. 2012;64(13):1394–1416.

Nel AE, Maedler L, Velegol D, Xia T, Hoek EMV, Somasundaran P, Klaessig F, Castranova V, Thompson M. Understanding biophysicochemical interactions at the nano-bio interface. *Nature Materials*. 2009;8(7):543–557.

Ni DL, Zhang JW, Bu WB, Xing HY, Han F, Xiao QF et al. Dual-targeting upconversion nanoprobes across the blood-brain barrier for magnetic resonance/fluorescence imaging of intracranial glioblastoma. *ACS Nano*. 2014;8(2):1231–1242.

Noh SH, Moon SH, Shin TH, Lim Y, Cheon J. Recent advances of magneto-thermal capabilities of nanoparticles: From design principles to biomedical applications. *Nano Today*. 2017;13:61–76.

Paz-Ares L, Ross H, O'Brien M, Riviere A, Gatzemeier U, Von Pawel J et al. Phase III trial comparing paclitaxel poliglumex vs docetaxel in the second-line treatment of non-small-cell lung cancer. *British Journal of Cancer*. 2008;98(10):1608–1613.

Pham E, Yin M, Peters CG, Lee CR, Brown D, Xu P et al. Preclinical efficacy of bevacizumab with CRLX101, an investigational nanoparticle-drug conjugate, in treatment of metastatic triple-negative breast cancer. *Cancer Research*. 2016;76(15):4493–4503.

Phillips E, Penate-Medina O, Zanzonico PB, Carvajal RD, Mohan P, Ye YP et al. Clinical translation of an ultrasmall inorganic optical-PET imaging nanoparticle probe. *Science Translational Medicine*. 2014;6(260):260ra149.

Rolfe BE, Blakey I, Squires O, Peng H, Boase NRB, Alexander C, et al. Multimodal polymer nanoparticles with combined F-19 magnetic resonance and optical detection for tunable, targeted, multimodal imaging in vivo. *Journal of the American Chemical Society* 2014;136(6):2413–2419.

Rosenfeld JV, Wong YT. Neurobionics and the brain-computer interface: Current applications and future horizons. *Medical Journal of Australia*. 2017;206(8):363–368.

Sadauskas E, Wallin H, Stoltenberg M, Vogel U, Doering P, Larsen A et al. Kupffer cells are central in the removal of nanoparticles from the organism. *Particle and Fibre Toxicology*. 2007;4:10.

Svenson S, Wolfgang M, Hwang J, Ryan J, Eliasof S. Preclinical to clinical development of the novel camptothecin nanopharmaceutical CRLX101. *Journal of Controlled Release*. 2011;153(1):49–55.

Tang F, Li L, Chen D. Mesoporous silica nanoparticles: Synthesis, biocompatibility and drug delivery. *Advanced Materials*. 2012;24(12):1504–1534.

Taylor DD, Gercel-Taylor C. MicroRNA signatures of tumor-derived exosomes as diagnostic biomarkers of ovarian cancer. *Gynecologic Oncology*. 2008;110(1):13–21.

Tenzer S, Docter D, Rosfa S, Wlodarski A, Kuharev J, Rekik A et al. Nanoparticle size is a critical physicochemical determinant of the human blood plasma corona: A comprehensive quantitative proteomic analysis. *ACS Nano*. 2011;5(9):7155–7167.

Tian X, Nguyen M, Foote HP, Caster JM, Roche KC, Peters CG et al. CRLX101, a nanoparticle-drug conjugate containing camptothecin, improves rectal cancer chemoradiotherapy by inhibiting DNA repair and HIF1 alpha. *Cancer Research*. 2017;77(1):112–122.

Tsao NH, Hall EAH. Enzyme-degradable hybrid polymer/silica microbubbles as ultrasound contrast agents. *Langmuir*. 2016;32(25):6534–6543.

Voss MH, Hussain A, Vogelzang N, Lee JL, Keam B, Rha SY et al. A randomized phase II trial of CRLX101 in combination with bevacizumab versus standard of care in patients with advanced renal cell carcinoma. *Annals of Oncology*. 2017;28(11):2754–2760.

Wang J. Can man-made nanomachines compete with nature biomotors? *ACS Nano*. 2009;3(1):4–9.

Wilhelm S, Tavares AJ, Dai Q, Ohta S, Audet J, Dvorak HF et al. Analysis of nanoparticle delivery to tumours. *Nature Reviews Materials*. 2016;1:16014.

Yu H, Sun J. Sweat detection theory and fluid driven methods: A review. *Nanotechnology and Precision Engineering*. 2020;3(3):126–140.

4 Nanotechnology

A Perspective from Quantum Biology

Adriana Marais

"We stand in the middle of infinity between outer and inner space, and there's no limit to either."

– Fantastic Voyage, 1966 (Fleischer, 1966)

CONTENTS

4.1 INTRODUCTION

Human imagination knows no bounds. Neither, it would seem, does reality. Our pursuit of knowledge has given rise to our capacity to describe the behavior of the smallest known constituents of reality via the most accurate scientific theory to date (quantum physics); as well as to investigate our position in the universe further than ever before, through unprecedented feats of engineering in space exploration. Despite this grand elucidation (from our perspective) of reality that spans vast scales, from trillionths of a meter to trillions of stars, there is a phenomenon that resides somewhere in between that eludes scientific understanding to date: life.

As far as we can tell from the cosmic microwave background, the Big Bang from which the universe purportedly emerged occurred ~13.8 billion years ago (Burbidge,1971). Hydrogen and some helium were produced, which combined to form the first molecules. The clumpiness of matter in the early universe, together with the force of gravity, subsequently resulted in the

DOI: 10.1201/9781003680093-4

formation of stars, from which the remainder of the elements were formed in ensuing supernova events. The atoms and molecules required for life already existed at the time the first stars in our Milky Way galaxy originated, ~13.5 billion years ago. Our Sun was formed around 9 billion years later.

Since the emergence of living systems on our planet, possibly as early as four billion years ago, much has transpired. According to the fossil record, Homo sapiens appeared relatively recently (only a few hundred thousand years ago), and in just the last few decades, we have experienced extremely rapid technological development. Consequently, we can now probe terrestrial life on scales smaller than the dimensions of atoms, as well as explore the heavens for evidence of extraterrestrial life and image galaxies as far as 13.5 billion light-years from Earth (Harikane et al., 2022). Planets beyond our solar system are difficult to see, just tiny disturbances in the light patterns of their stars; there may be as many as 200 billion of them in the Milky Way alone, with billions of trillions (~6×10^{22}) of stars in the universe (Manojlović, 2015). However, so far, Earth is the only planet we know of where life exists.

A significant breakthrough in our understanding of life was the discovery of the deoxyribonucleic acid (DNA) molecule. Initially imaged in 1952 (Franklin and Gosling, 1953) (Figure 4.1), this double-helical structure, just a few atoms (~2 nanometers (nm)) wide, contains not only the instructions but also the lineage of all terrestrial life (from spores to sperm whales); we all share significant portions of our genetic code. This is beautiful from the perspective of unity: life on Earth is one system; however, we are then left with just a single data point in terms of the study of life and its origins.

Nevertheless, it is a rich data point, since not only does Earth harbor life; it is teeming with it. The most abundant organisms are microscopic, with some researchers predicting the existence of one trillion species of microbes on Earth, of which the vast majority have yet to be discovered (Locey and Lennon, 2016). Some widespread species have the potential to alter the environment on a global scale. For example, the oxygenation of the atmosphere resulted when bacteria evolved the capacity to split water billions of years ago. Small organisms can also have profound effects in more localized environments; less than half of the cells in our bodies are human, while the rest belong to microbes that contribute to the maintenance of our state of being alive.

Currently, our human population is growing exponentially, and due to our resource usage, prolific associated industries, and culture of consumerism, we are having negative impacts on the environment on a planetary scale. One of the most fundamental challenges we face today is that of poor resource management and, as a result, being significantly out of balance with the finite environment on which we depend for life-support. We are already witnessing changes to our planet's climate and species diversity due to anthropogenic activities.

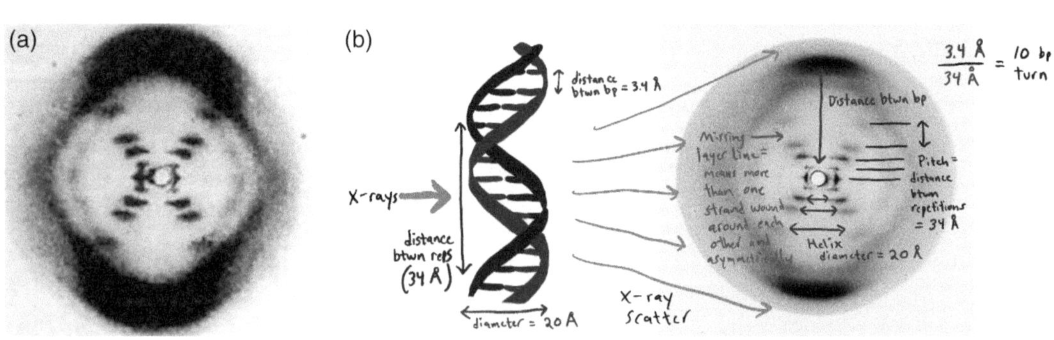

FIGURE 4.1 (a) *Photo 51* showing X-ray diffraction pattern of DNA (Franklin and Gosling, 1953). (b) Cartoon explaining how *Photo 51* obtained the double helix structure of DNA (Emily Willoughby) (Creative Commons Zero, Public Domain Dedication).

The contamination of the natural environment by industry has negative impacts on our health, while simultaneously, climate change is correlated with the increased likelihood of epidemics, conflicts, and poverty. Technological capability is a double-edged sword; the resource extraction and manufacturing required for producing technologies contribute to the harmful impacts we inflict on our ecosystems and ourselves. However, our technologies may also be harnessed to repurpose waste, regenerate the natural environment, and help us to live in a healthier and more sustainable way.

Resource cycling is not a new concept; life on Earth has been engaged in this for billions of years, operating at atomic and molecular scales for optimal efficiency. The transformation of sunlight, carbon dioxide (CO_2), and water to fuel the production of biomass via photosynthetic organisms is an example, and it turns out that this process harnesses quantum effects for optimization. Quantum physics describes how things behave at tiny scales, where atoms and molecules exist. Myriad phenomena at the quantum scale are counterintuitive and function differently than those in the macroscopic world, with which we are sensorially familiar. The strategic development of technologies that operate at these scales can utilize these unfamiliar effects to our advantage.

By understanding the basic components of matter and developing our capacities to manipulate these building blocks, we have witnessed remarkable progress in domains like computing and genetics, which in turn have enabled important advances in nanotechnology in areas such as materials science and healthcare. Since K. Eric Drexler's seminal book on nanotechnology in 1986, "The Engines of Creation" (Drexler, 1986), we also live in a different era with an additional three billion of us vying for livelihoods here on Earth.

An elucidation of matter at diminutive scales down to atomic and molecular resolution enables us to become smarter about how we manage our resources. This will be essential toward ensuring our strong and positive emergence from the maelstrom of uncertainties that we experience here on Earth, as well as our expansion to the Moon, Mars, and beyond, which is currently within reach. As we investigate the exceedingly small, we find that we are not the first to encounter it. Life on Earth has had a four-billion-year head start in the research and development of nanoworlds that dwell far beyond our visual perception. Recently developed technologies have enabled us to peer into life at increasingly minuscule temporal and spatial resolutions, as we attempt to further elucidate exactly what it is and where it came from. Although we have not yet understood life, in living systems we see a wealth of invaluable lessons on how to engineer at the scales of atoms and molecules. Moreover, by understanding how these astonishing feats of bioengineering are achieved, we learn not only about the nature of reality but also our place in it.

"There's plenty of room at the bottom." -

Richard Feynman, 1959 (Feynman, 1992)

4.2 ZOOMING IN: THE NANOWORLD

4.2.1 The Study of the Small

Our direct experience of reality is constrained by our senses. A micrometer, also referred to as a micron (μm), is a millionth of a meter. The human eye can distinguish objects down to ~50 μm, which is about the width of a fine hair; however, this has not stopped us from wondering what is going on at smaller scales. Philosophers have mused for thousands of years as to whether there might be indivisible elements from which matter is comprised, with the word "atom" derived from a Greek term meaning indivisible. With the advent of light microscopes in the early 1600s, our investigation of the small was extended beyond the visible world (Wollman et al., 2015).

Using a set of convex lenses to manipulate the light reflected from an object, magnifications of up to a few hundred times enabled us to observe the structures of microscopic biological systems

FIGURE 4.2 Large scanning tunneling microscope, London Centre for Nanotechnology. (Reproduced from O. Usher, UCL MAPS, 2014, Creative Commons Attribution 3.0 Unported license.)

(e.g., red blood cells, spermatozoa, bacteria, etc.) which are just a few μm wide. The resolution of a microscope using a light beam to illuminate a sample is limited by its wavelength. For visible light, this is a fraction of a μm (400 nm (violet) to 700 nm (red)) (Helmenstine, 2020). To "see" on even smaller scales, an alternate type of beam with shorter than visible light wavelengths is required.

Electron microscopes were developed in the 1930s (Haguenau et al., 2003). By using an electron beam to illuminate a specimen, images with a million-fold magnification can be generated. Since electrons are easily deflected by comparatively massive gas molecules, samples to be imaged need to be under vacuum. In 1981, the first scanning tunneling microscope (Figure 4.2) was demonstrated (Binnig and Rohrer, 1986). These instruments use the principle of quantum tunneling to enable visualization and manipulation of a wide range of biological and inorganic specimens (from cells to crystals) in environments not strictly requiring a vacuum, at the unprecedented resolution of individual atoms.

The conceptualization of the electron microscope, and in particular the application of quantum tunneling, was born from a new theory in physics (quantum mechanics) which was developed in the early 20th century to address the failures of classical physics to describe what was being observed on very small scales. Objects like molecules, atoms, and electrons are fundamentally explained by quantum theory. Just how small are these objects? A nm is a billionth of a meter, or a thousand times smaller than a μm. Most molecules are a few nm in size, while the atoms that comprise molecules are even smaller (0.1–0.5 nm). Quantum theory has revealed that the interior of an atom contains a cloud of negatively charged electrons that orbit around a tiny central positively charged nucleus. A nucleus is typically ~100,000 times smaller than its atom, and electrons (as particles) are so small; they are point-like ... with no size at all. Experiments have revealed that not only the dimensions but also the behaviors of quantum objects are fundamentally distinct from phenomena we are familiar with at larger scales.

Quantum entities possess unique properties such as spin and vibration; they can behave like particles or waves depending on the context and exist in more than one state simultaneously. Further, they can share correlations no matter how physically far apart they are. None of this is observed in the macroscopic world with which we are accustomed. For example, within the framework of classical physics, an object cannot spontaneously pass through a barrier without the additional energy to do so. However, in the quantum realm, a particle like an electron can also exist in a wavelike state, enabling just that. The capacity of an electron to tunnel through the space between the sharp conducting tip of an apparatus and a particular surface is the basis of how the scanning tunneling microscope can provide atomic level data on the structure of a surface being investigated.

The nanoworld exists at the interface between the macroscopic reality we are used to and the quantum regime, which provides an opportunity to explore the still mysterious transition between these worlds. Our nanoscale understanding of the molecules and atoms that comprise matter has led to the development of exciting new materials.

4.2.2 Materials

Nanomaterials are measured in unit sizes that range from one to 100 nm, which can occur naturally or be synthesized from carbon, metals, metal oxides, or other material composites. Carbon is the fourth most abundant element in the Universe, and owing to its electron configuration, it can bind with up to four other atoms simultaneously. Carbon can form polymers and diverse organic compounds under typical conditions on Earth and is a common element of all known life. Further, many different structures (some on the nanoscale) can be formed from carbon atoms that are bonded to each other.

Graphite, one of the oldest and most widely used naturally occurring materials, is a crystalline form of carbon that is stable under standard conditions with high electrical conductivity. As nanofabrication techniques have developed, graphite has become a starting material for the engineering of various types of carbon-based nanomaterials, including nanodiamonds, nanotubes, graphene, and diamond/graphene composites (diaphite) (Németh et al., 2020). Diamond is a solid form of carbon with its atoms arranged in a crystalline structure, which is formed when carbon deposits like graphite are subjected to high temperatures and pressures. It is the hardest and most thermally conductive natural material known. While typically we may be interested in diamonds we can see, nanodiamonds with dimensions below a micron can be synthesized inexpensively and have a wide range of applications.

Fullerenes are a class of carbon where the atoms are linked by single and double bonds in a closed or partially closed mesh, forming sheets, tubes, hollow spheres (C_{60}), cones, or other shapes (Bakry et al., 2007). Carbon nanotubes are tube-shaped fullerenes with diameters on the scale of nm, which can be single-walled or consist of multiple concentric structures with myriad applications (Raphey et al., 2019). The first clear electron microscope images of Ø50 nm tubes made of carbon were reported in 1952 (Radushkevich and Lukyanovich, 1952); however, our ability to synthesize carbon nanotubes may be much older. Unique black coatings with remarkable stability and adhesion were observed on pottery found in India dating back to the 6th century BC. Analysis revealed the presence of carbon nanotubes in the coating with diameters of less than a nm (Kokarneswaran et al., 2020).

Besides tubes, a mesh of carbon can form closed shapes. Closed form fullerenes had been predicted for some time, but only after the accidental synthesis in 1985 of C_{60} (Kroto et al., 1985) – a hollow sphere made of 60 carbon atoms (Figure 4.3) – were they detected in natural systems on Earth and then in outer space (Maier and Campbell, 2017). The remarkable symmetry and chemical properties of closed form fullerenes have led to a new understanding of the behaviors of sheet materials and their applications as nanomaterials.

Last but not least is the sheet or flat mesh form of fullerene: more commonly known as graphene. Graphene consists of a single layer of carbon atoms arranged in a two-dimensional lattice

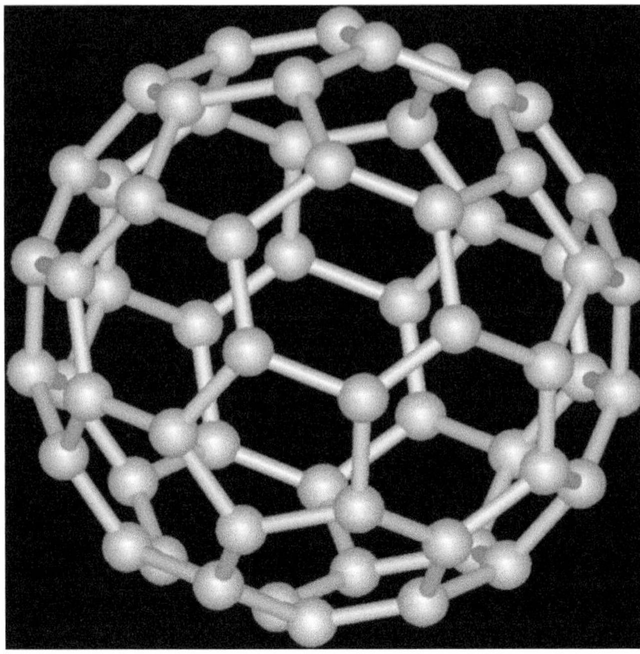

FIGURE 4.3 Ball and stick rendering of a C_{60} fullerene. (Reproduced from IMeowbot, 2005 Creative Commons Attribution-Share Alike 3.0 Unported license.)

nanostructure. Theorized for decades before, the term "graphene" was coined in the early 1960s (Boehm et al., 1962) for a hypothetical single-layer structure of graphite. When a simple method for the synthesis of graphene was published in 2004 (Novoselov et al., 2004), research rapidly evolved to explore the exceptional properties of this nanomaterial (Sattar, 2019).

Due to their unique molecular structures, fullerenes can exhibit extraordinary macroscopic properties with a wide range of applications, including high electrical conductivity, high heat conductivity, and semiconductivity. Because of the strength of the bonds between the carbon atoms, they form some of the strongest known materials. They are also relatively chemically inactive due to their mesh structure where no atoms are easily displaced. Fullerene class nanomaterials have complex nanoscale structures, while other common nanomaterials are composites that contain one or more types of nanoparticles.

Nanoparticles have dimensions between one and 100 nm and therefore possess an immense surface area to volume ratio. Consequently, the surface layers of nanoparticles can dominate and produce novel properties (often quantum mechanical in nature) compared to bulk amounts of the substance. In general, a range of nanoparticles or atoms can be introduced as impurities to various nanomaterials to modulate their structural, electronic, and optical attributes. For example, modern automobile tires consist of rubber that is strengthened by reinforcers such as nanoparticles of silicon dioxide (Giftson and SivaKumar, 2014) or carbon nanotubes (Shao et al., 2018). Further, so-called doping with a range of different nanoparticle species is widely employed in the semiconductor industry to produce solar cells and transistors for computers.

More specifically, titanium dioxide nanoparticles have useful light interaction attributes. As additives, they can modulate the light scattering that occurs in the coatings of light-emitting diodes (LEDs) or solar cells; mitigate the degradation of materials or prevent sunburned skin caused by ultraviolet light; or (through the absorption of light) catalyze chemical reactions such as the production of hydrogen from water or wastewater treatment. Platinum group metals possess unique physical and chemical properties, including being used as catalysts. Since platinum can drive the removal or transfer of hydrogen or the addition of oxygen, nanoparticles comprised thereof have applications across a vast array of industries, from improving power conversion in fuel cells, batteries, and solar cells to removing pollutants from the environment.

To make large-scale use of the remarkable properties of nanomaterials, efficient strategies for their production are required. Some nanoparticles occur naturally; for example, nanodiamonds can be produced by volcanic eruptions or meteoritic impacts, and in general the natural processes of fire and erosion continuously generate a range of nanoparticles. This type of top-down approach where bulk material is broken down to nanostructures is one way of producing nanomaterials. Another approach is from the bottom up, where nanostructures are built up through the assembly of individual atoms and molecules. The detection of fullerenes in stars and interstellar dust, as well as in rock sediments and near meteorite impact sites, demonstrates that under specific conditions, complex nanostructures may form spontaneously. How close are we to synthesizing nanomaterials with nanostructures built from scratch?

Our current utilization of nanomaterials is largely limited to passive applications like material coatings, nanocomposites, and semiconductor doping. In other words, we mostly incorporate rudimentary nanostructures (e.g., nanoparticles), which are typically produced top-down, into existing products or processes to improve efficiency or some other characteristic. We have had a look at the matter at the nanoscale; the next level of sophistication will involve precise spatial positional control at molecular and atomic resolution. For this, nanomachines may be indispensable.

4.2.3 Machines

Quantum theory has enabled us to understand semiconductivity. Based on this knowledge, the first transistor was demonstrated in 1947 (Bardeen and Brattain, 1948). A transistor is an electronic component that contains a semiconductor material (often silicon) that can both conduct and insulate current, meaning that it can be employed as a switch in electrical circuits. Transistors are the fundamental building blocks of computers and one of the most significant technologies ever developed. The number of transistors in computer processors has been increasing exponentially since the invention of computing. From dimensions of ~1 cm, the current commercial transistors fabricated by Taiwan Semiconductor Manufacturing Co. Ltd. are 3 nm in size, which is on track to produce 2 nm chips by 2025, with 1 nm chips under development (Ahmad, 2022).

Moore's Law tracks the associated increase in computing power as a function of time: the speed and capabilities of computers can be expected to double every two years. However, we are approaching a fundamental limit in terms of the number of transistors that may be packed into a chip, not due to technological limitations, but rather because the laws of physics will prevent transistors from functioning in the conventional sense if they are any smaller and closer together.

Beyond this limit, quantum computers offer further acceleration to address specific types of problems, including the capacity to solve certain open problems. Rather than employing transistors, quantum computers encode data directly into entities such as particles of light (photons), electrons, or atoms. Potential applications include rapid data processing, drug design, and the modeling of complex phenomena (e.g., protein folding or climate patterns). It has been demonstrated that a universal quantum computer would be able to crack traditional encryption systems that are used to keep data private. However, in the drive toward the universal quantum computer, we have only recently achieved so-called quantum supremacy, which means that state-of-the-art quantum computers can only just outperform our best supercomputers, and only for certain specific tasks (Harrow and Montanaro, 2017).

Our ability to process and share information using computing in conjunction with communication systems has developed rapidly and culminated in the revolution of connectivity, manifest as the Internet. However, what about the control of matter itself on the level of individual atoms and molecules? In the last few decades, we have witnessed impressive progress in 3D printing, the process of building objects layer by layer via the deposition of material according to digital 3D design data. However, this is not considered nanotechnology, since bulk feedstocks are utilized as input materials, with each layer typically being on the order of millimeters.

Using scanning tunneling microscopy, individual atoms can be identified and repositioned, although not terribly quickly. In 1989, the ultrasharp tip of a scanning tunneling microscope was used to pick up 35 xenon atoms and spell "IBM" in 5 nm tall letters (Eigler and Erhard, 1990), which required 22 hours. New proposed methods to enhance the speed of scanning tunneling microscopy might achieve rates of just over 100 atoms moved per second (Celotta et al., 2014; Serrate et al., 2010). Further, optical tweezers (Barredo et al., 2018) and electron beams (Dyck et al., 2018) have been employed to facilitate the construction of three-dimensional structures atom-by-atom, layer-by-layer. However, these examples are still millions of times slower than standard techniques using lasers and solvents that are used to produce the tiniest electronic structures, which on the most advanced computer chips are currently ~3 nm (Ahmad, 2022).

There remains ample room for the sophistication of techniques to manipulate matter on the scale of atoms and molecules via macroscopic devices like the tips of scanning tunneling or atomic force microscopes (Custance et al., 2009). However, what if we instead built tiny machines to do the job? Nanorobotics is an emerging field of engineering that focuses on the design and building of devices that act on their environment, constructed of nanoscale components. A simple example of such a device is a molecular switch, which is a molecule that can be reversibly shifted between two or more conformations in response to environmental stimuli (Fahrenbach et al., 2013).

For example, a pH indicator alters color as a function of pH level (Wang and Chen, 2018). A molecular shuttle is a molecular machine capable of moving molecules or ions from one location to another (Yao et al., 2022). Further, molecular and atomic resolution fabrication might also be achieved by exploiting the base-pair recognition capacities of oligonucleotides (Boehm, 2004; Krishnan and Seeman, 2019). A fun example of the current state-of-the-art is the Nanocar Race (Nanocar-Race II, 2022), where molecular machines comprised of hydrogen and carbon atoms, measuring between 1 and 3 nm, roll along a 100 nm wide track. The nanocars are propelled without direct physical contact via a current of tunneling electrons that originate from the tip of a scanning tunneling microscope, with a speed record of 95 nm per hour (Figure 4.4).

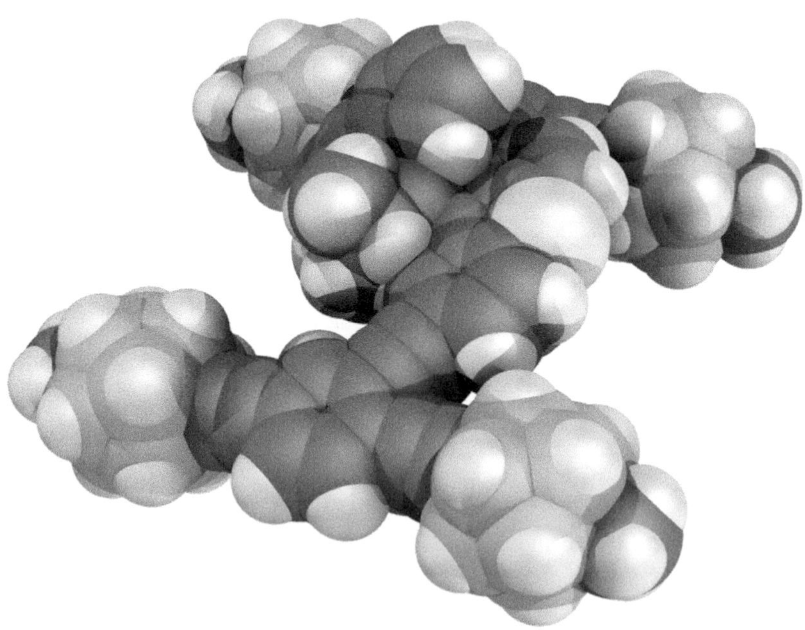

FIGURE 4.4 Space-filling model of the "nanocar 9." (Reproduced from ChemSpider by Qoheletl2. Creative Commons Zero, Public Domain Dedication.)

What if, rather than physically interacting with matter, we could program it? If we could transmit a set of instructions to a swarm of nanobots, we could achieve macroscopic outputs with microscopic complexity and precision. We could manufacture products from raw materials or remove contaminants from our environment. To create such collective activity, we would need only to engineer a single so-called "universal constructor" (Pesavento, 1995; von Neumann and Burks, 1966) with the ability to self-replicate *under human control*. Scaling this idea: every human being on Earth could have access to such nanoreplicators.

This could spell the end of scarcity, since people could in theory produce all the resources, materials, and devices they required from the abundant elements around them. Everyone could have access to safe shelter, nutritious food, clean water and air, communication and computing devices, healthcare, and much more. Conversely (which is where *under human control* comes in), errors in replication could propagate uncontrollably and, in a worst-case scenario, convert all material on Earth into "grey goo," a scenario initially coined by Drexler (Drexler, 1986). This scenario has been investigated in detail and refuted, with multiple countermeasures being proposed by Freitas (Freitas, 2000).

While our vision for controlling matter at atomic and molecular levels remain beyond the reach of our current technical capabilities, in fact, ingenious feats of nanoengineering are already taking place every moment within our own bodies, and in all the living systems around us. Life on Earth has undergone four billion years of trial and error in the nanoworld, which provides a rich source of knowledge and inspiration for our own endeavors at these scales.

4.2.4 Biology

4.2.4.1 Biological Nanotechnology

Living systems are extraordinarily complex open systems. Every moment in the human body, there are a plethora of nanoscale interactions that occur internally, where a host of bionanomachines proceed with the ultraprecise business of maintaining life within each of our cells. Further processes occur between organisms and their environment where there is a continuous exchange of energy and matter.

Even the simplest of living systems we see around us are mind-boggling. The level of complexity challenges not only our minds but also our computational power. The precise modeling of the ~100 billion atoms in a single simple bacterial cell (the order of magnitude of the stars that comprise the Milky Way Galaxy) is a feat that is currently beyond our computing capabilities. A molecular system almost 100 times smaller that contained 1.6 billion atoms was modeled using the world's most powerful supercomputer in 2021, Japan's Fugaku (TOP500, 2021), with standard approximations that left some data out. One second of activity would require more than 300,000 years to computationally simulate (Jung et al., 2021). Fortunately, researchers are also interested in events that occur on much shorter timeframes (e.g., nanoseconds or billionths of a second), which can be computed in hours.

Multicellular life engages in even more intense activities. For example, motor proteins convert chemical energy from adenosine triphosphate (ATP) (an organic fuel compound that provides energy in cells) into mechanical force (Tafoya and Bustamante, 2018). These motor proteins include myosin, which is responsible for muscle contraction and other motility functions, as well as kinesin and dynein that transport molecular cargos like protein complexes or messenger RNAs (mRNAs) within cells (Pollard and Goldman, 2018; van den Heuvel and Dekker, 2007). Motile cilia are found in the lungs, respiratory tract, and middle ear, and are so named due to their rhythmic waving motion, which among other functions clear the airways of mucus and dirt. Motile cilia are composed of hundreds of proteins, many of which also function independently as nanomachines; for example, the dynein motor protein generates their movement (Klena and Pigino, 2022).

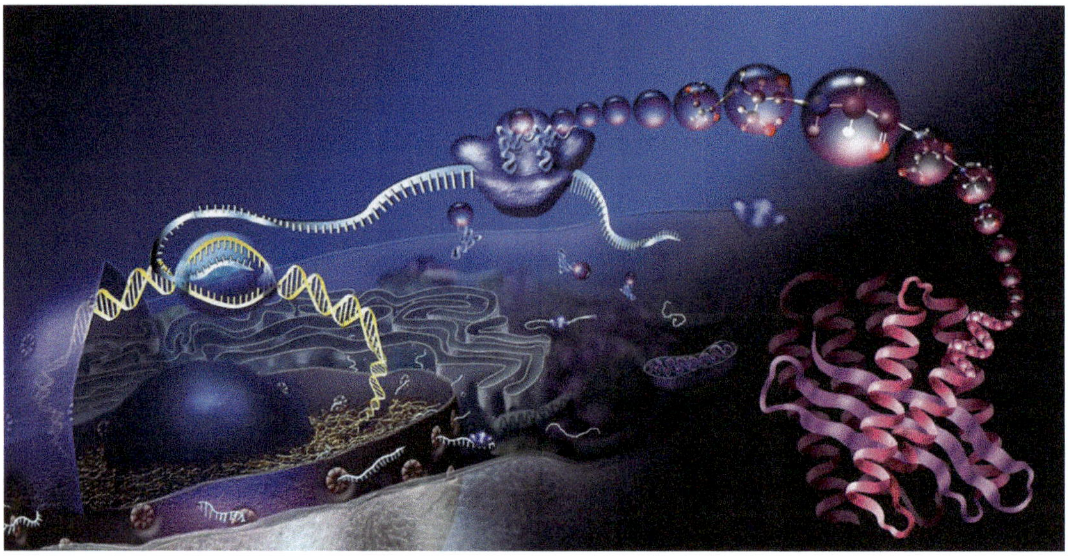

FIGURE 4.5 Depiction of ribosome translating DNA into an amino acid strand that folds to form a functional protein. (Reproduced from Nicolle Rager, National Science Foundation. Public domain.)

Some bionanomachines generate energy; for example, ATP synthase drives a turbine-like motion to produce ATP fuel (Guo and Rubinstein, 2022). Others are responsible for gene expression, including DNA polymerases (Jain et al., 2018) for replicating DNA and RNA polymerases (Girbig et al., 2022) for producing mRNA (ribonucleic acid is akin to DNA but single-stranded). The ribosome is essentially a nanofactory with dimensions of Ø30 nm, serving as the site for protein synthesis within the cell (Khatter et al., 2015) (Figure 4.5). DNA encodes sequences of amino acids as instructions to produce specific proteins.

These are transcribed to an mRNA chain to which the ribosome binds, determines the sequence, and generates the protein. Although artificial ribosome-inspired protein synthesis systems have been devised, natural ribosomes are so good at what they do that repurposing them to accept a wider variety of building blocks beyond the 20 amino acids that make up life on Earth is an intense focus of investigation. The goal is to produce new types of proteins with entirely new properties and functionalities (d'Aquino et al., 2018).

Living nanomachines and their dynamics are far more complex than any molecular machine we have yet conceived of, designed, or created in the lab. However, progress in our understanding of one specific biomolecule has led to one of the fastest growing fields of human knowledge: genetics. The basis of activities in the cell resides in the DNA that encodes it, which is a double helical shaped molecule only a few nm wide (but when laid out straight stretches up to several meters long) that contains four bases: cytosine (C), guanine (G), adenine (A), and thymine (T) (Klug, 2004; Watson and Crick, 1953). Since the vast majority of DNA for each species is essentially the same, it is possible to determine the code for the entire sequence, or genome, for these species.

The first bacterial genome was sequenced in 1995 (Fleischmann et al., 1995), and by 2001 Craig Venter, and subsequently the Human Genome Project, announced the first drafts of the base pair sequences that comprise the genes that constitute human DNA (Lander et al., 2001; Venter et al., 2001). The total data required to store the 2.9 billion base pair sequence of the human genome required over 700 megabytes. Work continues to fill in unresolved gaps, as well as to repair the errors revealed in quality checks.

Humanity's ability to sequence the human genome has led to the elucidation of correlations between nanoscale segments of the genetic sequence with macroscopic characteristics, which can be manifest as anything from physical appearance to risks for certain diseases. This provides us with the ability to predict, for example, that an individual may have a high probability of susceptibility to a certain disorder, which enables preventative actions or early treatments. We have also learned to manipulate the genetic code at molecular and even atomic resolution, enabling, for example, genetically edited embryos.

The combined CRISPR-Cas9 technology was initially developed in 2012 (Jinek et al., 2012), which operates at a nanoscale to remove, modify, or add segments to a genetic sequence. CRISPR is a homing device that guides molecular scissors (the Cas9 enzyme) to a targeted section of DNA to perform cutting and pasting. Although human genomic editing remains controversial, particularly as relates to unborn children, the genetic modification of microbes to perform certain tasks may have important applications for environmental remediation/rejuvenation here on Earth, as we prepare for humanity's propagation among the stars.

Might gene editing also shed light on life itself? One line of investigation involved an existing simple organism, such as a bacterial cell, into which a string of DNA extracted from another bacterium was inserted (one that was minimized and optimized), after which the cell was observed to survive (Hutchison 3rd et al., 2016). Though such exercises may instruct us as to the functionality of different fragments of DNA, they cannot convey how to engineer a cell membrane, or any of the myriad internal organelles and other components of a cell from scratch, nor impart insights into how these complex systems emerged in the first place.

Humanity must admit that we have not yet even come close to a full understanding of life, cognition, sentience, nor explain how they managed to emerge from physics and chemistry. We have neither synthesized an organism from scratch nor satisfactorily defined this phenomenon or the processes involved. In other words, we have yet to pinpoint exactly what it is that distinguishes an aggregate of molecules from a functional life form. Albert Einstein commented that, "one can best feel in dealing with living things how primitive physics still is" (Rosen, 2000). All living systems fundamentally consist of atoms and molecules, the properties of which may be described by quantum mechanics. However, the question burns: must the prerequisite be that we delve all the way down to the quantum domain to elucidate exactly what is going on in biology?

Some aspects of biology may be adequately described by classical physics and chemistry. However, as we peer deeper into the functionality of living things, into the realms of the very atoms and molecules that make life possible, we encounter an increasing number of processes that require quantum physics for their explanation: quantum biology is the field that deals with such cases (Marais et al., 2018).

4.2.4.2 Quantum Biology

4.2.4.2.1 Photosynthesis

For over three billion years, a prolific class of organisms, including bacteria, algae, and plants, have been utilizing sunlight and abundant terrestrial resources (e.g., carbon dioxide and water) for the generation of energy-rich compounds to power cellular activities. Photosynthesis is one of the oldest known living process and its study the most well-established area in quantum biology (van Grondelle and Novoderezhkin, 2011). The initial stages of photosynthesis occur in less than a nanosecond with near-perfect quantum efficiency, where an electron is transferred for every photon of sunlight absorbed by pigment molecules in the light-harvesting antennae of organisms. The electrons leave positive holes behind and add negative charges to their destination sites.

This separation of positive and negative charges forms an effective battery that powers the subsequent production of molecules like glucose and ATP. In a nutshell, photosynthesis employs intricate nanoscale machinery powered by sunlight to convert input resources into fuel. A comprehensive understanding of how nature harnesses sunlight (the primary renewable energy resource on Earth) with near-perfect efficiency has important implications for our own energy generating systems.

Living systems continually update their internal processes based on information obtained from their sensing of the environment, where even subtle changes may translate to significant adaptations. Proposed biological sensing mechanisms so sensitive as to detect changes in the environment at quantum resolution include navigation in the Earth's magnetic field, vision, smell, and cognition.

4.2.4.2.2 Magnetoreception

Over the last few decades, we have come to rely heavily on global positioning satellites that keep accurate time and orbits to allow us to trilaterate our positions on Earth with accuracies of a few meters. Imagine if humans were inherently physiologically endowed with such navigation capabilities? Magnetoreception is a sense that enables an organism to detect a magnetic field for navigation. A range of organisms, from bacteria to birds, have this capacity (Xie, 2022), with some claiming that humans do as well; however, until recently we have not had a good theory of how this might be the case (Wang et al., 2019).

Although the Earth's magnetic field is powerful enough to protect us from the stream of charged particles that are blasted at us from the Sun, it was thought not strong enough to have an impact on the chemical reactions typical of life. Cryptochrome proteins are found in plants and animals (including humans) and are sensitive to blue light, from which magnetically sensitive quantum states are produced. Recent research has revealed that these proteins found in the human retina are promising candidates for enabling the ability to sense the Earth's magnetic field (Foley et al., 2011; Wiltschko and Wiltschko, 2019).

4.2.4.2.3 Vision

Quantum mechanics reveals that while light may be manifest as a wave, it is simultaneously comprised of discrete particles called photons. In the 1930s, single-photon detectors were invented with sensitivities for individual quanta of light,, which have advanced considerably since then (Xu et al., 2021). A few years after their invention, initial experiments were performed, which concluded that human retinal rods could respond to a single photon during low light vision (Hecht et al., 1942). While much about the functionality of rhodopsin (the protein responsible for the conversion of light into optic nerve impulses) is not yet clearly understood, more recent experiments have revealed that quantum coherence plays a role in the ultrafast processes involved with the initialization of vision (Polli et al., 2010; Tscherbul and Brumer, 2015).

4.2.4.2.4 Olfaction

The conventional explanation for human olfaction involves a lock-and-key mechanism, wherein uniquely shaped odorants match specific receptors that trigger the recognition of smells. However, this notion remains woefully inadequate for elucidating how we manage to distinguish a trillion scents (Bushdid et al., 2014), yet appear to possess only several hundred odorant receptors (Fukutani et al., 2023). A theory initially proposed almost a century ago suggests that our sense of smell detects not the morphologies, but the quantum vibrations of molecules (Gane et al., 2013). More recently, experiments have indicated that different molecules with the same shape are easily distinguished by fruit flies, while molecules with similar quantum vibrational frequencies elicit similar reactions despite being unrelated in shape (Franco et al., 2011). The engineering of biologically inspired room-temperature molecular sensing devices that operate on the principle of quantum vibration detection could validate this theory. They could also be applied for the detection of viruses, toxins, or other substances in the air, even at extremely low concentrations.

4.2.4.2.5 Consciousness

The question of whether quantum physics might play a role in explaining the still unresolved issue of how the physical structure of the brain accounts for things like the acquisition of knowledge via the senses is as old as quantum physics itself. The phenomenon of consciousness (the simultaneous awareness of external and internal existence) is poorly understood. It is not clear that brain processes like neural signal transmission and synaptic connectivity are sufficient to explain the full functionality of the brain. Through the field of anesthetics, one thing about consciousness was discovered; we know how to turn it off (Turin et al., 2014). Chloroform, xenon, and an array of molecules that appear to be quite disparate, so-called anesthetics, can reliably achieve this. A compelling area of consciousness research involves quantum spin models of brain processes to explain the efficacy of anesthetics, and work continues to clarify the biological sites of proposed quantum effects in the brain (Adams and Petruccione, 2020; Kurtev, 2022; Melkikh and Khrennikov, 2015).

Recent research has revealed that microtubules have the capacity to enable conductivity, signaling, and communications, while theoretical models allude to their ability to perform classical and quantum information processing. Electron mobility and the potential for quantum tunneling and superconductivity between the aromatic amino acids of tubulins (which comprise microtubules), in conjunction with the alignment of internal tubulin pathways with helical microtubule lattice patterns, give rise to topological quantum effects that resist quantum decoherence (Hameroff et al., 2002; Kalra et al., 2023). The role of microtubules in brain function and potentially consciousness itself remains to be better understood.

4.2.4.2.6 DNA Replication

Since the issue of whether we might fully understand the brain using the brain itself remains unresolved, let us get back to basics. DNA replication is one of life's most fundamental processes, which involves seemingly random mutations. We may wonder then whether quantum effects are somehow at play, being themselves probabilistic in nature. The unpredictable "errors" that occur when strands of DNA are being copied (for example, the substitution of one base for another) are in fact the fundamental mechanism that allows life to trial variations on the level of individuals. This enables adaptations and eventually the emergence of new species in response to changing environments. The ongoing investigation of life's most basic functionality (replication) reminds us of how much we still have to learn (Epum and Haber, 2022; Prioleau and MacAlpine, 2016; Slocombe et al., 2021).

On a more pragmatic level, since quantum effects play essential roles in certain biological processes to enable complex systems, which have been perfected over billions of years, they offer us a wealth of information from which to draw inspiration for our own technologies. Whether quantum or not, through our comprehensive investigation of life and its core processes, we can identify and mimic the ingenious feats of nanoengineering that commonly occur in living systems, ranging from bacteria to brains.

> Every mammal on this planet instinctively develops a natural equilibrium with the surrounding environment, but you humans do not. You move to an area and you multiply and multiply until every natural resource is consumed. The only way you can survive is to spread to another area. There is another organism on this planet that follows the same pattern. A virus. – The Matrix, 1999.
>
> *(The Wachowskis, 1999)*

4.3 ZOOMING OUT: NANOTECHNOLOGY APPLICATIONS

4.3.1 Basic Resources

The biomolecules that comprise the machinery of life are typically nanometric entities; thus, the knowledge and tools that have been developed via nanotechnology may also be applied to healthcare for the prevention and treatment of disease. The discipline of nanomedicine involves the use of nanoscale materials, such as biocompatible nanoparticles, and in the future, devices like autonomous nanorobots, for applications that encompass diagnostics, drug delivery, and biomedical engineering, which are explored in detail in other chapters of this book. Zooming out further, a prerequisite for fundamental human rights, such as the inalienable right to access healthcare, is the availability of basic resources: life requires constant inputs of specific resources to sustain living. For many terrestrial life forms, including humans, the most fundamental of these are warmth, water, air, and food.

We have harnessed nanotechnology to increase the efficacy of the manufacturing and functionality of consumer products, which has evolved into a multi-billion-dollar industry. Earth is currently the only life-support system we have. Can we harness and exploit the positive benefits of nanotechnology to reverse some of the devastating impacts we have inflicted on our natural environment, as well as learn to live in a more resource-efficient way, whether on this planet or beyond?

4.3.1.1 Energy

Energy is required for life and for literally anything interesting to happen. Over four billion years, life on Earth has absorbed sunlight and stored it in organic compounds. A million years ago, humans learned to control fire, which extracted energy from the dead biomass of plants. Subsequently, we discovered an even more energy-rich fuel formed from fossils exposed to intense heat and pressure beneath the Earth's surface. Burning these organic compounds is an indirect means of harnessing energy from the Sun. However, current conflicts over the control and use of dwindling supplies of fossil fuels are frequent, while the carbon dioxide generated via their combustion is reaching levels that are unprecedented in human history, which plays a major role in global climate change.

More direct strategies for accessing energy include the conversion of sunlight to electricity; converting fuel to electricity through chemical reactions rather than combustion; and at an even more fundamental level emulating the processes of the Sun itself by accessing the nuclear energy stored in the cores of atoms. Might nanotechnology advance humanity's acumen in the production of energy?

Conventional solar cells have conversion efficiencies of up to ~20% and typically employ silicon as a basic semiconductor material, which is the second most abundant element on the Earth's surface. Nanomaterials and their enhanced properties can facilitate higher solar energy conversion efficiencies while reducing the requirement for those materials that are rare or expensive to produce. For example, black silicon is a nanomaterial characterized by needle-shaped single-crystal silicon structures on its surface, which can result in a ~20% higher light absorption when used in solar cells, due to its far larger surface area (Roy et al., 2016; Zhang et al., 2022).

While rare metals (e.g., indium) may be employed to reduce the thickness of the absorption layer, the emergence of organic solar cells that utilize conductive organic molecules (e.g., hydrogen, carbon, and oxygen) is more relevant for resource efficiencies, in terms of light absorption and charge transport. Organic solar cells have attractive properties that include flexibility and visible light transparency. With the integration of metal nanomeshes, graphene, and various nanoparticles into these technology platforms, their efficiencies are likely to improve significantly beyond today's best traditional solar cells.

While solar power is often thought to be the solution to renewable energy, solar cells and the batteries required for energy storage/supply in the absence of sunlight have limited lifespans. Further, their fabrication requires metals and other materials whose extraction, transport, and processing have detrimental impacts on the environment. Rather than produce electricity directly, might it be possible to store solar energy as an organic chemical fuel to biomimetically enable photosynthetic ATP production?

Fuel cells are devices that convert the energy of aqueous fuels into electricity through chemical reactions rather than by combustion. In an archetypal hydrogen fuel cell, hydrogen and oxygen are combined to produce energy and water. Microbial fuel cells convert chemicals into electrical energy during wastewater treatment and can also produce hydrogen. Despite our existing in a universe that is primarily comprised of hydrogen and residing on a planet whose surface is 71% covered by water, the techniques we have developed thus far for the production of hydrogen remain unbelievably primitive. We extract hydrogen from fossil fuels and natural gas, using electricity that is typically generated by burning coal.

However, recently developed artificial photosynthesis systems employing nanostructures and bioengineered microbes to split water molecules into hydrogen and oxygen using sunlight have demonstrated several fold higher solar energy conversion efficiencies than what were previously possible (Dogutan and Nocera, 2019; Zeng et al., 2021). Chemical fuels are typically comprised of atomic and nanoscale entities, ranging from hydrogen atoms (Ø0.1 nm) to hydrocarbon molecules (tens of nm wide). A range of nanotechnological processes may be employed to optimize the generation of such fuels from the most abundant materials in the universe: hydrogen, oxygen, and carbon.

The Sun exists as the ultimate source of energy in our Solar System. Since the feasibility of generating power as the Sun does remains elusive for now (we have yet to attain the stable and scalable nuclear fusion (Figure 4.6) of hydrogen), nuclear power generated via the cleaving of atomic nuclei (nuclear fission) remains the best option available.

While fears surrounding nuclear energy poignantly linger long after the deployment of weaponized uranium on Japanese citizens during World War II, as well as the Chernobyl and Fukushima disasters, nanotechnology also has applications in this domain. Whether finding use in traditional nuclear power stations or in radioisotope thermoelectric generators, an active area of research involves the effective radiation shielding properties of nanomaterials, particularly graphene. Containers for radioactive materials can be made of thousands of layers of graphene, replacing the toxic lead or immense volumes of concrete that are typically employed (Huang et al., 2019; Singh et al., 2023). Graphene oxide can also remove toxic and radioactive materials from contaminated water (Gao et al., 2022; Zhang et al., 2020), which brings us to the next topic.

FIGURE 4.6 ITER Tokamak fusion reactor cross section mockup, 2022. (Reproduced by Johannes Reimer, Creative Commons Attribution 4.0.)

4.3.1.2 Water

Water is essential to all known life, and Earth is the only planet in the solar system that currently supports large bodies of liquid water on its surface. From the perspective of modern science, aspects of the bulk properties and phase transitions of water remain to be fully elucidated. Even its origins are mysterious, with isotopic analysis revealing that much of Earth's water is in fact older than the Sun itself (Gibney, 2014). A liquid substrate is likely a prerequisite for the type of complex chemistry associated with the emergence of life, as well as its sustenance; with bodies up to 60% water by mass, humans can survive for only a few days without water. Despite living on a planet that is mostly covered with oceans, we face the reality of running out of freshwater in many regions in the coming decades, as a result of urbanization and shifting rainfall patterns, coupled with poor water management. Furthermore, our supplies of this precious liquid are contaminated by our industrial activities, with pollutants consisting of a complex mixture of metals, bulk and microplastics, petroleum, pesticides, fertilizers, pharmaceutical chemicals, and raw sewage.

While these contaminants impact our living environment on a global scale, the pollutants themselves consist of molecular and atomic scale particles. Our capacity to manipulate matter at the nanoscale can play essential roles in the remediation and restoration of the natural ecosystems that serve as our life-support systems. Nanoparticles such as titanium dioxide can improve the efficacy of water disinfection (Chen and Selloni, 2014), while membranes made of various nanomaterials (Singh et al., 2021) can be used to filter out larger contaminant particles from water, since H_2O molecules are extremely compact (~2.75 Å or 0.27 nm) in contrast to most contaminants.

Graphene-based materials and carbon nanotubes, as well as combinations of the two, have applications in nanofilters to remove impurities in wastewater management systems (Lingamdinne et al., 2019). Seawater constitutes ~97% of all water on the planet. The main limitation to its global utilization is the energetic cost of up to 10 kWh/m^3 for the removal of the salts. Desalination using graphene oxide membranes as filters could be achieved with up to 50% less energy than traditional techniques (Cohen-Tanugi and Grossman, 2012; Rabiee et al., 2023).

At the heart of current wastewater management strategies is our collaboration with the advanced nanomachinery found in bacteria and other microbes. Microorganisms do the heavy lifting in existing wastewater treatment systems by removing pollutants and pathogens, recovering nutrients and energy, as well as producing clean water. The initial challenge is to identify them; using DNA-based technologies, we are beginning to categorize the range of microbe species that are present in wastewater plants and digesters in different climates worldwide. Next, we need to identify the specific tasks that each performs, to enhance or minimize their populations to more efficiently achieve specific processing goals. A further step might be to genetically modify these microbes to optimize their accomplishment of these outcomes. For example, in cases of contamination by oil spills or radioactive material leaks, designer/engineered microbes could do cleanup jobs faster and more efficiently than naturally occurring bacteria. However, we revert to the grey goo scenario if we imagine what could happen if certain robust genetically modified superbugs make their way out of treatment areas and into the ambient environment.

4.3.1.3 Air

Both of our planetary neighbors (Venus and Mars) once held bodies of water on their surfaces, where we now observe average temperatures of ~500°C and <-60°C, respectively. These dramatic changes in surface conditions are due to influences that include changes in their atmospheric compositions. The runaway greenhouse effect: a feedback loop that results from the accumulation of insulating gases in the atmosphere is thought to be the reason why Venus has heated up so significantly. Conversely, on Mars the temperature drop might be explained by the cessation of the global magnetic field, which resulted in its insulating atmosphere being stripped away by the solar wind.

The vital air we breathe today has not always existed on Earth. Roughly 2.5 billion years ago, photosynthetic cyanobacteria with the ability to split water began to produce oxygen on a large scale. One of the greatest extinction events in Earth's history was the result of reactive oxygen species, which were toxic to living cells. However, simultaneously, the conversion of oxygen into ozone by ultraviolet light in the upper atmosphere provided a protective layer beneath which life with the capacity to harness oxygen flourished. Multicellular life emerged as a result, which eventually evolved into modern humans ~200,000 years ago.

The emergence of the Industrial Revolution had direct impacts on Earth's atmosphere. CO_2 is certainly not the only molecule that is introduced into the ambient atmosphere on a massive scale. While humanity has previously taken clean air for granted, today billions of people breathe air that is contaminated with myriad contaminants (e.g., PM2.5, PM10, carbon monoxide, methane, lead, mercury, nitrogen oxides, nitrogen dioxide, ozone, black carbon, sulfur dioxide, forever chemicals, microplastics, etc.), exposure to which translates to a host of deleterious health effects (Klimkaite et al., 2023; Krittanawong et al., 2023; Mousavibaygei et al., 2023; Sunyer and Dadvand, 2023). Due to the increasing amounts of greenhouse gases in the atmosphere, along with our continued destruction of the natural habitats that serve to regulate the atmosphere, Earth retains double the Sun's heat that it did two decades ago (Loeb et al., 2021), which results in global scale climate change, extreme weather events, an increased likelihood of epidemics, as well as a mass extinction event, the so-called Anthropocene extinction (Ceballos et al., 2015; Cowie et al., 2022; Turvey and Crees, 2019).

Air filtering systems rely on the molecular analysis of airborne contaminants to facilitate their removal. Furthermore, as with water management, nanomaterials may assist with the absorption or filtering of these pollutants, and our collaboration with microbes can play an important role. Platinum (Pt) nanoparticles are utilized to reduce unwanted exhaust gases from vehicles. By installing a platinum-containing catalytic converter into the exhaust system of the engine, unwanted emissions including nitrogen oxides, carbon monoxide, and hydrocarbons can be reduced. Ironically, however, Pt nanoparticles have since been found to be released into roadside environments with their own damaging downstream effects on ecosystems and human health (Taskula et al., 2022). Thus, better filtering is still required.

Nanotechnologies make a host of strategies possible for the utilization of atmospheric resources. The conversion of CO_2 in the ambient atmosphere to useful products like fuels or materials can be energy-intensive, and ironically, again, fossil fuels are typically burned to generate pure CO_2. To overcome these energy barriers, among other things, the efficient separation of multiple components from CO_2 including CO and other carbonyl products is required, for which nanoscale catalysts can play a crucial role. For example, batteries have been developed that suck in air or other gas streams, which passes over coated carbon nanotube electrodes as they charge up to subsequently release CO_2 in its pure form upon discharge (Voskian and Hatton, 2019). This pure CO_2 may be utilized for many applications including the carbonation of beverages or for greenhouses to facilitate plant growth. Alternately, nanotechnology can be applied to strip the CO_2 of its oxygen and convert the resulting carbon into hydrocarbon fuels such as methanol. Techniques include the use of copper nanowires with specifically designed surfaces to convert CO_2 from the ambient atmosphere to ethylene fuel (Choi et al., 2020), which is one of the most-used organic molecules for a range of industrial applications, as well as a precursor for the production of graphene (Wang et al., 2017).

4.3.1.4 Food

Food is the primary source of nutrients for the support of life. Not only do we rely on photosynthesis for our fuel and air, but also our food, whether directly or indirectly. Agriculture enabled the emergence of the first urban areas ~10,000 years ago. With rapid rates of urbanization, which means more people in one place, albeit fewer producing food in rural areas, our agricultural strategies need to advance to be more efficient. Precision indoor hydroponic farming, where soil can be replaced by moist air, uses 95% less water than traditional agriculture, removes the need for pesticides, and can be implemented vertically in urban areas to dramatically decrease the space required for production and reduce transportation distances. In such controlled environments, AI/data can be used to optimize nutrient ratios, as well as the light and air conditions for each species, for more nutritious food grown faster from fewer resources (Jacquez et al., 2023; Wang et al., 2023; Yegül, 2023).

In 2020, nearly one in three people in the world did not have access to adequate food (FAO, 2021). By 2050, the global population may reach 10 billion people, meaning that if we continue to produce food in the same way, an additional land area about the size of Australia will be required to feed us all. Genetically engineered plants, vertical hydroponic farming as mentioned above, and the generation of cultured meats derived from various animal cells (Aminian-Dehkordi et al., 2023; Ge et al., 2023; KhokharVoytas et al., 2023; Louis et al., 2023) could increase the amount of food we produce, while significantly reducing the land and resource requirements to produce them.

Nevertheless, while physics and chemistry may provide insights into the foods we eat and how they nourish our bodies, our understanding of biology remains incomplete. A range of unanticipated effects and potential health risks posed by the consumption of genetically engineered food have been predicted and can already be observed (Delaney et al., 2018; Kumar et al., 2020; Taylor and Hefle, 2002; National Research Council (US), 2004). Animals kept as livestock currently take up nearly 80% of global agricultural land, yet constitute less than 20% of consumed calories (FAO, 2009).

One solution might be cultured meat, which was first approved for commercial sale by the Singapore government in 2020 (Lucas, 2020). Cultured meat is produced from stem cells extracted from living animals, does not require their slaughter, is antibiotic-free, and identical to conventional meat at the cellular level (Broucke et al., 2023). Our expanding knowledge of life at the nanoscale has enabled us to think about food in this way, and even more recently to grow meat under chemically defined conditions; an important milestone for real food produced with laboratory efficiency.

4.3.2 Space Exploration

Our deepest enquiries into matter and life have shown us that all in the Universe is made of the same fundamental constituents of reality: energy, atoms, and molecules. However, in terms of attempting to understand what life is from our perspective on Earth, we have just a single data point: terrestrial life. Can life emerge elsewhere? Would we recognize it? These are queries that are difficult to address from Earth. The discovery of definitive evidence of extraterrestrial life, whether ancestral or completely distinct from life here on Earth, would be an immense leap toward understanding ourselves and our place in the Universe. One motivation for exploring Mars is that it may provide a crucial second data point toward elucidating who we are and where we come from.

Given that water is vital for all life here on Earth and that Mars had oceans four billion years ago, while considering the significant volume of debris that has been exchanged between Earth and Mars over millennia, it is not inconceivable that life on Earth was seeded by a Martian meteorite. Such an object may have already been found. The so-called Alan Hills meteor fragment was found in Antarctica in 1984 (The Meteoritical Society, 1984). Subsequently, President Clinton announced that "it speaks of the possibility of life" (The White House, 1996); and the debate continues as to whether it may contain fossilized evidence of extraterrestrial life from Mars.

The largely white surface of Antarctica lends itself to the search for rocks arriving from space. It is also arguably the most remote place that humans have visited, since it is completely inaccessible for months at a time during winter. No one lives there permanently, and overwinter crews rely on much anticipated resupply missions each summer. Our activities in space stations 400 km up in Earth's orbit and even our three-day journeys to the Moon a half-century ago have also relied heavily on the proximity of Earth. Arguably, at the heart of our challenges here on Earth is our lack of focus on the efficient repurposing of waste. There is a place, just within reach, that is both remote and extreme enough to teach us some important lessons on this topic: Mars.

The technology required to inhabit Mars already exists; living in the ISS or even on the Moon is more challenging from an engineering perspective than setting up camp on a rocky planet with significant gravity and an atmosphere. However, unlike Earth orbit or the Moon, Mars is on average over 200 million km away from Earth. Its remoteness will define approaches to infrastructure choices focused on circular resource utilization and increasing levels of independence from Earth. The human settlement of our neighboring planet will be a critical demonstration of community resilience under resource-constrained conditions. The tools and mindset to achieve this enterprise are just as critical for survival on Mars as for a more sustainable future here on Earth, where a growing population that competes for Earth's finite resources is facing increasingly uncertain times. Thus, nanotechnology, our ability to manipulate matter on the level of atoms and molecules, will play a critical role.

Elon Musk established SpaceX with the goal of expanding beyond Earth and has designed the Starship's propulsion system to run on fuel that can be produced on Mars from local resources like atmospheric CO_2. Assuming we get there, after securing shelter, the most fundamental step toward establishing a community on Mars will be power production. The availability of energy will enable all other functionalities of the settlement. If power systems fail, within minutes, so does life; thus, the establishment of multiple robust protocols for power redundancy is essential and indeed a matter of survival.

Radioisotope thermoelectric generators fueled by radioactive materials with appropriate shielding (e.g., graphene) are reliable sources of heat and energy that could be functional upon arrival at Mars. Solar irradiance on its surface is at best around half of that on Earth. While establishing a solar array and connecting batteries brought from Earth is a good start, the limited lifetimes of solar energy components under the harsh surface conditions of Mars in view of the challenges involved with manufacturing such technologies locally means that supporting or replacement systems will be required.

Fuel cells that can utilize local resources are good candidates for a Martian settlement not only for their energy generation and storage capacities but also because of their byproducts. For example, a reversible solid oxide fuel cell system (Gómez and Hotza, 2016) can store energy via the nanoscale electrolysis of water to produce molecular oxygen and hydrogen, with the energy generation phase generating water via a reaction between these gases.

While water can be extracted from surface sand using rovers and heating procedures, CO_2 can be directly pumped from the Martian atmosphere, and from either molecule, oxygen can be produced by electrolysis. However, pressurized breathable air also requires an inert gas like nitrogen, which makes up only a few percent of the Martian atmosphere. Another source of nitrogen is wastewater, managed through the controlled use of specific microbes employed to retrieve both pure water and resources like fuels and nutrients from the digestion of organic waste. Critical for the transformation of resources is the oldest living process on Earth, photosynthesis. Food for the community can be grown within agricultural units and photobioreactors optimized for biomass production. A lively ecosystem can be supplemented with insects and fish that are conveniently transportable via eggs, and sources of agricultural nutrients and dietary protein, such as algae, which are rich with a range of essential vitamins and minerals. Further, an array of medicinal plants, fungi, and bacteria, among other biomaterials, as sources of biomolecules for healthcare applications will be invaluable.

Over the longer term, a Mars settlement (Figure 4.7) would need to locally manufacture its infrastructure to generate power and cycle resources as outlined above, where 3D printing would serve as a critical starting point for such manufacturing capabilities.

In situ manufacturing and the incorporation of nanomaterials and technologies into the toolbox may lead to unexpected paradigm shifts. For example, new methods to mass produce graphene-like materials that combine hydrocarbon gas, oxygen, and a spark plug inside a chamber (Sorensen et al., 2016), or methods of producing hydrogen efficiently as already discussed, could mean that the impetus to innovate using graphene as a material and hydrogen as a fuel may be far greater on Mars than Earth. This is because profitable legacy systems and their custodians here on Earth tend to impede progress.

The establishment of a viable healthcare system 200 million km from Earth would be a challenge unlike anything we have ever faced. While telemedicine may have its place terrestrially here on Earth, emergency surgery from 10 light minutes away would not be practical. The first inhabitants of

FIGURE 4.7 Conceptual Mars habitat, 2021. (Reproduced from Spongy101010 Creative Commons Attribution-Share Alike 4.0 International.)

Mars will rely on local expertise and pharmaceutical production for treatment. Essential medicines like antibiotics, as well as anesthetics, painkillers, and perhaps cancer treatment drugs, will eventually need to be produced locally. Here, our understanding of biomolecules produced in greenhouses, and nanotechnological strategies for the extraction (and the eventual molecular manufacturing) of the active ingredients for such drugs, as well as their optimized delivery in the body, will be critical.

This is where nanomedicine may assist, as advances in this discipline concomitant with our arrival on Mars may allow for a healthcare system that emulates a planetwide model envisaged for here on Earth, as Global Health Care Equivalency (GHCE) (Boehm, 2025). Molecular manufacturing (the progeny of today's 3D printing technologies) (Dinu et al., 2007; Drexler, 1995) may enable us to generate autonomous nanomedical devices atomic layer by atomic layer, which may eventually have the capacity to address practically any condition (including aging) afflicting the inhabitants of a Mars colony (Boehm, 2013).

While deep space missions pose the biggest challenges imaginable to efficient resource usage (when explorers and space settlers may be travelling between stars for generations), planning to establish communities on the Moon, Mars, and other appealing locations in the Solar System will certainly be enough to occupy us for now. The nanomachinery of life has already mastered the efficient utilization of abundant resources to thrive on this planet. Once we deepen and more fully immerse ourselves in collaboration with Nature, further secrets of how to achieve this level of mastery at the nanoscale using the building blocks of life here on Earth, or in even more extreme environments off-world, promise to be revealed.

4.4 CONCLUSION

It has taken billions of years for the conditions in which we live to be established. In a relatively extremely short period of time, we are having significantly destabilizing global scale impacts on this environment. The drive to understand and control our environment has resulted in rapid technological development, which has taken us down to the subatomic level as well as to the interstellar space beyond our Solar System. The industries that support this development and the technology-enabled lifestyles of a portion of our society are having increasingly deleterious impacts on our life-support systems. This, while much of Earth's population struggles in relative poverty without access to clean water, fresh air, nutritious food, or the reliable power and communication systems that are essential to participate meaningfully in this era. It is abundantly clear that for an ever-expanding population inculcated in a culture of perpetual greed-based "profit above all" growth (which is decoupled from reality), and over-the-top materialism confined to a single planet, things do not end well.

Our understanding of the nanoworld at the level of atoms and molecules comes with the responsibility to apply this knowledge for the regeneration of our natural environment. Further, to restore good health for all humanity through access to basic resources and a robust, massively distributed global healthcare system (e.g., GHCE), such that we may survive this tumultuous transitional era and emerge into a new 21st century world paradigm of abundance and peace. Our knowledge of the very small can enable us to secure a more sustainable future in harmony with the natural environment and each other on our home planet Earth, and to expand beyond Earth toward an even more advanced understanding of reality and our role in it.

Author Biography

Adriana Marais has a background in theoretical physics; her PhD and postdoctoral work in quantum biology focused on photosynthesis and the origins of the building blocks of life in space. She is currently developing a blockchain-based economic system for extreme and resource-constrained environments toward a second PhD. Adriana is a Director at the Foundation for Space Development Africa, developing Africa's first mission to the Moon, the Africa2Moon Project. She is a Scientific Moderator on Space Resources with the Geneva Science and Diplomacy Anticipator, Chair of Space Sciences at Tod'Aérs Aeronautics and Space Research, a member of the South African Government task team on the 4th Industrial Revolution, and Faculty at the Singularity University and Duke Corporate Education. In 2019, Adriana left her position as Head of Innovation at SAP Africa to found Proudly Human, and she is currently leading the organization's Off-World Project: a series of off-grid habitation experiments in the most extreme environments on the planet, in preparation for life on the Moon, Mars and beyond, as well as a sustainable future on Earth.

REFERENCES

Adams B, Petruccione F. Quantum effects in the brain: A review. *AVS Quantum Science*. 2020;2(2): 022901.

Ahmad M. TSMC approaching 1 nm with 2D materials breakthrough. *EDN Aspencore Network*, November 3, 2022. https://www.edn.com/tsmc-approaching-1-nm-with-2d-materials-breakthrough/ (accessed June 19, 2023).

Aminian-Dehkordi J, Rahimi S, Golzar-Ahmadi M, Singh A, Lopez J, Ledesma-Amaro R et al. Synthetic biology tools for environmental protection. *Biotechnology Advances*. 2023;68:108239.

Bakry R, Vallant RM, Najam-ul-Haq M, Rainer M, Szabo Z, Huck CW et al. Medicinal applications of fullerenes. *International Journal of Nanomedicine*. 2007;2(4):639–649.

Bardeen J, Brattain WH. The transistor, a semiconductor triode. *Physical Review*. 1948;74(2):230–231.

Barredo D, Lienhard V, de Léséleuc S, Lahaye T, Browaeys A. Synthetic three-dimensional atomic structures assembled atom by atom. *Nature*. 2018;561(7721):79–82.

Binnig G, Rohrer H. Scanning tunneling microscopy. *IBM Journal of Research and Development*. 1986;30(4):355–369.

Boehm FJ. Nanotechnology, Nanomedicine, and AI: Toward the Dream of Global Health Care Equivalency. 2025. https://www.nanoappsmedical.com/nanotechnology-nanomedicine-and-ai-toward-the-dream-of-global-health-care-equivalency/ (accessed March 28, 2025).

Boehm FJ. An Investigation of Nucleic Acid/DNA-Based Manufacturing. *Research contribution to the Center for Responsible Nanotechnology: Thirty Essential Nanotechnology Studies – #10.* August 2004. https://www.scribd.com/document/183051565/DNA-MANUFACTURING-pdf# (accessed June 20, 2023).

Boehm FJ (Ed.). Chapter 16 "Nanomedical Device and Systems Design in Space Applications." *Nanomedical Device and Systems Design: Challenges, Possibilities, Visions* (1st ed.). CRC Press, 2013.

Boehm HP, Clauss A, Fischer GO, Hofmann U. Surface properties of extremely thin graphite lamellae. In: *Graphite and Precursors.* Amsterdam: Elsevier; 1962. p. 73–80.

Broucke K, Van Pamel E, Van Coillie E, Herman L, Van Royen G. Cultured meat and challenges ahead: A review on nutritional, technofunctional and sensorial properties, safety and legislation. *Meat Science.* 2023;195:109006.

Burbidge G. Was there really a Big Bang? *Nature.* 1971;233:36–40.

Bushdid C, Magnasco MO, Vosshall LB, Keller A. Humans can discriminate more than 1 trillion olfactory stimuli. *Science.* 2014;343(6177):1370–1372.

Ceballos G, Ehrlich PR, Barnosky AD, García A, Pringle RM, Palmer TM. Accelerated modern human-induced species losses: Entering the sixth mass extinction. *Science Advances.* 2015;1(5):e1400253.

Celotta RJ, Balakirsky SB, Fein AP, Hess FM, Rutter GM, Stroscio JA. Invited article: Autonomous assembly of atomically perfect nanostructures using a scanning tunneling microscope. *Review of Scientific Instruments.* 2014;85(12):121301.

Chen X, Selloni A. Introduction: Titanium dioxide (TiO_2) nanomaterials. *Chemical Reviews.* 2014;114(19):9281–9282.

Choi C, Kwon S, Cheng T, Xu M, Tieu P, Lee C et al. Highly active and stable stepped Cu surface for enhanced electrochemical CO_2 reduction to C_2H_4. *Nature Catalysis.* 2020;3:804–812.

Cohen-Tanugi D, Grossman JC. Water desalination across nanoporous graphene. *Nano Letters.* 2012;12(7):3602–3608.

Cowie RH, Bouchet P, Fontaine B. The sixth mass extinction: Fact, fiction or speculation? *Biological Reviews of the Cambridge Philosophical Society.* 2022;97(2):640–663.

Custance O, Perez R, Morita S. Atomic force microscopy as a tool for atom manipulation. *Nature Nanotechnology.* 2009;4(12):803–810.

d'Aquino AE, Kim DS, Jewett MC. Engineered ribosomes for basic science and synthetic biology. *Annual Review of Chemical and Biomolecular Engineering.* 2018;9:311–340.

Delaney B, Goodman RE, Ladics GS. Food and feed safety of genetically engineered food crops. *Toxicological Sciences.* 2018;162(2):361–371.

Dinu CZ, Chrisey DB, Diez S, Howard J. Cellular motors for molecular manufacturing. *Anatomical Record (Hoboken).* 2007;290(10):1203–1212.

Dogutan DK, Nocera DG. Artificial photosynthesis at efficiencies greatly exceeding that of natural photosynthesis. *Accounts of Chemical Research.* 2019;52(11):3143–3148.

Drexler KE. *Engines of Creation.* Anchor Books, 1986.

Drexler KE. Molecular manufacturing: Perspectives on the ultimate limits of fabrication. *Philosophical Transactions of the Royal Society of London. Series A: Physical and Engineering Sciences.* 1995;353(1703):323–331.

Dyck O, Kim S, Jimenez-Izal E, Alexandrova AN, Kalinin SV, Jesse S. Building structures atom by atom via electron beam manipulation. *Small.* 2018;14(38):e1801771.

Eigler DM, Erhard KS. Positioning single atoms with a scanning tunnelling microscope. *Nature.* 1990;344(6266):524–526.

Epum EA, Haber JE. DNA replication: The recombination connection. *Trends in Cell Biology.* 2022;32(1):45–57.

Fahrenbach AC, Warren SC, Incorvati JT, Avestro AJ, Barnes JC, Stoddart JF et al. Organic switches for surfaces and devices. *Advanced Materials.* 2013;25(3):331–348.

FAO, IFAD, UNICEF, WFP, WHO. 2021. The State of Food Security and Nutrition in the World 2021. Transforming food systems for food security, improved nutrition and affordable healthy diets for all. Rome, FAO. 2021. https://doi.org/10.4060/cb4474en

Feynman RP. There's plenty of room at the bottom. *Journal of Microelectromechanical Systems.* 1992;1(1): 60–66. (A transcript of the talk given by Richard P. Feynman on December 26, 1959).

Fleischer R. (Director), *Fantastic Voyage.* Brandywine Productions, 1966.

Fleischmann RD, Adams MD, White O, Clayton RA, Kirkness EF, Kerlavage AR et al. Whole-genome random sequencing and assembly of Haemophilus influenzae Rd. *Science.* 1995;269(5223):496–512.

Foley LE, Gegear RJ, Reppert SM. Human cryptochrome exhibits light-dependent magnetosensitivity. *Nature Communications.* 2011;2:356.

Food and Agricultural Organization (FAO) of the United Nations. The State of Food and Agriculture Livestock in the Balance. Rome, 2009. https://www.fao.org/3/i0680e/i0680e.pdf (access August 31, 2023).

Franco MI, Turin L, Mershin A, Skoulakis EM. Molecular vibration-sensing component in Drosophila melanogaster olfaction. *Proceedings of the National Academy of Sciences of the United States of America.* 2011;108(9):3797–3802.

Franklin RE, Gosling RG. The structure of sodium thymonucleate fibres. I. The influence of water content. *Acta Crystallographica.* 1953;6(8–9):673–677.

Freitas RA Jr. Some Limits to Global Ecophagy by Biovorous Nanoreplicators, with Public Policy Recommendations. *Zyvex Corp. report.* April 2000. http://www.rfreitas.com/Nano/Ecophagy.htm (accessed June 20, 2023).

Fukutani Y, Abe M, Saito H, Eguchi R, Tazawa T, de March CA et al. Antagonistic interactions between odorants alter human odor perception. *Current Biology.* 2023;33(11):2235–2245.e4.

Gane S, Georganakis D, Maniati K, Vamvakias M, Ragoussis N, Skoulakis EM et al. Molecular vibration-sensing component in human olfaction. *PLoS One.* 2013;8(1):e55780.

Gao Y, Qin Y, Zhang M, Xu L, Yang Z, Xu Z et al. Revealing the role of oxygen-containing functional groups on graphene oxide for the highly efficient adsorption of thorium ions. *Journal of Hazardous Materials.* 2022;436:129148.

Ge C, Selvaganapathy PR, Geng F. Advancing our understanding of bioreactors for industrial-sized cell culture: Health care and cellular agriculture implications. *American Journal of Physiology-Cell Physiology.* 2023;325(3):C580–C591.

Gibney E. Earth has water older than the Sun. *Nature News.* 2014. https://doi.org/10.1038/nature.2014.16011 (accessed June 05, 2023).

Giftson F, SivaKumar G. Nano particles in automobile tires. *IOSR Journal of Mechanical and Civil Engineering (IOSR-JMCE).* 2014;11(4):2320–2334.

Girbig M, Misiaszek AD, Müller CW. Structural insights into nuclear transcription by eukaryotic DNA-dependent RNA polymerases. *Nature Reviews Molecular Cell Biology.* 2022;23(9):603–622.

Gómez SY, Hotza D. Current developments in reversible solid oxide fuel cells. *Renewable and Sustainable Energy Reviews.* 2016;61:155–174.

Guo H, Rubinstein JL. Structure of ATP synthase under strain during catalysis. *Nature Communications.* 2022;13(1):2232.

Haguenau F, Hawkes PW, Hutchison JL, Satiat-Jeunemaître B, Simon GT, Williams DB. Key events in the history of electron microscopy. *Microscopy and Microanalysis.* 2003;9(2):96–138.

Hameroff S, Nip A, Porter M, Tuszynski J. Conduction pathways in microtubules, biological quantum computation, and consciousness. *Biosystems.* 2002;64(1–3):149–168.

Harikane Y et al. A search for *H*-dropout Lyman break galaxies at *z* ~12–16. *Astrophysical Journal.* 2022;929(1).

Harrow AW, Montanaro A. Quantum computational supremacy. *Nature.* 2017;549(7671):203–209.

Hecht S, Shlaer S, Pirenne MH. Energy, quanta, and vision. *Journal of General Physiology.* 1942;25(6):819–840.

Helmenstine AM. "The Visible Spectrum: Wavelengths and Colors." *ThoughtCo.* 2020. https://www.thoughtco.com/understand-the-visible-spectrum-608329 (accessed June 10, 2023).

Huang ZW, Li ZJ, Zheng LR, Wu WS, Chai ZF, Shi WQ. Adsorption of Eu(III) and Th(IV) on three-dimensional graphene-based macrostructure studied by spectroscopic investigation. *Environmental Pollution.* 2019;248:82–89.

Hutchison 3rd CA et al. Design and synthesis of a minimal bacterial genome. *Science.* 2016;351(6280): aad6253.

Mendoza Jacquez JA, Tiwari DK, Pérez González R. Lettuce production using nano nutrient-based hydroponic system: vertical plant factory production. *Microsc Microanal.* 2023;29(Suppl 1):72–73.

Jain R, Aggarwal AK, Rechkoblit O. Eukaryotic DNA polymerases. *Current Opinion in Structural Biology.* 2018;53:77–87.

Jinek M, Chylinski K, Fonfara I, Hauer M, Doudna JA, Charpentier E. A programmable dual-RNA-guided DNA endonuclease in adaptive bacterial immunity. *Science.* 2012;337(6096):816–821.

Jung J, Kobayashi C, Kasahara K, Tan C, Kuroda A, Minami K et al. New parallel computing algorithm of molecular dynamics for extremely huge scale biological systems. *Journal of Computational Chemistry.* 2021;42(4):231–241.

Kalra AP, Benny A, Travis SM, Zizzi EA, Morales-Sanchez A, Oblinsky DG et al. Electronic energy migration in microtubules. *ACS Central Science.* 2023;9(3):352–361.

Khatter H, Myasnikov AG, Natchiar SK, Klaholz BP. Structure of the human 80S ribosome. *Nature.* 2015;520(7549):640–645.

KhokharVoytas A, Shahbaz M, Maqsood MF, Zulfiqar U, Naz N, Iqbal UZ et al. Genetic modification strategies for enhancing plant resilience to abiotic stresses in the context of climate change. *Functional & Integrative Genomics*. 2023;23(3):283.

Klena N, Pigino G. Structural biology of cilia and intraflagellar transport. *Annual Review of Cell and Development Biology*. 2022;38:103–123.

Klimkaite L, Liveikis T, Kaspute G, Armalyte J, Aldonyte R. Air pollution-associated shifts in the human airway microbiome and exposure-associated molecular events. *Future Microbiol*. 2023;18:607–623.

Klug A. The discovery of the DNA double helix. *Journal of Molecular Biology*. 2004;335(1):3–26.

Kokarneswaran M, Selvaraj P, Ashokan T, Perumal S, Sellappan P, Murugan KD et al. Discovery of carbon nanotubes in sixth century BC potteries from Keeladi, India. *Scientific Reports*. 2020;10(1):19786.

Krishnan Y, Seeman NC. Introduction: Nucleic acid nanotechnology. *Chemical Reviews*. 2019;119(10):6271–6272.

Krittanawong C, Qadeer YK, Lavie CJ. Air pollution, physical activity, and lifespan. *Mayo Clinic Proceedings*. 2023;98(8):1113–1115.

Kroto HW, Heath JR, O'Brien SC, Curl RF, Smalley RE. C_{60}: Buckminsterfullerene. *Nature*. 1985; 318:162–163.

Kumar K, Gambhir G, Dass A, Tripathi AK, Singh A, Jha AK et al. Genetically modified crops: Current status and future prospects. *Planta*. 2020;251(4):91.

Kurtev S. Wave-like patterns in parameter space interpreted as evidence for macroscopic effects resulting from quantum or quantum-like processes in the brain. *Scientific Reports*. 2022;12(1):18938.

Lander ES et al. International human genome sequencing consortium. Initial sequencing and analysis of the human genome. *Nature*. 2001;409(6822):860–921.

Lingamdinne LP, Koduru JR, Karri RR. A comprehensive review of applications of magnetic graphene oxide-based nanocomposites for sustainable water purification. *Journal of Environmental Management*. 2019;231:622–634.

Locey KJ, Lennon JT. Scaling laws predict global microbial diversity. *Proceedings of the National Academy of Sciences of the United States of America*. 2016;113(21):5970–5975.

Loeb NG, Johnson GC, Thorsen TJ, Lyman JM, Rose FG, Kato S. Satellite and ocean data reveal marked increase in Earth's heating rate. *Geophysical Research Letters*. 2021;48:e2021GL093047.

Louis F, Furuhashi M, Yoshinuma H, Takeuchi S, Matsusaki M. Mimicking Wagyu beef fat in cultured meat: Progress in edible bovine adipose tissue production with controllable fatty acid composition. *Materials Today Bio*. 2023;21:100720.

Lucas A. "Singapore issues first regulatory approval for lab-grown meat to Eat Just." CNBC. December 2020 (accessed June 01, 2023).

Maier JP, Campbell EK. Fullerenes in space. *Angewandte Chemie International Edition*. 2017;56(18):4920–4929.

Manojlović LM. Photometry-based estimation of the total number of stars in the universe. *Applied. Optics*. 2015;54:6589–6591.

Marais A, Adams B, Ringsmuth AK, Ferretti M, Gruber JM, Hendrikx R et al. The future of quantum biology. *Journal of the Royal Society Interface*. 2018;15(148):20180640.

Melkikh AV, Khrennikov A. Nontrivial quantum and quantum-like effects in biosystems: Unsolved questions and paradoxes. *Progress in Biophysics and Molecular Biology*. 2015;119(2):137–161.

Mousavibaygei SR, Bisadi A, ZareSakhvidi F. Outdoor air pollution exposure, bone mineral density, osteoporosis, and osteoporotic fractures: A systematic review and meta-analysis. *Science of the Total Environment*. 2023;865:161117.

"Nanocar-Race II." March 2022. https://memo-project.eu/flatCMS/index.php/Nanocar-Race-II (accessed June 01, 2023).

National Research Council (US) Committee on Identifying and Assessing Unintended Effects of Genetically Engineered Foods on Human Health. *Safety of Genetically Engineered Foods: Approaches to Assessing Unintended Health Effects*. Washington (DC): National Academies Press (US), 2004.

Németh P, McColl K, Smith RL, Murri M, Garvie LAJ, Alvaro M et al. Diamond-graphene composite nanostructures. *Nano Letters*. 2020;20(5):3611–3619.

Novoselov KS, Geim AK, Morozov SV, Jiang D, Zhang Y, Dubonos SV et al. Electric field effect in atomically thin carbon films. *Science*. 2004;306(5696):666–669.

Pesavento U. An implementation of von Neumann's self-reproducing machine. *Artificial Life*. 1995; 2(4):337–354.

Pollard TD, Goldman RD. Overview of the cytoskeleton from an evolutionary perspective. *Cold Spring Harbor Perspectives in Biology*. 2018;10(7):a030288.

Polli D, Altoè P, Weingart O, Spillane KM, Manzoni C, Brida D et al. Conical intersection dynamics of the primary photoisomerization event in vision. *Nature*. 2010;467(7314):440–443.

Prioleau MN, MacAlpine DM. DNA replication origins—where do we begin? *Genes & Development*. 2016;30(15):1683–1697.

Rabiee N, Sharma R, Foorginezhad S, Jouyandeh M, Asadnia M, Rabiee M et al. Green and sustainable membranes: A review. *Environmental Research*. 2023;231:116133.

Radushkevich LV, Lukyanovich VM. O strukture ugleroda, obrazujucegosja pri termiceskom razlozenii okisi ugleroda na zeleznom kontakte (On the structure of carbon formed by the thermal decomposition of carbon monoxide to the contact with iron). *Zh. Fis. Khim. (Russ. J. Phys. Chem.)*. 1952;26:88–95.

Raphey VR, Henna TK, Nivitha KP, Mufeedha P, Sabu C, Pramod K. Advanced biomedical applications of carbon nanotube. *Material Science & Engineering C-Materials for Biological Applications*. 2019;100:616–630.

Rosen R. *Essays on Life Itself*. Columbia University Press, 2000.

Roy AB, Dhar A, Choudhuri M, Das S, Hossain SM, Kundu A. Black silicon solar cell: Analysis optimization and evolution towards a thinner and flexible future. *Nanotechnology*. 2016;27(30):305302.

Sattar T. Current review on synthesis, composites and multifunctional properties of graphene. *Topics in Current Chemistry (Cham)*. 2019;377(2):10.

Serrate D, Ferriani P, Yoshida Y, Hla SW, Menzel M, von Bergmann K et al. Imaging and manipulating the spin direction of individual atoms. *Nature Nanotechnology*. 2010;5(5):350–353.

Shao HQ, Wei H, He JH. Dynamic properties and tire performances of composites filled with carbon nanotubes. *Rubber Chemistry and Technology*. 2018;91(3):609–620.

Singh BK, Mahzan NS, Abdul Rashid NS, Isa SA, Hafeez MA, Saslow S et al. Design and application of materials for sequestration and immobilization of ^{99}Tc. *Environmental Science & Technology*. 2023;57(17):6776–6798.

Singh NB, Susan MA, Guin M. Applications of green synthesized nanomaterials in water remediation. *Current Pharmaceutical Biotechnology*. 2021;22(6):733–761.

Slocombe L, Al-Khalili JS, Sacchi M. Quantum and classical effects in DNA point mutations: Watson-Crick tautomerism in AT and GC base pairs. *Physical Chemistry Chemical Physics*. 2021;23(7):4141–4150.

Sorensen C, Nepal A, Singh GP. Process for high-yield production of graphene via detonation of carbon-containing material. U.S. Patent No. 9,440,857. September 13, 2016. https://patents.google.com/patent/US20140335010A1/en (accessed June 02, 2023).

Sunyer J, Dadvand P. Air pollution and COVID-19 severity. *European Respiratory Journal*. 2023;62(1):2300818.

Tafoya S, Bustamante C. Molecular switch-like regulation in motor proteins. *Philosophical Transactions of the Royal Society of London. Series B, Biological Sciences*. 2018;373(1749):20170181.

Taskula S, Stetten L, von der Kammer F, Hofmann T. Platinum nanoparticle extraction, quantification, and characterization in sediments by single-particle inductively coupled plasma time-of-flight mass spectrometry. *Nanomaterials (Basel)*. 2022;12(19):3307.

Taylor SL, Hefle SL. Genetically engineered foods: Implications for food allergy. *Current Opinion in Allergy and Clinical Immunology*. 2002;2(3):249–252.

The Meteoritical Society. "Meteoritical Bulletin Database: Allan Hills 84001." December 1984. https://www.lpi.usra.edu/meteor/metbull.php?code=604 (accessed June 01, 2023).

The Wachowskis (Directors). *The Matrix*. Warner Bros, 1999.

The White House. "Clinton WJ. Statement on the NASA discovery of possible evidence of life on Mars." 1996 Aug 7. https://clinton.presidentiallibraries.us/items/show/36218 (accessed August 11, 2025).

TOP500. The List "November 2021." November 2021. https://top500.org/lists/top500/2021/11/ (accessed June 01, 2023).

Tscherbul TV, Brumer P. Quantum coherence effects in natural light-induced processes: cis-trans photoisomerization of model retinal under incoherent excitation. *Physical Chemistry Chemical Physics*. 2015;17(46):30904–30913.

Turin L, Skoulakis EK, Horsfield AP. Electron spin changes as a general mechanism for general anesthesia? *Proceedings of the National Academy of Sciences*. 2014;111(35):12598–12603.

Turvey ST, Crees JJ. Extinction in the Anthropocene. *Current Biology*. 2019;29(19):R982–R986.

van den Heuvel MG, Dekker C. Motor proteins at work for nanotechnology. *Science*. 2007 Jul 20;317 (5836):333–6.

van Grondelle R, Novoderezhkin VI. Quantum effects in photosynthesis. *Procedia Chemistry*. 2011; 3(1):198–210.

Venter JC et al. The sequence of the human genome. *Science*. 2001;291(5507):1304–1351.

von Neumann J, Burks AW. *Theory of Self-Reproducing Automata*. University of Illinois Press, 1966.

Voskian S, Hatton TA. Faradaic electro-swing reactive adsorption for CO_2 capture. *Energy & Environmental Science*. 2019;12(12):3530–3547.

Wang B, König M, Bromley CJ, Yoon B, Treanor M, Torres JA et al. Ethene to graphene: Surface catalyzed chemical pathways, intermediates, and assembly. *Journal of Physical Chemistry C*. 2017; 121:9413–9423.

Wang C, Chen Y, Sun M, Wu J. Potential of technological innovation to reduce the carbon footprint of urban facility agriculture: A food-energy-water-waste nexus perspective. *Journal of Environmental Management*. 2023;339:117806.

Wang CX, Hilburn IA, Wu DA, Mizuhara Y, Cousté CP, Abrahams JNH et al. Transduction of the geomagnetic field as evidenced from alpha-band activity in the human brain. *eNeuro*. 2019;6(2):ENEURO.0483-18.2019.

Wang L, Chen Y. Lanthanide doped carbon dots as a fluorescence chromaticity-based pH probe. *Mikrochimica Acta*. 2018;185(10):489.

Watson JD, Crick FH. Molecular structure of nucleic acids; a structure for deoxyribose nucleic acid. *Nature*. 1953;171(4356):737–738.

Wiltschko R, Wiltschko W. Magnetoreception in birds. *Journal of the Royal Society Interface*. 2019; 16(158):20190295.

Wollman AJ, Nudd R, Hedlund EG, Leake MC. From Animaculum to single molecules: 300 years of the light microscope. *Open Biology*. 2015;5(4):150019.

Xie C. Searching for unity in diversity of animal magnetoreception: From biology to quantum mechanics and back. *Innovation (Camb)*. 2022;3(3):100229.

Xu Y, Kuzmin A, Knehr E, Blaicher M, Ilin K, Dietrich PI et al. Superconducting nanowire single-photon detector with 3D-printed free-form microlenses. *Optics Express*. 2021;29(17):27708–27731.

Yao B, Sun H, Yang L, Wang S, Liu X. Recent progress in light-driven molecular shuttles. *Frontiers in Chemistry*. 2022;9:832735.

Yegül U. Development of an embedded software and control kit to be used in soilless agriculture production systems. *Sensors (Basel)*. 2023;23(7):3706.

Zeng G, Pham TA, Vanka S, Liu G, Song C, Cooper JK et al. Development of a photoelectrochemically self-improving Si/GaN photocathode for efficient and durable H_2 production. *Nature Materials*. 2021;20(8):1130–1135.

Zhang Q, Gao Y, Xu Z, Wang S, Kobayashi H, Wang J. The effects of oxygen functional groups on graphene oxide on the efficient adsorption of radioactive iodine. *Materials (Basel)*. 2020;13(24):5770.

Zhang X, Liu Y, Yao C, Niu J, Li H, Xie C. Facile and stable fabrication of wafer-scale, ultra-black *c*-silicon with 3D nano/micro hybrid structures for solar cells. *Nanoscale Advances*. 2022;5(1):142–152.

5 Development of Universal Medical Millirobots

Sook Kuan Goh, Sumeet Chaudhary, Griffin Wagner, Nicholas Lalley, John Yin, Sarah Pixley, Kevin J. Haworth, Maham Rahimi, Marc Cahay, Keshav Alagarsamy, Zhongyun Dong, Frank J. Boehm, and Mark Schulz

CONTENTS

5.1 OVERVIEW OF MEDICAL GOALS AND ROBOT DESIGN CHALLENGES

Nanomedicine is a term used to imply the diminutive dimensions of structures and mechanisms involved in the medical field. The scale range of biomedical components (e.g., molecular, organellar, cellular, organs, tissues) represents significant challenges for medical research (Dogangil et al.,

DOI: 10.1201/9781003680093-5

2010; Carstensen and Parker, 2014; Engh et al., 2006; Frasson et al., 2012; Hambley et al., 1988; Lissandrello et al., 2017; Stanford University, Lung Cancer, 2025; Moreira and Misra, 2015; Perica et al. 2014; Taylor and Anderson, 2000; Wang and Gao, 2012). Envisaged nanomedical devices, potentially ~1000 nanometers (nm) (1 micron (μm)) in diameter and larger, and often remotely operated, will be required to diagnose and treat diseases in the future.

In 1999, Robert A. Freitas Jr. published *Nanomedicine, Volume I: Basic Capabilities*, which was the first book to describe the myriad technical challenges involved in the medical applications of molecular nanotechnology and nanodevice design (Freitas, 1999). Conceptual autonomous nanodevices and nanorobots, preferentially comprised of diamondoid or sapphire nanomaterials with dimensions in the nanometer range, are described in great detail. Nevertheless, ~25 years later, the required prerequisite technology (e.g., molecular manufacturing (MM)) (Ballard et al., 2015) that will enable the atomically precise fabrication of practical electromechanical nanodevices has yet to materialize.

However, technologies are now emerging with the capacity to fabricate and mass-produce devices at the millimeter scale with sub-micron features through the use of nanomaterials. These robots with nanoscale features/materials will have the ability to operate at the cellular level and work on larger structures. Though this chapter is currently speculative, the goal is to inspire and expedite research toward the development of microscale machines/devices to address myriad medical conditions. We survey the primary challenges that may limit progress in the creation of millimeter sized robots (millirobots), as well as new microtechnologies that might assist with overcoming these barriers. Further, we provide concepts for numerous nanorobotic devices that may be feasible to manufacture in the relatively near-term, which may assist in addressing specific medical maladies for which there are currently no cures.

Conventional medical robots for minimally invasive soft tissue surgeries confront significant challenges (Dogangil et al., 2010), as current technologies for the construction of small robotic devices lack the precision or actuation required (Ali et al., 2019; Carstensen and Parker, 2014; Dhar et al., 2024; Engh et al., 2006; Moreira and Misra, 2015; Taylor and Anderson, 2000; Wang and Gao, 2012). Since multi-steerable catheters and ultrasound guided needles have difficulty with accessing remote physiological sites (Engh et al., 2006; Moreira and Misra, 2015), the diminutive sizes and dynamic in vivo mobility of the millirobots proposed herein are novel.

Earlier efforts (Taylor and Anderson, 2000) in this area did not produce as small a robot as proposed here, partially because the leveraging technology was not yet available. At this level of miniaturization, the scientific problems to be overcome (e.g., power and mobility) have traditionally been insurmountable when employing conventional materials and manufacturing processes (Frasson et al., 2012; Hambley et al., 1988; Lissandrello et al., 2017; Stanford University, Lung Cancer, 2025; Perica et al., 2014). We addressed these challenges by proposing designs for a new class of robot that incorporates carbon nanotube (CNT) biomedical wire, hydraulics, miniaturized pull cables, microtubing, and 3D microprinting.

Relevant to state-of-the-art research and practice, one of the major hurdles in robot design is power (Freitas, 1999, 2019), as miniaturization does not favor traditional chemical battery technologies. Wireless power transfer (e.g., utilizing externally transmitted radio waves to generate electricity within the body) also becomes problematic at small scales. To harvest energy, a nano- or microrobot would need to incorporate an antenna that is larger than the robot itself, which would have to stay in proximity to the energy source.

In terms of propulsion, options include the creation of microchemical rockets, which can react with innate substances in vivo (e.g., gastric acid) to traverse the human body. Further, bio-hybrids may be devised, for which bacteria are harnessed to provide the propulsive force (swimming) to pursue a target. This would be based on changes in the concentration gradients of particular molecules, or microswimmers that follow changes in pH or temperature.

Existing approaches for the actuation and control of 2D mobile microrobots include magnetically induced crawling and rotation (Lee et al., 2021), thermally driven, electrically driven, acoustically powered (microswimmers), piezoelectric-magnetic, electrowetting, and potentially many others (Abdelmohsen et al., 2014; Ahn et al., 1993; Büttgenbach, 2014; Garcia-Gradilla et al., 2013; Garimella and Eltorai, 2016; Golestanian et al., 2005; Ibele et al., 2009; Magdanz et al., 2013; Manesh et al., 2010; Mei et al., 2008; Mirkovic et al., 2010; Mozaffarian et al., 2016; Nelson et al., 2010; Paxton et al., 2004; Peng et al., 2016; Tang et al., 2012; Takato et al., 2016; Trimmer and Gabriel, 1987; Zygomalas et al., 2015).

The diminutive scales of micro/nanoscale robots require the use of propulsion strategies that are significantly unlike those employed in the macroscale world. Considerable additional efforts are required toward the development of high-performance nanomotor propulsion to enhance the force required for tissue penetration, as well as to add new functionalities (e.g., actuation, imaging) to these nano/microscale machines. This, while ensuring the use of non-toxic, biocompatible, and environmentally benign materials. External guidance or power will be required to operate more advanced microdevices within the body.

In the submillimeter domain, options for millirobot propulsion and navigation become limited. In our preceding work, we relied on flexible tethers such as a cable or multi-lumen tubes that can be pushed or pulled; however, this limited robotic locomotion due to tether flexibility and friction issues. Moving a microrobot with magnetic fields is possible, albeit this requires six actuators for a full six degrees of freedom. The simplest robots may be steered magnetically; however, since the magnetic flux drops off with the square of distance, the magnetic force is difficult to control. Benchtop testing that involved the use of a simple tube revealed that magnetic propulsion was jerky due to a combination of robot friction against the tube (vessel) wall and nonlinear magnetic force over a distance.

A Micro-Pharmaceutical-Delivery-Vehicle was envisioned as a tethered magnetically navigated vehicle to deliver drugs/nanoparticles directly to a target through a hollow tether. More complex diagnostic and therapeutic functions will require onboard actuators that cannot be controlled via external magnetic fields. It is anticipated that future medical demands will require the development of various classes of microrobots/nanorobots with dedicated functionalities. Some of these will utilize chemical reactions or ultrasound for propulsion, while others employ pliable braided nanotube yarn tethers to power electric motors/actuators, and micro-lumen tubing to provide fluid for the operation of hydraulic motors, communications, and the retrieval of robots. A solenoid actuated tool is a further concept, for which tethered robots with electrical and hydraulic power are considered the most practical approach; thus, magnetic, wireless, battery, and chemical power approaches are not investigated here. New technologies that will enable the design/development of millirobots are discussed next.

5.2 NEW TECHNOLOGIES

New materials and instrumentation will enable the development of microrobots and nanotools to operate within the human body for the improved diagnosis and treatment of diseases for which no cure is currently known. Thanks to microextrusion technologies (e.g., MicroLumen, Inc.), customized microtubing is available in sizes that range from only a Ø0.1 mm (Ø100 μm) interior lumen with a wall thickness of ~0.05 mm (50 μm). Extruded tubing such as microcatheters and peripheral intravenous cannulas are applied in microfluidics, pediatrics, and microanalytics. Millirobot devices will be comprised of miniaturized diagnostic instruments, innovative imaging technologies, and advanced therapeutic systems that provide far greater functionality, dexterity, and accuracy than conventional devices.

This will enable specialist clinicians to perform minimally invasive and percutaneous medical procedures to address an extensive array of medical conditions, to drastically reduce inconclusive outcomes and medical errors. This need-driven medical device technology can be quickly translated from the bench to the bedside to enhance the daily lives of patients. Millirobot development is inspired by the demand for treatments that can improve patient outcomes through increasingly significant advances, such as the invention of ever smaller and less invasive devices. The day will come when millirobots and future autonomous nanorobotic devices will eliminate most threats to human health, quickly and efficaciously repair our bodies, and dramatically reduce the cost of healthcare. On this point, the premise of an envisioned equitable and cost-effective Global Health Care Equivalency (GHCE) healthcare system would be that sophisticated nanomedical devices may be molecularly manufactured (MM) domestically via Factory@Home (F@H) systems (Domschke and Boehm, 2014). These nanodevices might address virtually any human condition, including the disease of aging.

Initially, millirobot devices might be developed to address cardiovascular disease (the most common cause of death in the U.S., where ~703,000 people die each year (CDC, 2024a)) and cancer (which claims the lives of more than ~608,000 Americans each year (CDC, 2024b)). As medicine transitions further away from major surgeries to minor outpatient procedures, and from systemic to localized, millirobots and nanorobots will assist clinicians with seeing more, reaching further, and treating more intelligently. The family of innovations described here will equip clinicians with new tools and techniques that will establish research and educational standards for others to follow, while influencing what happens next in medicine.

5.2.1 Micromotor Applications in Biology and Medicine

Lightweight micromotors will likely have myriad applications in biological and medical domains, with their primary application being in microrobotics. The development of microrobotics will allow physicians/surgeons to diagnose and treat diseases much earlier and more quickly than ever before. Microrobots will be deployed to remedy disorders that are currently untreatable. Nelson and team (Nelson et al., 2010) determined that biomedical microrobotic applications might encompass targeted therapies, material removal, controllable structures, and telemetry. Biomedical microrobotics employed for targeted therapies will enable drug therapy, brachytherapy, hyperthermia, and stem cell development. Applications for material removal include ablation and biopsy, whereas those for controllable structures include scaffolding, stents, occlusions, and electrodes. Biomedical microrobotics for telemetry will be utilized for sensing and marking/tagging (Nelson et al., 2010).

Garimella and Eltorai (2016) discussed the use of nanotechnology in orthopedics and its benefits for the treatment of osteosarcoma. Further, micromotors may facilitate advanced imaging strategies. A team at Shenzhen University designed and fabricated an electromagnetic micromotor for intravascular ultrasound imaging (IVUS) (Peng et al., 2016). The creation of IVUS is important, as according to the American Heart Association (Virani et al., 2021), cardiovascular disease (as mentioned above) is the leading cause of death worldwide. Through the use of micromotors and IVUS, doctors will be able to diagnose and treat conditions with far greater precision.

Additionally, micromotors and microrobotics will revolutionize surgical techniques. Zygomalas et al. (2015) discussed miniaturized surgical robotics for use in natural orifice transluminal endoscopic and single site laparoendoscopic surgeries. Utilizing the natural orifices of the body or only laparoendoscopic single sites enables minimally invasive surgical processes. Further, minimally invasive surgery allows for expedited healing, thus reducing or negating hospitalization stays, which is less time-consuming and at a lower cost. The use of microrobotics will allow for these experimental surgical processes to be realized. Since the development of micromotors remains in its infancy, various strategies have been employed to fabricate them. Distinct classes of micromotors include (but are not limited to) catalytic, electrostatic, electromagnetic, as well as those that are driven by light, organisms, and ultrasound.

Catalytic micromotors catalyze H_2O_2, H_2O, glucose, acids, or bases, which can be utilized as fuels. Abdelmohsen et al. (2014) articulated the use of catalytic micromotors and nanomotors for biomedical applications, with the most common mechanisms encompassing bubble, diffusiophoresis and self-electrophoresis propulsion. Bubble nanomotor and micromotor propulsion exploits the "spontaneous decomposition of a fuel by a catalyst into micron sized gas bubbles whose release from the nano-motor surface drives the motion of the motors in the direction away from the catalyst". Bubble propulsion micromotors/nanomotors are the most promising catalytic motors, as they have the capacity operate independently of the ionic content of the media they traverse (Manesh et al., 2010; Mei et al., 2008). Thus, they will be able to function in blood and serum (Lissandrello et al., 2017).

Diffusiophoresis utilizes the asymmetric properties of catalytic motors, which results in the "decomposition product of the fuel preferentially on one side of the motor. A concentration gradient of the decomposition products is thus generated along the surface of the rod. Subsequently, as the accumulated decomposition products reach a critical point, where the local concentration is high and the product starts to diffuse away, this diffusion results in the movement of the motor" (Abdelmohsen et al., 2014; Golestanian et al., 2005; Paxton et al., 2004). "Self-electrophoresis propulsion is usually attributed to an electric gradient generated inside asymmetric conductive bimetallic rods. In principle the system acts as a self-contained electrochemical cell whereby one metallic end acts as the anode and the other as the cathode" (Abdelmohsen et al., 2014). These catalytic micromotors have strong potential for applications in the biological, environmental, and medical domains. However, catalytic micromotors and nanomotors have very fundamental shortcomings, including a dependency on fuel and power outputs. Consequently, they require further development and functional improvements.

Electromagnetic micromotors were initially introduced by Ahn et al. (1993) (Trimmer and Gabriel, 1987) with their applications including microfluidic systems, x-y positioning systems, biomedicine, pharmaceutics, biotechnology, and microrobotics (Büttgenbach, 2014). Tang et al. (2012) developed an omnidirectional wall-climbing microrobot with magnetic wheels that were directly integrated with electromagnetic micromotors. Uchikoba and team developed an electromagnetic MEMS motor without winding wire (Takato et al., 2016). Consequently, electromagnetic micromotors have shown great potential in the biological and medical domains toward the development of microrobotics. However, the greatest shortcoming of electromagnetic micromotors will be the scaling the motors to the milli and micro levels due to Ampere's Law.

Light activated micromotors utilize photons as their power source. Sen and co-workers developed a light activated micromotor that utilized the photosensitive properties of AgCl/inert silica particles driven by UV-induced dissolution (Ibele et al., 2009). A shortcoming of light driven micromotors includes the natural absorption characteristics of the human body, which make in vivo applications nearly impossible. Additional shortcomings include that there is currently extremely limited research in the scientific community in this area.

Organism driven micromotors attempt to mimic and exploit the biomolecular micromotors that are commonly found in nature. Magdanz and team developed a sperm flagella-driven micro-biorobot. This strategy, which coupled a biological power source with a microdevice, is an intriguing approach toward the development of microrobotics (Magdanz et al., 2013). The shortcomings of this strategy include that there is limited research being done in this area by the scientific community, and the biorobots require an external magnetic field for guidance.

Ultrasound propelled motors utilize ultrasound, which can significantly enhance biological and medical disciplines. Garcia-Gradilla et al. (2013) developed ultrasound propelled nanomotors that were magnetically guided. These nanomotors were used to capture and transport *E. coli* and *S. aureus* bacteria and attained a significant velocity induced by ultrasonic waves. The shortcomings included that they required an external magnetic field for guidance and could only operate in low Reynolds number fluids. Further, improvements are required on current models to improve functionality. Although diverse types of micromotors continue to be developed, each has its benefits and shortcomings and all require functional improvements.

Currently, electromagnetics show promise toward the development of microrobotics for biological and medical applications. A commercial electric motor based on a rotor and stator design that might be suitable for microrobots is the Ø2.0 mm × 15.5 mm long gear motor from Namiki. However, small motors require high RPM to generate adequate power, as well as a gearbox to produce the lower RPM and higher torque required for propulsion, or for the cutting of tissues. Nevertheless, these low voltage DC motors are too large in diameter and too long for some millirobot applications and are also expensive to manufacture. Electrostatic motors can be smaller; however, higher

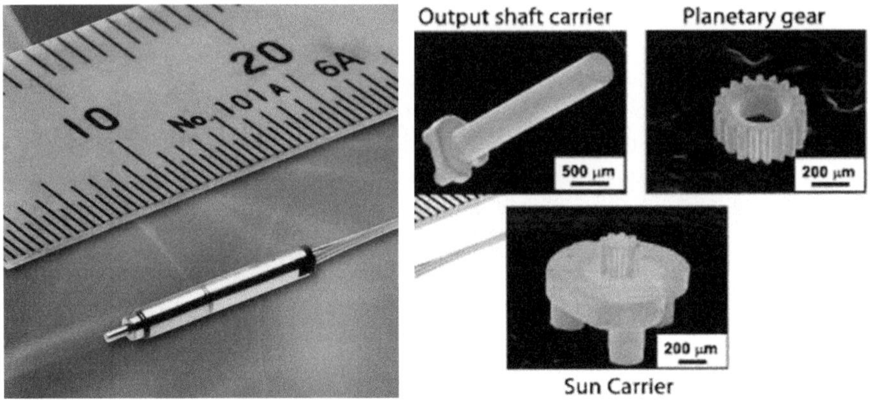

FIGURE 5.1 Tiny motor. Namiki developed a Ø1.5 mm micro-geared motor that was the smallest diameter electromagnetic motor in the world, which is expected to be utilized in various areas, such as in advanced medical equipment (catheters and endoscopes) and microrobots. The Namiki miniature motor will satisfy the demand for a motion control actuator that can operate in small spaces (figure from Wikipedia page; Namiki Precision Jewel Co, Creative Commons Attribution-Share Alike 3.0 Unported license).

voltages are necessary, which is undesirable for use in the human body. Electrostatic motors are typically not employed at the macroscale level; however, electrostatic micromotors show promise for multiple microscale applications, and critical issues such as pinhole free dielectrics and intrinsic lubrication must be resolved (Trimmer and Gabriel, 1987). Currently, there are few applications for electrostatic micromotors.

Several features of the Namiki DC motor include its diminutive size (Ø2 mm and Ø4 mm), low inertia, and quick response. Since the development of the smallest coreless motor in 1973, Namiki has successfully commercialized the DC micro Ø2 mm brushless motor. Subsequently, it has been utilized in medical equipment, analytical, physics, and chemical industries, which require small and precise motion control functions. Namiki developed the smallest and lightest geared motor in the world featuring the world's smallest metallic glass micro gearhead (Figure 5.1). This motor is well-suited for applications in medical and other miniaturize devices. The development of tiny motors is ongoing, and a literature survey should be performed to identify the latest advances.

5.2.2 3D Microprinting

Medical research facilitated by 3D microprinters will involve the development of new devices for diagnostics and therapeutics. Many illnesses cannot be treated efficaciously as current medical instruments are simply too large. A 3D microprinter developed by Nanoscribe (Nanoscribe, 2025) demonstrated the capacity to fabricate diminutive robot components for use within the human body, as well as print dissolvable medical implants. These microrobots and implants will be smaller and more accurate than existing devices, thereby ushering in a new era of precision medicine. Dynamic 3D microprinters will have the ability to fabricate medical micro/nanodevices for myriad diagnostic, therapeutic, and health monitoring applications. Further, through sequentially advanced iterations, the emergence of mature MM (which will be significantly expedited by AI and downstream AGI/ASI) may make possible the atomic resolution fabrication of autonomous nanomedical devices with the capacity to treat virtually all disease.

The 3D printing technology of Nanoscribe (Figure 5.2a) provides a number of advantages for the additive manufacturing of polymer components with dimensions that range from several microns up to the millimeter scale. The high printing resolution enables sub-micron features and design freedom for very complex parts with integrated internal features that are otherwise impossible to

FIGURE 5.2 3D microprinters. (a) The Nanoscribe Quantum X shape is an example of a new Quantum X platform printer. The endomicroscope shown is an example of a complex 3D-printed structure based on the new printing technology, 3D printing by 2GL®, which enables the highest level of shape accuracy, surface quality, and significantly increased throughput. 3D-printed microneedles are also shown. (The source of the images with permission is: Nanoscribe Company, nanoscribe.com). (b) MicroArch® is the first commercial high resolution, 3D micro-fabrication printer based on PμSL technology. This ultra-high resolution 3d printer can be used as an example for Micro 3D Printing of Disposable Medical Devices. (The image is from Boston Microfabrication Company: https://bmf3d.com/product/s230/ with permission.)

produce using conventional manufacturing techniques. These benefits support research and development tasks across a broad range of academic disciplines and industries.

Ultra-high resolution 3D printers based on Projection Micro Stereolithography (PμSL) technology, such as microArch® S230, provide down to 2 μm resolution (Figure 5.2b). Boston Microfabrication Company (https://bmf3d.com/) and different users provide prototyping services that employ this printer, which enables initial designs of biomedical devices (Vinnakota, 2024a, 2024b, 2024c, 2025) to be 3D printed and investigated.

5.2.3 Carbon Nanotube Biomedical Wire

CNTs are ideal materials for biosensing and other medical applications. Individual CNTs have high surface area to volume ratios, high mechanical strength, excellent chemical stability, high aspect ratios (length/diameter), and exceptional electrical and thermal conductivity. CNTs may be woven as macroscale yarns and sheets, albeit they have fewer unique properties in contrast to individual CNTs. CNT yarns (~Ø50 μm) are flexible, pliable, and have moderate strength and electrical conductivity; thus, they are suitable candidates for use as bio-removable wires in biomedical applications.

To create biomedical wires, CNT yarns were coated with a biodegradable polymer coating. Following characterization, a novel approach for the use of CNT-based wires for transcutaneous power and communication in the body was investigated. The extraction of the wires, biocompatibility observations, and histology analyses were performed using a mouse model in a sterile environment (Alagarsamy, 2017). The mouse model testing indicated that the CNT wires were biocompatible with tissues and the removal of the wires did not harm the mouse. However, the wire crossing the skin was chewed by the mice, which indicated that it caused irritation. Thus, it is likely that smaller diameter wires may reduce this irritation. Long-term testing will also be required to assess the risk of infection using the wire. This study revealed, perhaps for the first time, that tiny flexible CNT wires may be a novel strategy for the provision of power and communication across the skin for a range of potential medical applications, such as the powering sensors, valves, pacemakers, and artificial organs.

CNT fibers (Figure 5.3) are ideal materials for medical wires since they are pliable, lightweight, flexible, strong, biocompatible, and electrically conductive. This eliminates the difficulties associated with the electrical leads and connections that are currently used in the human body.

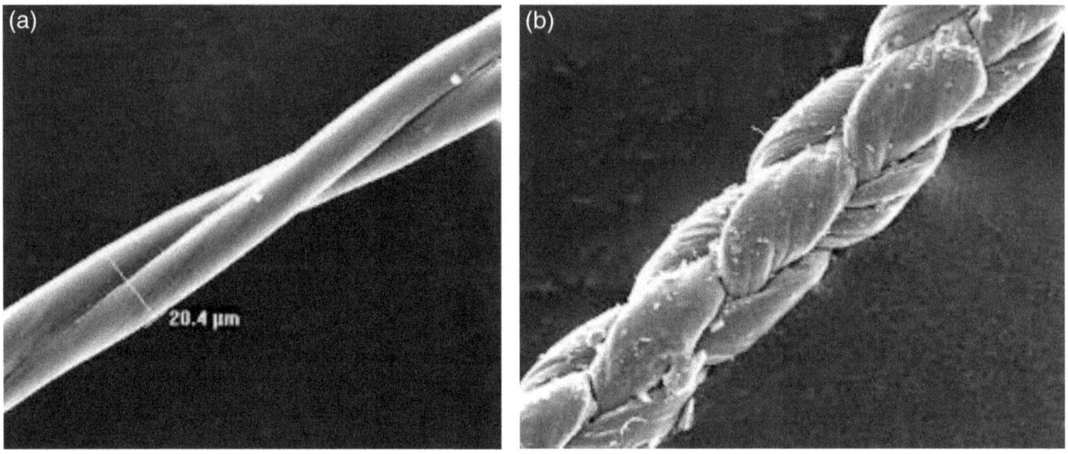

FIGURE 5.3 Carbon nanotube wire: (a) two ply yarn; (b) four-ended braid. The braiding was performed by Atkins and Pearce Company, Covington, KY, http://www.atkinsandpearce.com/about-us/. (This figure was generated by the authors.)

CNT fibers have also been employed to traverse the skin to provide electrical power, and for sensing the impedance and corrosion of implants in animals, which may be plied to create tougher larger diameter (Figure 5.3a) or braided (Figure 5.3b) wires. However, the electrical conductivity of CNT wires is lower than those comprised of Cu, which limits certain applications. Hybridized wires with increased electrical conductivity are being developed (Alagarsamy, 2017), which can maintain the flexibility of nanotube wires.

5.2.4 Micro- and Multi-Lumen Tubing

High-performance micro- and multi-lumen medical and surgical tubing can be employed as robot tethers (Figure 5.4). The primary materials that comprise this tubing include polyimide, PTFE, polyurethane, Pebax, and nylon. Medical tubing has excellent mechanical, thermal, and chemical attributes, which are held to tighter tolerances than conventionally extruded plastic tubing. Many medical devices such as surgical electronics and instrumentation require small diameter tubing. Another requirement is lubricity, where applications such as cardiology and neurology require increased sensitivity of movement and the ability to push the tubes without friction that can damage vessels or tissues.

Critical minimally invasive applications that require such tubing include cardiovascular catheters, stent delivery systems, urological retrieval devices, and drug delivery. Secondary operations involved in the fabrication of this tubing include laser machining, etching, composite tube constructions, novel assembly methods, and braid/coil reinforced shafts. Tubes comprised of extruded fluoropolymers, nylons, LCP, other engineered plastics with multi-lumens and fabricated with

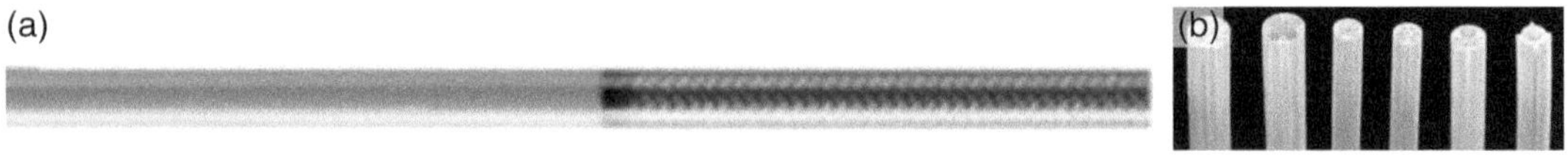

FIGURE 5.4 Tubing for robot tethers: (a) Microlumen braided polyimide tube with select length laser ablation and nylon reflow over the ablated area for variable flexibility, see *Introduction to Carbon Hybrid Materials*, 2018 (image with permission from Microlumen, http://www.microlumen.com/). (b) Zeus Multi–Lumens are unique tubular extrusions which provide multiple lumens or passages. Absorv Bioabsorbable Extrusions are also available that dissolve and disappear in the body over a period of months to years (image with permission from Zeus, https://www.zeusinc.com/).

secondary operations are employed in almost every medical niche. Extensive customization is also available for the development of tethers for robots, as well as a wide range of temporary or permanent implantable and bio-absorbable devices.

5.3 DEVELOPMENT OF UNIVERSAL MILLIROBOTS

Medical requirements and robot development encompass an extensive range of topics (3D MicroPrint GmbH, 2025; ASRM, 2024; Bakar et al., 2014; Barbot et al., 2014; Boston Scientific, 2025; BTG Vascular, 2025; CareFusion, 2025; CDC, 2024a, 2024b; Choi and Kim, 2010; CTCA, 2021; Dogangil et al., 2010; Erni et al., 2013; Espalin et al., 2014; G.O.B.A. Surgery, 2025; Galt Medical, 2019; Gao and Wang, 2014; Garcia-Gradilla et al., 2013; Ghosh and Fischer, 2009; Gou et al., 2014; Haworth et al., 2008; Hengsbach and Lantada, 2014; Hogg and Sretavan, 2005 ;Hogg, 2014, 2016; Hogg, 2016; Hohmann et al., 2015; Hong et al., 2006; Huang et al., 2015; *IJINT*, 2013; Jain et al., 2012 ;*Journal of Micro-Bio Robot*, 2025; Kagan et al., 2012; Karachaliou et al., 2015; Kim et al., 2013; Leshansky et al., 2007; Ma et al., 2007; Mahar et al., 2007; Martel, 2013; Mathieu et al., 2005; Mimee et al., 2018; Moreira and Misra, 2015; Mushtaq et al., 2015; Nanoscribe, 2025; Nanoworld Research Laboratories, 2025; QIAGEN, 2025; Qiu et al., 2015; Repasky et al., 2018; Schulz et al., 2009a; Schulz et al., 2009b; Schulz et al., 2013; Seeger, 2013; Sitti et al., 2015; Sivasankar and Durairaj, 2012; Thamphiwatana et al., 2015; Tottori et al., 2012; UCSF Health, 2025; Wang and Gao, 2012; Xiong et al., 2012; Yu et al., 2007; Yun et al., 2007; Yun et al., 2009; Zeeshan et al., 2014; Zeus, 2025; Zhang et al., 2015; Zhu et al., 2015). From an overall perspective (in view of current state-of-the-art technologies), micro- and nanoparticle entities are still too small to perform common medical procedures. Catheters in the mm size domain are appropriate for performing medical procedures; however, catheters are not steerable through various tissues and have limited tip functionalities.

Since the robotic manipulators of da Vinci surgical systems are currently too large to access many small in vivo sites (Zygomalas et al., 2015), more functional devices are required that can operate at the intermediate (mm resolution) range. The University of Cincinnati Nanoworld Laboratory (Nanoworld Laboratories, 2025) is developing mm size devices (millirobots) for medical applications. Minimally invasive millirobots are envisaged to be fluoroscopically guided to traverse tissues (QIAGEN, 2025), possessing functional hydraulically or electrically powered tips, which will likely be 3D printed using surgical grade polymeric or metallic materials.

5.3.1 Millirobot Design

Medical robotics for minimally invasive surgical procedures for soft tissues face significant challenges (Dogangil et al., 2010; Wang and Gao, 2012) including a reliable means for precision actuation (Büttgenbach, 2014; Manesh et al., 2010). Preceding work in the literature on multi-steerable catheters and ultrasound guided needles revealed that they were too large or lacked controllable access (Ali et al., 2019; Carstensen and Parker, 2014; Dhar et al., 2024; Engh et al., 2006; Frasson et al., 2012). These limitations indicated that the small dimensions and mobility of millirobots in biological tissues were unique. An earlier effort by Taylor and Anderson (2000) did not produce as small a robot as proposed herein, in part because 3D microprinters (Nanoscribe, 2025; Boston Microfabrication, 2025) and other leveraging technologies were not available at the time.

Medical and engineering teams worldwide have sought to design miniaturized devices that can be directed to specific remote and otherwise inaccessible sites to perform a series of diagnostic and therapeutic tasks (Freitas, 1999; Takato et al., 2016; Tang et al., 2012; Zygomalas et al., 2015). Just as laparoscopic and catheter-based interventions have revolutionized medicine, millirobots will lay the groundwork for another revolution in precision medicine toward the development of sophisticated autonomous nanomedical robots. This section will address the two major threshold design challenges for the miniaturization of medical devices, which are (i) power and (ii) mobility.

We propose that millirobots might be electrically powered using CNT wires and hydraulically energized using pressurized micro-lumen tubing (MicroLumen, 2024; Zeus, 2025). The delivered power will enable the operation of devices at the robot tip. Mobility will be achieved by controlling a microrobot body at the tip of a flexible tether using tension cables for steering and oscillating hydraulic pressure to generate tip axial vibration. This will be required to shear through tissue spaces with minimal damage, or the hydraulic pressure might be used to operate a cutting device. Axial force transferred through the tether will be used to advance and remove the millirobot.

With these functionalities, millimeter sized robots will be movable and steerable over extensive distances from the operator. A tethered millirobot is being developed as a testbed to evaluate navigation through simulated tissues to normally inaccessible sites, perform medical techniques, and be safely removed. Novel tether designs will be critical to facilitate the traversal of tissues with minimal damage. A unique variable durometer, spiral wire multi-lumen tether is being designed with high radial stiffness for pressurization, bending flexibility for steering, and axial flexibility for the oscillation of the tip. Various robot tips have been proposed to be 3D printed using a microprinter (Espalin et al., 2014; Nanoscribe, 2025) (Boston Microfabrication, 2025).

5.3.2 Medical Impact

Controlling the fine movements of tethers that drive millirobots will revolutionize the functionalities of multiple existing medical devices including robotic arms and catheters. Further, it will introduce a platform for the development of an extensive armamentarium of mini-surgical tools to improve healthcare and reduce costs (Erni et al., 2013; Hogg, 2014, 2016; Hogg and Sretavan, 2005; Leshansky et al., 2007; Martel, 2013; Sivasankar and Durairaj, 2012; Sitti et al., 2015; Zhu et al., 2015). Further, microscale tube fittings can be 3D printed to use with micro-lumen tubing that currently requires larger Tuohy-Borst fittings. Qosina Co., Microlumen Co., or other companies might commercialize the micro-fittings developed for use by the millirobot.

A ~Ø5 mm robot will have difficulty manipulating a bladed robot tip that proceeds straight through simulated tissues using hydraulic pressure, as friction will likely induce vibrations at the tip. Smaller stiffer tethers will increase the force required to move through tissues. Potential applications might include lung cancer screening, in vivo subarachnoid space chemical sensing, pressure, flow, glaucoma, tubes, needles, surgery, smart implants, communications, and control. Other small devices for the traversal of various tissues are also under development.

5.4 MILLIROBOT DESCRIPTION

The millirobot will be manifest as an advanced miniaturized technology platform toward the development of mobile steerable and powered functional robots that can perform precise manipulations at remote locations from the operator. This direct access will significantly expand current capabilities for minimally invasive medical operations or engineering repairs. The diminutive size and remote operation of the millirobot (Figure 5.5a) lead to two fundamental design challenges and basic science impediments, which include (i) power delivery and (ii) mobility. Power is required for millirobot functionality and the cutting of various tissues; however, this is currently constrained as electric motors are too large and require a gearbox to generate sufficient torque. Furthermore, batteries and chemical power are not practical due to their being too bulky or possessing insufficient power.

Mobility will be critical for enabling finely controlled millirobot movement through tissues, which accounts for the natural heterogeneity of the in vivo environment, and to guide and position the millirobot at/within target tissues. However, these capacities have been limited due to the difficulty of movement without tissue maceration. We addressed these issues through the engineering of a concept for a novel tether (Figures 5.5b–d) that incorporates four miniaturized Dyneema pull

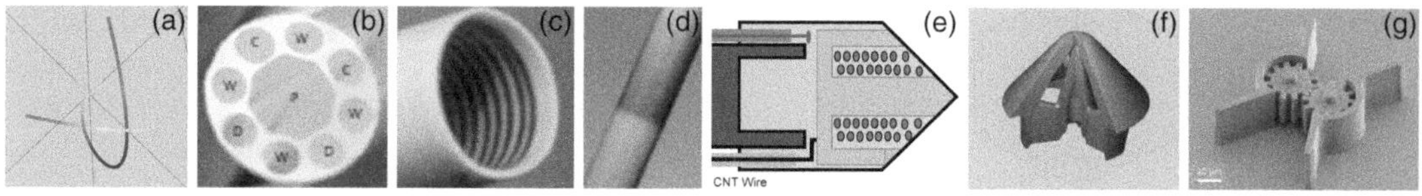

FIGURE 5.5 Design of millirobot components (tube diameters in a–e are Ø1 mm): (a) Graphic representation of millirobot navigating through and supported by tissue, and steered by multiple pull wires; (b–d) multi-lumen coil reinforced variable durometer tether with lumen for pull wires (W), drug/nanoparticle delivery (D), pressure (P), and CNT wire (C); (e) robot body side view cross-section showing CNT wire (and coil circles) and pull wire (red top) in tube. Conventionally, only 1% of any nanoparticle (NP) dose gets into tumors, as in the bloodstream, proteins bind weakly to the NP and form a cloud around it, covering tumor-binding molecules. The millirobot will deliver NPs directly into the tumor tissue/vasculature and then electromagnetically heat or remove the NPs. The delivery of drugs deep within the brain using this strategy might enable the treatment of glioblastoma; (f) Nanoscribe 3D printed robot tip for fluid/NP delivery; (g) example of 3D printed microhydraulic gear motor for surgery (blades not shown). (f, g) from Nanoscribe (Nanoscribe, 2025). (This figure was generated by the authors. Two subfigures f and g are from Nanoscribe (Nanoscribe.com) are with permission.)

wires (W), hydraulic pressure (P), drug delivery/biopsy extraction tubes (D), and electrically conductive CNT wire (C). This one-piece extruded tether with tailored directional stiffness is novel in robot design.

The multipurpose millirobot is designed for universal applications. Future efforts will employ the basic technologies described in this chapter to develop specific novel multifunctional tooltips with capabilities such as microactuators for surgery, thermal deposition/tissue ablation, diagnostic sensing, and therapeutic nanoparticle-based drug delivery. Relevant efforts in robot development include millirobot fabrication at the NSF Sponsored University of Cincinnati (UC) Pilot Microfactory for Nanomedicine Devices.

5.4.1 Methodology

The strategies being investigated toward the development of the millirobot are described below. A first iteration robot is in development with the initial aim of producing a larger sized prototype and evaluating its performance, after which it will be dimensionally scaled down to enable further applications.

5.4.1.1 Millirobot Power

Power will be delivered to the millirobot through tethered electrical and hydraulic means. In general, electric motors might be employed in larger robots, while hydraulic motors could be used in smaller robots.

5.4.1.1.1 Electrical Power via CNT Wire

Custom-made flexible Ø50 μm hybrid CNT-Ag-Cu wires will be fabricated in-house to provide current to power the millirobot at DC or RF. Coiled CNT wires at the robot tip can generate an electromagnetic field (Figure 5.5e) for the ablation of tumor tissues or the delivery/collection of magnetic nanoparticles for tumor therapies. The robot tip with CNT coils (Figure 5.5e) will be 3D printed (Figure 5.5f) and metallized, whereas CNT wires will be cooled by the water that flows through the micro-lumen tubes.

5.4.1.1.2 Hydraulic Power via Pressurized Lumen

The envisaged millirobot tether (Figures 5.5b–d) will provide high pressure fluid, with significant power and safe operation, to vibrate the robot tip to shear (micro-sliding) through tissue, or (by switched tips) operate hydraulic motors (Figure 5.5g) with surgical blades for shard cutting, fascia/hard tissue penetration, or alternating from sharp to blunt dissection. Testing will verify the modeling of hydraulic pressure, friction, and axial tip shear (Carstensen and Parker, 2014) as it traverses tissue. Tip vibration will enable a reduction in the required penetration force to minimize pain and trauma.

5.4.1.2 Millirobot Mobility

5.4.1.2.1 Tether Design for Mobility through Tissue

Conventional tubing is stiffer in the axial extruded direction and circumferentially soft; thus, it is unsuitable as a tether. The proposed tether design (with an internal volume of $V = \pi R^2 L$) employs a spiral wire-reinforced multi-lumen variable durometer tube that enables oscillating high pressure (water or perfluorocarbon) to axially vibrate the tip to allow a robot (for the first time) to go around and between organs (e.g., liver and intestine) and safely push tissue aside to access targeted sites. Taking the differential into account reveals that the tube volume is 100x more sensitive to changes in radius in contrast to its length. Thus, the one conceptual design is where the tether is radially stiff, flexible/bendable, and axially compliant to enable the hydraulic vibration of the tip at a 100 μm amplitude to traverse tissues. Tether stiffness (k) and tip mass (m) are selected to resonate at the hydraulic pressure frequency $f = \left(\frac{1}{2\pi}\right)\sqrt{k/m}$ to increase the tip amplitude. If the movement of the tip is blocked, a hydraulic gear motor (Figure 5.5g) with a surgical blade can cut through scar tissue.

5.4.1.2.2 Robot Steering

The principle behind millirobot steering is that the tether should be flexible, particularly near the tip (Figure 5.5a). Steering is achieved through the manipulation of hydraulic stress in the tube wall using robust pull cables (braided flexible Dyneema) within the lumens (W in Figure 5.5b) attached to the tip and within the microlumen tubes. A computer-control system will automatically manipulate four pull cables to steer the robot. The specific flexibility of the tubes will be designed to be modeled using COMSOL software (Figures 5.5b–d).

5.4.1.2.3 Navigation

Fluoroscopy imaging will be used for navigation, which provides millimeter resolution. Alternatively, ultrasound guidance may be used to provide non-ionizing real-time imaging (Szabo, 2014). Physicians and collaborators will evaluate the effects of the robot on tissues, including soft tissue analysis using micro-CT (Hildebrand et al., 2024).

5.4.1.2.4 Manufacturing

Scientific developments include the customization of the tether axial/bending stiffness, use of braided Dyneema cabling for steering, hydraulic oscillation (e.g., using an air pressurized cylinder up to 18,000 cycles/min to move through tissue), CNT wiring, and a 3D microprinted robot tip. With precise control over directionality, various classes of millirobots may be designed for specific procedures.

5.4.1.2.5 Cost

The cost of individual medical millirobots is estimated to be moderate (in the range of balloon catheters), which can be discarded after use.

5.4.2 Initial Millirobot Component Testing

Initially, for a conceptual millirobot, the materials used will be larger in scale (~5 mm), rather than the envisioned final small scale (1 mm) prototype design. The size of the tether tube

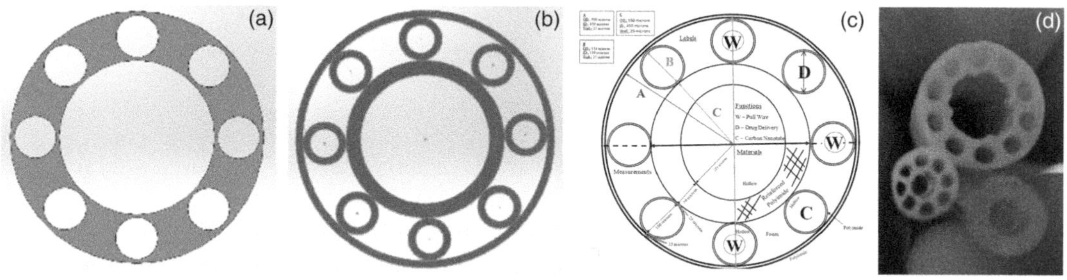

FIGURE 5.6 Tether tube design: (a) Microlumen tube with solid fill between tubes; (b) microlumen tube with space between tubes; (c) design with materials and specifications; (d) multi-lumen and single-lumen tube shapes. (This figure was generated by the authors.)

(comprised of a latex material) is 1600 μm OD, while the final expected diameter of the tube will be 900 μm OD and consist of a polyimide material. The first iteration employs commercial microlumen tubing comprised of eight smaller peripheral tubes (150 μm OD), and a single central tube (550 μm OD), giving an overall tube dimension of Ø900 μm (Figure 5.6). The rationale behind the central tube having a larger diameter is to ensure that the overall tube can withstand the pressure required for the actuation force to traverse through tissue or operate devices at the millirobot tip.

Each small microlumen tube has a dedicated function, which includes a pull wire for steering, a CNT wire for power, and a guideway for the delivery of drugs (Figure 5.6c). Reinforced polyimide is used for the central tube to increase stiffness for enabling pushing and mobility. A compression coil spring external to the tube enables the pushing of the robot using the tether, while facilitating bending for steering. The coil spring may also be utilized to extend and increase the elongation of the tube for vibrating and moving through tissues. The end tip of the tube contains a lower durometer rubber for connecting to the robot tip. This allows the axial extension vibration of the end of the tether, enables the robot body to move through tissue, and allows bending to permit steering. The tip is sharp and small to facilitate the cutting of tissues in parallel, while, if possible, minimizing tissue damage. The pull wire is made of Dyneema fiber, which is very strong, albeit susceptible to cutting on sharp edges. The multi-lumen and single-lumen tube shapes are shown in Figure 5.6d.

For initial testing and proof of concept, a larger scale will be used for the millirobot as the required materials are easier to obtain and are low cost. The polyimide tube will be replaced with latex, whereas the pull wire used is a Ø0.04 mm nano-filament braid. A sewing needle is used to attach the pull wire to the latex tube and to sew the thread into the tube. Using a syringe, the tube is filled with water for the oscillating motion created by an air oscillating tool to demonstrate navigation and steering. Once the tube is filled with water, the tip is inserted at one end to prevent water leakage. Subsequently, an equally long extension spring is placed over the tube. At the opposite end, a barbed fitting is connected to the syringe to the tube (Figures 5.7a, b). During testing under hydraulic pressure, issues included tip loosening and leakage through the latex tube.

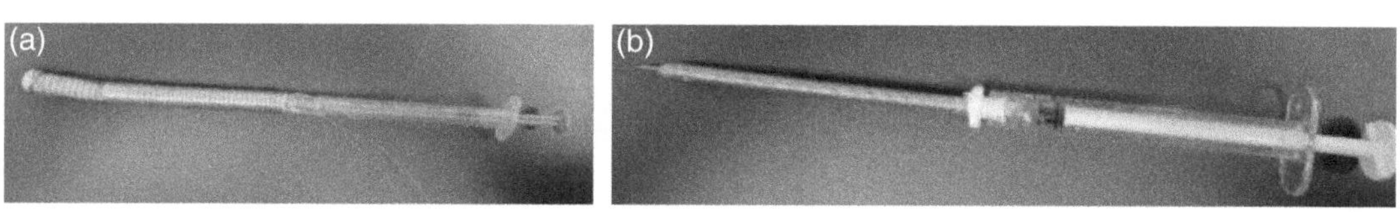

FIGURE 5.7 Prototype to test mobility: (a) Prototype using multi-lumen tube, and open coil spring with a Dyneema thread for steering; (b) simple prototype using single-lumen tube and a contracting coil spring for pushing through tissue. (This figure was generated by the authors.)

5.4.2.1 Larger Scale Prototype

Two experiments were conducted using various types of tubes, needles, and syringes:

Steering

Steering for the prototype was tested using Dyneema fibers as pull wires. Initially, a Dyneema thread was inserted as a pull wire into a multi-lumen tube using a microscope. Due to its diminutive size, installing the Dyneema thread pull wire was difficult and time-consuming (Figure 5.7) using a needle. The steering of the multi-lumen tube was actuated when a pulling force was applied to the Dyneema thread (Figure 5.7a).

Mobility

Mobility was tested by pressurizing the prototype robot tether with hydraulic fluid (water) for penetrating a mass of gelatin, which served as a human tissue analog.

Mobility was tested using different types of tubes (single- and multi-lumen) and materials. A single-lumen tube was comprised of latex, whereas multi-lumen tubes were made of various polymeric materials. The single-lumen latex tube was filled with water via a probe needle tip inserted at its end. A syringe was connected to the other end of the tube to form a simple actuation system. An oscillating tool was attached to the plunger of the syringe to produce tip vibration by pressurizing the fluid stored in the barrel of the syringe. A high hydraulic pressure was generated by the oscillating tool. Barbed fittings were employed to connect the tip, tube, and syringe to prevent the separation of the components due to the large hydraulic pressure. In this configuration, which included a spiral spring on the exterior of the latex tube, the tube primarily expanded axially as it was constrained from expanding radially due to the spiral coil.

The body of the tube should extend only axially when hydraulic pressure is applied for pushing and cutting tissue to facilitate movement within the human body. To tailor the flexibility of the tether, an external coil spring was employed to allow for the extension of the tether but with negligible radial expansion. Epoxy was used to attach both ends of the tube that were connected to the barbed connector and tip to prevent water leakage. These steps were repeated for multi-lumen tubes for testing.

5.4.2.2 Testing of Hydraulically Actuated Robot Tip

Since the main focus at this juncture was mobility, a single-lumen latex tube was initially tested rather than the multi-lumen tube. The millirobots are envisaged to operate in blood vessels, or they can move within tissues. Thus, the robot design was universal, while the design of the tip was application specific. Figure 5.8 reveals that the penetration of the robot tip into a gelatin tissue analog was challenging.

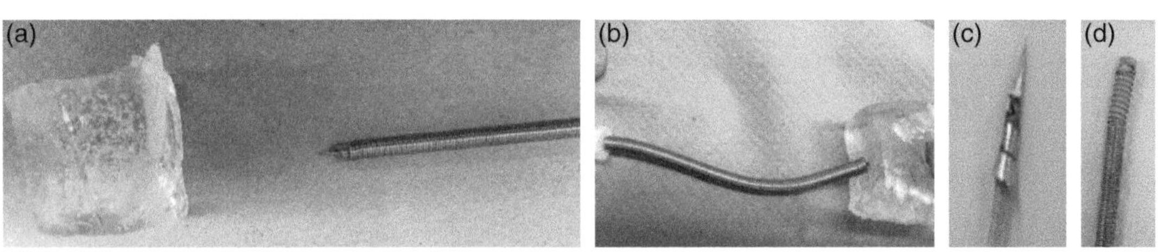

FIGURE 5.8 Difficulty with penetration of the robot tip into gelatin tissue analog: (a–b) The tether bending due to lack of lateral support when piercing through the gelatin (tissue); further, the elasticity of the gelatin resisted cutting; (c) a sharper tip was fabricated. The tip should be redesigned with a lateral (parallel to tissue) cutting motion, as the elasticity of the gelatin resisted incisions normal to the surface; (d) external compression coil spring over the inner tether. (This figure was generated by the authors.)

FIGURE 5.9 Millirobot testbed consists of the robot tip with a cutting blade, the tether with a coil spring for constraining radial vibration, a syringe for providing hydraulic pressure to the tether, and a pneumatically actuated piston. (This figure was generated by the authors.)

Figure 5.9 depicts the millirobot testbed with a syringe and oscillating pneumatic piston to oscillate the pressure within the syringe. The tether latex tube extends axially along with the spiral extension spring. When the oscillating force is applied, the displacement of the robot tip (blade) is ~0.5 in.

Figure 5.10 depicts the further mobility testing of the millirobot prototype through the simulated tissue. The tether was guided to prevent lateral vibration, whereas in operation, tissues or a vessel would provide lateral support (manually provided here). The sharp, pointed tip did penetrate the gelatin once a sufficient force was applied. The lubrication of the tip reduced friction as well as the required applied force. Since traversing the blade through the analog tissue was difficult, it was determined that a new design would be required for the cutting or separation of tissues.

The results of the testing revealed that the mobility and tip design required further modifications, including a sharper blade for the tip. Lubrication was applied to the blade and gelatin to

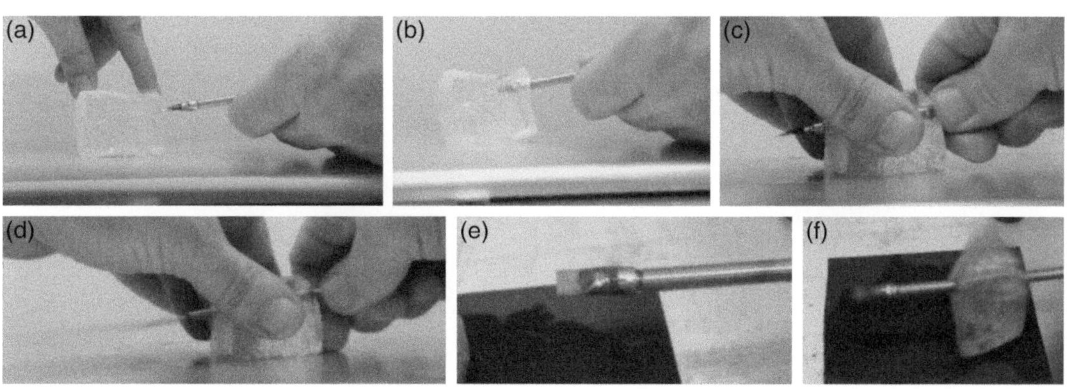

FIGURE 5.10 Large millirobot prototype shows difficulty with moving through the gelatin tissue analog: (a–c) Hand guided mobility through the gelatin (human tissue) analog; (d) the high pressure induced oscillation of the tip; (e–f) a wide tip, which requires significant pressure to move through gelatin. Normal cutting was not effective (very slow) due to the elasticity of the gelatin. (This figure was generated by the authors.)

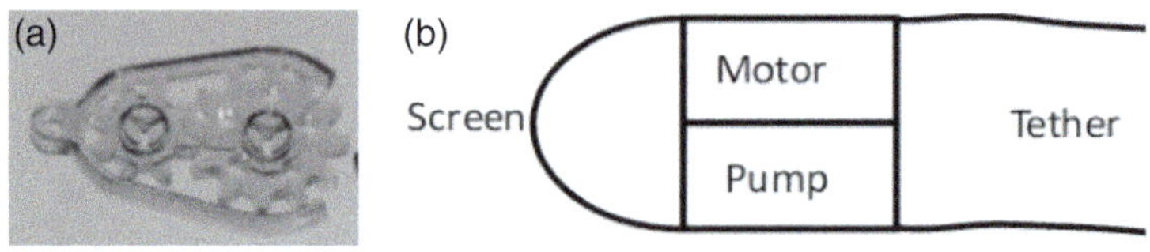

FIGURE 5.11 Millirobot with fluid power: (a) AI generated image of 3D printed gears that may be used as a hydraulic motor or pump in millirobots or micromachines (image from Nanoscribe with permission, Nanoscribe.com). (b) Millirobot design for the sorting of particles in fluids. (This figure was generated by the authors.)

decrease the friction of the surface area. As a strategy for improvement, a stiffer tube with a suitable tip was required. Alternatively, a tip that can separate or cut tissue may be necessary, where a blade powered by a hydraulic motor may be employed to cut tissue. Hydraulic motors for millirobots are discussed below.

5.4.2.3 Future Millirobots with Hydraulic Motors

Hydraulic pressure (water within the central lumen of the tether) may also be employed to operate a hydraulic motor, which can drive a pump. The pump can sample/deliver fluids, or sort biological elements or other particles. Particle filters using screens or posts were designed to capture Ø10 μm cells, while the millirobot particle sorter will be a pump that continuously removes particles. The particle sorter is envisaged to remove Ø100 nm nanoparticles (e.g., viruses) and anything larger. Manufacturing the gears for this hydraulic pump may be achieved using a 3D printer, whereas a stamp or mold of the motor components may be formed using Nano-Imprint Lithography (NIL) for mass production. An example of 3D printed gears is shown in Figure 5.11a using a Nanoscribe microprinter. A schematic of the millirobot using hydraulic motors in its tip is shown in Figure 5.11b.

Figure 5.12 depicts a hydraulic motor for the millirobot, which would facilitate the collection of nanoparticles from the bloodstream. Microprinting using the Nanoscribe printer is an enabling technology for the development of gears that are small enough to capture Ø100 nm nanoparticles (predicted). Many other types of robot tips are possible, which might also be powered by hydraulic motors.

5.4.3 Conceptual Hydraulic Motor Propelled Millirobot

A concept for a larger sized robot that is propelled by four hydraulic motors on wheels (Figure 5.13), rather than being propelled by pushing a tether, is outlined below.

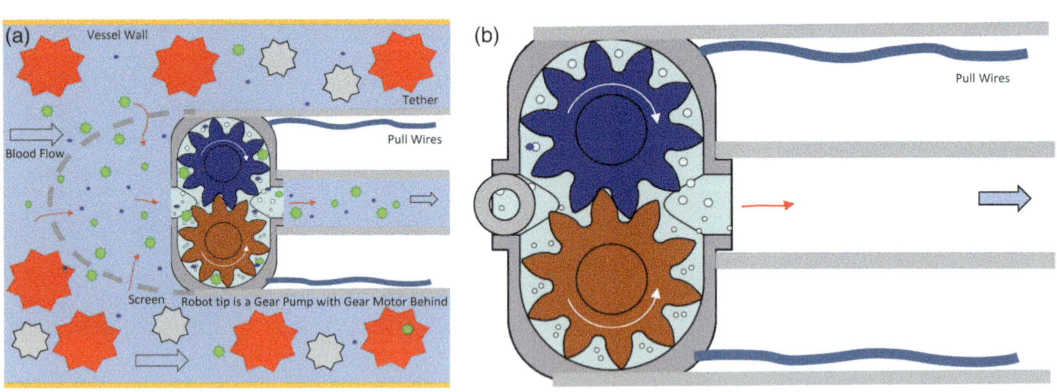

FIGURE 5.12 Robot hydraulic motor to drive the pump and the pump tip to collect bloodborne nanoparticles. (This figure was generated by the authors.)

FIGURE 5.13 Conceptual millirobot with four hydraulic wheel motors. (This figure was generated by the authors.)

5.4.3.1 Self-Propelled Millirobot Design

A conceptual tethered millimeter sized millirobot device is described here that traverses the vascular system/through tissues to deliver drugs, perform sensing, or collect biopsy samples. The millirobot is guided by driving the wheels at different speeds. A flexible tube at the tip of the robot enables hydraulic pressure (water) to drive the gear motors/pumps. A tether with spiral wire reinforcement is the lifeline of the millirobot. This tether tube arrangement allows for the application of a small axial force to move the millirobot, as pulsed pressure or heating actuates the tip, which transmutes the tissue ahead of the device into particulate mass. This enables its movement within a minimally invasive passageway. RF ablation can also be employed to remove tissue, and the millirobot can be extracted by pulling the tether.

Once the millirobot has reached its target, drugs can be delivered through the microlumen tether to small orifices at the tip. Sensing can be accomplished using CNT wires within the device. Suction can be employed to sample fluids or to facilitate the extraction of a liquid biopsy. Initially, the robot's outer dimension will be ~Ø5 mm, with the central tube being ~Ø3 mm. Hydraulic actuation is anticipated to provide the highest power per volume of actuation in contrast to a larger electric motor, which requires a gear reducer to increase torque, or electromagnetic actuators that are larger in size and generate heat. The use of external magnetics is too complex and expensive, and the motion control is too inaccurate for the precise movements required for surgical procedures. Further, the hydraulic actuator motor can be 3D printed (Espalin et al., 2014; Hengsbach and Lantada, 2014; Huang et al., 2015; 3D 3D MicroPrint GmbH, 2025; Nanoscribe, 2025; Xiong et al., 2012; Zeeshan et al., 2014).

5.4.3.1.1 Self-Propelled Millirobot Power

The powering of the millirobot will be accomplished via electrical and hydraulic means. Novel Ø50 μm flexible hybrid CNT-Ag wires will be manufactured to provide current at a greater density than Cu wires to power the millirobot via DC or RF. Coiled CNT wires will generate an electromagnetic field for ablating tumor tissues or delivering/collecting magnetic nanoparticles for tumor therapy. Millirobot tips with CNT coils may be 3D printed and metallized, whereas CNT wires can be cooled by water flowing through the microlumen tubes. The tether will supply a high-pressure fluid (having significant power and safe operation) to operate the hydraulic motors.

5.4.3.1.2 Self-Propelled Millirobot Mobility

The design of tethers to facilitate millirobot mobility through the blood or tissues has been considered. Conventional plastic tubing is stiffer in the extruded axial direction and circumferentially soft, which is unsuitable for use as a tether. A proposed tether (Figure 5.13) used a spiral wire-reinforced multi-lumen variable durometer tube to enable an oscillating high hydraulic pressure (e.g., water or perfluorocarbon) to power the four wheels for propulsion and steering. The tube volume was 100 times more sensitive to changes in radius compared with length. Thus, our tether was radially stiff, flexible, and axially compliant to enable hydraulic fluid delivery to the gear motors.

5.4.3.1.3 Self-Propelled Millirobot Fabrication

Scientific developments will include tailoring the tether axial/bending stiffness, the use of Dyneema braided cabling for steering, hydraulic actuation to traverse tissues, CNT wiring, and 3D micro-printing of the millirobot tip. With precise directional control, millirobots can be designed for specific procedures. The conceptual millirobot possesses hydraulic motors and wheels for propulsion (Figure 5.13), a tether to supply pressure for the operation of the hydraulic motors, as well as hollow lumen tethers for tissue debris removal and drug delivery, with an electrosurgical tip.

If necessary, laser drilled holes through the wheels can employ tether suction to increase traction. First-generation millirobots will employ a tether with a CNT wire for material delivery and retrieval, as well as communications. A nanotube tether would be a breakthrough in itself as it would be soft, pliable, strong, electrically conductive, and overcome most of the issues that plague metallic or polymeric tethers. A metamorphic crawler would also have the capacity to change size. The end result is that the millirobot would have high power density, be able to travel against blood flow, and push through tight access locations, all of which will be required to realize their maximal potential. The bifurcating human vascular system has close to 60,000 miles of potential pathways, which offer an immense range of opportunities for in vivo microrobotic submersibles.

Gaining access to previously inaccessible blood vessels would also make it possible to sample and test blood in situ in real time, in contrast to conventional blood draws, which sample systemic blood circulation. The accurate early diagnosis of disease will enable physicians and surgeons to pursue far more extensive targeted treatments. Gaps in knowledge and technology, as well as technical, regulatory, and societal barriers that prevent the achievement of this proposed vision, might conceivably negatively impact healthcare system economics and costs on a global scale. Procedures that are far less invasive with the capacity to treat more diseases will significantly reduce costs. An example is a medical robotic submarine that may emerge as a new segment in the catheter market, which was valued at $18.7 billion in 2020 and estimated to reach $30.2 billion by 2026 with a CAGR of 8.3% from 2021 to 2026 (Research and Markets, 2022).

5.5 APPLICATION OF FUTURE MILLIROBOTS TO DIAGNOSE AND TREAT LUNG CANCER

Future millirobot applications may encompass lung cancer screening, in vivo subarachnoid space (chemical, pressure, and flow) sensing, glaucoma detection (tubes, needles, and surgery), as well as facilitating the communications between and control of smart implants. Lung cancer is the leading cause of cancer-related deaths in the U.S., where ~226,650 new cases are estimated for 2025 (American Cancer Society, 2025). Exposure to cigarette smoke causes ~90% of lung cancer cases; however, it also increases the risk of cancer in non-smokers due to secondhand smoke; worsening ambient air pollution is also a significant risk factor. A cancerous tumor is shown in Figure 5.14. Thoracic surgeons work with oncologists, pathologists, and radiologists to diagnose and provide

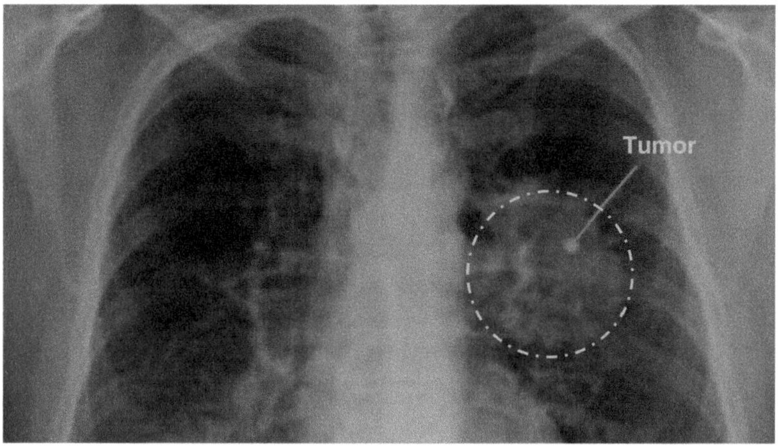

FIGURE 5.14 Lung tumor shown on a chest X-ray (figure from Wikipedia page; Creative Commons Attribution-Share Alike 3.0 Unported license, https://en.wikipedia.org/wiki/Lung_tumor).

therapy for lung cancer, and surgeons may perform various types of lung cancer resections. Millirobotics may assist with the diagnosis and treatment of lung cancer, which is discussed below.

5.5.1 Lung Cancer Diagnosis

A chest X-ray or CT scan is typically performed to scan for nodules or masses within the lung (Figure 5.15a). If nodules or a mass are present, more detailed tissue diagnosis may be performed via bronchoscopy or through the use of a transthoracic needle biopsy prior to further staging and therapy. In some cases, the physician may proceed directly to surgical procedures based only on radiology studies to prevent unnecessary delays in providing surgical therapy. It is also critical, prior to therapy being initiated, to accurately assess the status and stage of cancer to determine whether it has metastasized to other parts of the body. A chest CT scan will identify the location and size of the lung tumor and any enlarged lymph nodes within the chest (Figure 5.15b).

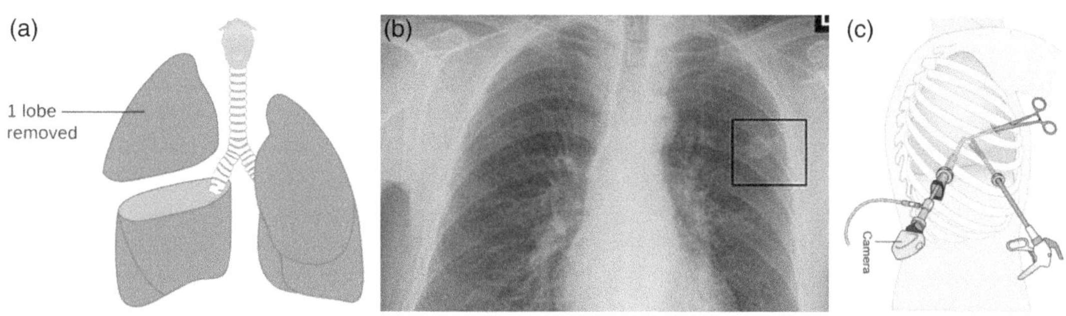

FIGURE 5.15 Illustration of lung tumor and therapeutic procedures. (a) Illustration of lung showing cancer and current lobectomy technique of upper right lung, https://en.wikipedia.org/wiki/Lung_lobectomy; (b) lung nodes, https://en.wikipedia.org/wiki/Lung_nodule (permission is granted to copy, distribute, and/or modify this document under the terms of the GNU Free Documentation License); (c) schematic of video-assisted thoracoscopic surgery (VATS). The millirobot may enable the removal of the tumor, thus avoiding lobectomy. (Images a and c have been released as part of an open knowledge project by Cancer Research UK and are attributed to Cancer Research UK/Wikimedia Commons, diagram showing video assisted thoracoscopy (VATS) CRUK 378 – Video-assisted thoracoscopic surgery – Wikipedia.)

PET scans provide additional information regarding the potential spread of a tumor to the lymph nodes and other sites in the body. However, a brain MRI (magnetic resonance imaging) may be required to determine if the cancer has spread to the brain, as PET is not a reliable indicator of brain tumors. Biopsies of the lymph nodes in the chest may also be performed using a procedure referred to as mediastinoscopy. In some cases, endobronchial ultrasound may be employed to biopsy lymph nodes from within the airway. If the lymph nodes in the chest contain a tumor, chemotherapy or chemoradiation may be given prior to surgery. Once the diagnosis of lung cancer is made and the staging process is completed, cancer therapy can proceed.

5.5.2 Lung Cancer Treatment

Patients with uncomplicated early-stage tumors may be treated by thoracic surgeons alone, with cure rates of ~80%. Patients with more advanced tumors may be evaluated and treated by a multidisciplinary team. Typical stage-based therapies include:

- Stage I – Surgery followed by chemotherapy if the tumor is large.
- Stage II – Surgery followed by chemotherapy.
- Stage III – Chemotherapy or chemoradiation prior to and possibly following surgery.
- Stage IV – Chemotherapy alone; however, surgery is possible at single sites if the cancer has spread to the brain or adrenal glands.

Different types of surgery or lung resections include lobectomy, which is the anatomic removal of a complete lobe of lung (Figure 5.15a). This is the most common type of operation, which has very low complication and mortality rates, as there are three lobes in the right lung and two in the left lung. Sleeve lobectomy involves the removal of a complete lobe of the lung, as well as a portion of the airway that conducts air to the remaining lobe, and then reconnecting that airway and the remaining lobe. This is a more complex procedure that may circumvent the requirement of a pneumonectomy. Pneumonectomy involves the complete removal of the lung on one side, which may cause early complications and long-term disability due to shortness of breath. The overarching goal is to completely remove the cancer without pneumonectomy. Segmentectomy involves the removal of only a segment of a lobe, whereas wedge resection removes an even smaller part of the lung. Segmentectomy and wedge resection are employed when patients have small tumors or severe lung disease.

Two additional complex thoracic surgical procedures provide the lowest possible complication and highest cure rates following lung cancer surgery. These are video-assisted thoracic surgery (VATS) and sleeve lobectomy. In most hospitals, lobectomies are performed by thoracotomy where the ribs are spread apart to provide access to the chest. The large incision, rib spreading, and cutting of muscle tissues cause significant discomfort following surgery. In contrast, VATS lobectomy allows surgeons to carry out the identical procedure within the chest that is performed by thoracotomy; however, it is done using only three to four small incisions (Figure 5.15c). The surgeon employs a small video camera inserted through one of the incision sites to see inside the chest. The surgical procedure itself is carried out using long instruments that are passed through the other incisions. Patients who undergo VATS lobectomies have less pain and expedited recovery. In general, a VATS lobectomy is an option only for patients with Stage I lung cancer.

Sleeve lobectomy is an option when a tumor is located at the origin branch of an airway that supplies a lobe of the lung. This procedure allows for the total removal of the tumor without the complete removal of the lung. This is made possible by removing a small portion of the airway that conducts air to the remaining lobes, and subsequently reattaching the airway and the remaining lobes such that they continue to function in the usual manner. Thus, sleeve lobectomies provide the same chance of cure as pneumonectomy with fewer operative complications and improved quality of life.

Although the envisaged millirobots are not yet a clinical reality, they are being developed to be used in different universal ways. Larger size millirobots may be employed to perform VATS and other types of minimally invasive surgeries. Smaller sized millirobots may enter the airways or blood vessels to extract biopsies, deliver drugs, or perform simple surgeries.

5.6 CONCEPTUAL MILLIROBOT PROTOTYPE: LOCALIZED DIAGNOSTICS AND HYPERTHERMIC CANCER THERAPY

Glioblastoma multiform, or GBM, is a primary tumor that arises in the brain or spinal cord. This type of cancer remains in place and rarely travels to other organs; however, symptoms appear in regions of the body that are controlled by the areas of the brain afflicted with GBM. GBM tends to grow rapidly within the brain and is deadly. Other tumors, such as breast tumors that travel to the brain, are "secondary" brain tumors. Glioblastoma is a major condition that is currently profoundly inadequately addressed in medicine. US Senators Ted Kennedy and John McCain died of glioblastoma nine years apart, despite receiving the best medical care available (Begley, 2018; CTCA, 2021; Glioblastoma (Wikipedia), 2025; Jain et al., 2012; Mimee et al., 2018; Repasky et al., 2018). Their deaths brought renewed attention to a disease that affects nearly 700,000 Americans.

Glioblastoma is a complex, drug-resistant, and adaptive cancer, for which there is no early detection or prevention. One thing to consider within this context is that glioblastoma has traditionally been difficult to treat for several reasons: (1) the delivery of drugs across the blood-brain barrier (BBB) is extraordinarily difficult, which limits the effectiveness of chemotherapeutic drugs; (2) glioblastoma can be in close proximity to regions of the brain that are essential for life or substantially impact the quality of life; thus, there needs to be exceptional control to ensure that these areas of the brain are not adversely damaged; (3) blastomas typically have tendrils that radiate outward into the brain; thus, the removal of the core bulk tumor might slow the progress of the tumor; however, it is not curative. The latter issue might make it quite difficult for the millirobot to be considered a curative device; however, there is ample room and an urgent need for the development of devices that can at least extend the quality of life for patients for the remaining time that they have. A focus on addressing some of the fundamental issues involved with the movement of the robot tip will be a key step toward the introduction of millirobots to clinical use.

Understanding GBM is an essential first step in developing millirobotic therapies. Most importantly is the fact that a single glioblastoma can contain cells with many different mutations. This is why the cancer almost always recurs following the first round of treatment, which typically encompasses surgery, radiation, and chemotherapy. The remaining cells not eradicated by the initial treatment can regrow the tumor (Weathers, 2019). Blocking one pathway is not effective as the tumors are too heterogeneous, and the cancer takes an alternative route. Thus, a variety of drugs may be required to treat GBA.

Since the heterogeneity of drug resistance is a significant issue, drug combinations must be thoroughly investigated to elucidate which might result in optimal efficacy. Further, many cancer drugs (~90%) are incapable of crossing the BBB (Wen, 2019). If it becomes possible to deliver drugs directly to the tumor, drugs with some level of efficacy against glioblastoma that cannot penetrate the BBB may be used in combination (Butowski, 2019). Millirobots may be useful for the direct delivery of drugs to the tumor site, as well as to perform precision surgery by burrowing through the GBM spindles that are difficult to resect using conventional strategies.

The manipulation of the immune system to attack cancer cells is another promising approach, which has transformed the treatment and prognosis of types of cancer that were nearly as unmanageable as glioblastoma. Since many patients do not respond to new immuno-oncological drugs, researchers are creating vaccines that prime the immune system to attack specific molecules that are present on the exteriors of cancer cells. It remains unknown whether glioblastoma will eventually defeat the immune system as it does everything else. The creation of electric fields in the brain is

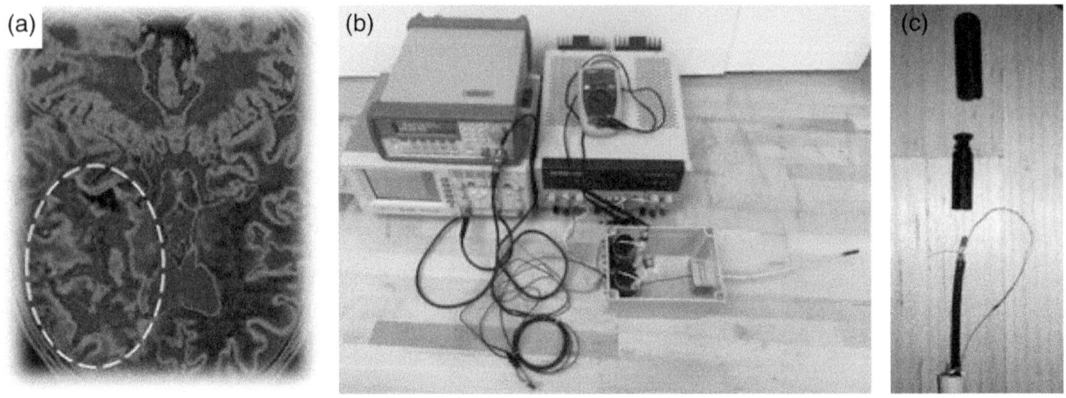

FIGURE 5.16 Possible GBM application of millirobots: (a) Exemplar GBM brain tumor (white dashed region); (b) large size prototype millirobot to demonstrate robot technology. To treat GBM, the cancerous cells must be identified and treated. This robot has three degrees of freedom, side-to-side translation, up and down translation, axial motion of the tether, and further extension of the robot tip. A mm sized robot is required to be minimally invasive; (c) robot tip with thermocouple and CNT wire for RF localized hyperthermia. (This figure was generated by the authors.)

another newer approach that buys glioblastoma patients extra months of life. Figure 5.16 illustrates a prototype of a large sized millirobot to treat GBM by delivering drugs, or by burrowing through and removing spinel tumor tissue via hyperthermia.

Figure 5.17 depicts the design of a millirobot tip that contains a thermocouple and CNT wire for RF heating. The millirobot may have multiple modalities to treat the tumor. Navigating the robot tip by controlling the tether, as well as the extension of the tip, must be highly precise, whereas the tumor tissue can potentially be identified using fluoroscopy.

A tip with a hydraulic motor and blade for cutting that can be 3D printed is suggested as a future improvement for traversing tissue.

Glioblastoma is particularly difficult to treat for two reasons. First, it is not known exactly why glial cells turn cancerous. Hypotheses range from genetics to environmental risk factors to some combination of the two. Second, the precise demarcation of the location of the cancer is problematic. Since glioblastoma is typically lodged deep within the brain, it is nearly impossible to cleanly separate the tumor from this delicate organ. Malignant tumors are a complex mixture of various species of cancer cells with different behaviors (Yung, 2019). Glioblastoma is challenging to treat

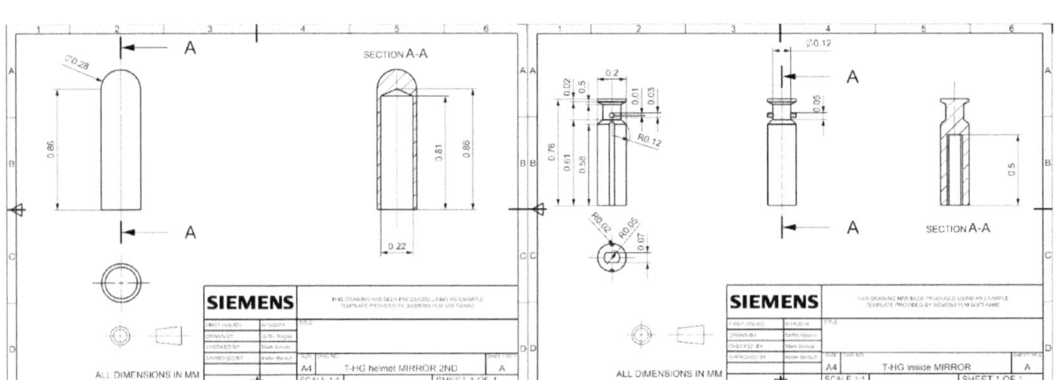

FIGURE 5.17 Design of the millirobot tip to contain a thermocouple and CNT wire for RF heating. (This figure was generated by the authors.)

and currently impossible to cure as it affects the glial cells, which are the sticky supportive membranes that surround the neurons of the brain.

Glioblastoma is also distinguished from other types of brain cancer due to its rapid growth. There are many types of brain cancers, which range from slow growing, to benign brain tumors, to those that derive from glial cells. Glioblastoma is currently incurable as it can multiply and take over the organ, which is central to the operation of the entire body, and due to the challenges involved with the identification and removal of cancerous cells and tumors (Li, 2019). Due to the dense folds of the brain, even small errors in treatment using a laser or radiation can be harmful. Even if the primary tumor is removed, the cancer almost always returns. Glioblastoma contains microtumor cells that can spread throughout the brain; thus, the removal of only one portion of the brain will not be curative.

The survival rate for glioblastoma is ~5%; thus, researchers must determine what makes the glial cells become cancerous. This might involve strategies for earlier diagnosis to facilitate the earlier removal of the cancer, or the development of improved contrast agents to enable enhanced tumor imaging toward which oncologists and researchers have investigated a variety of techniques. The complexity and recalcitrance of glioblastoma mean that a combination of drugs and therapies is required to treat it, which can be harmful to the patient due to significant side effects. This is because researchers do not yet fully understand the precise biology and genetics of the tumors, which would provide useful guidance as to the specific types and dosages of drugs to use, as well as radiation levels.

Further, the BBB prevents toxic therapeutic drugs in the bloodstream from accessing the internal cells of the brain. Immunotherapy is an additional area of research that has been shown to be effective for cancer treatment. One bottleneck is that glial cells do not contain many T-cells; however, antibodies and gene therapies may be employed to modify glial cell functionalities. MRI and various radiographic tracers can be utilized to image most of the tumor, which assists physicians with the preemptive application of radiation and chemotherapies. Advanced non-magnetic imaging might help to precisely guide millirobots (with the proviso that they are MRI compatible) for the removal of tumors. Further, an improved elucidation of genetic markers might translate to improved treatments.

There is a rapidly growing interest in nanoscale technologies across multiple scientific disciplines and industries; however, the medical community still needs robots that have the capacity to perform specific tasks at a larger scale. The purpose of the research in this chapter is to describe the development of a Ø1 mm robot, required for tasks that are still considered to be intractable in the medical community today. Although many robots have already been developed for a variety of tasks, very few have been created for the sole purpose of localized cancer and tumor treatment. Some of the challenges that millirobots have encountered in the past include the ability to fabricate and operate mechanical components at the mm scale. Nevertheless, millirobots are still potentially capable of performing complex tasks that might not otherwise be achieved at the nano- or macroscale. A few of the tasks that millirobots can perform include:

1. In vivo body temperature sensing using thermocouples scaled down to the size of a tenth of a millimeter (Ø100 μm) (Goldman et al., 1987; He et al., 2002).
2. Tumor cell separation using the individual lumen of the tubing to drive a motor at the head of the device.
3. Localized hyperthermia induction via electrical resistance or direct gold nanoparticle (GNP) implementation (Jain et al., 2012).
4. Potential for specific drug delivery and nanoparticle delivery to relatively restricted regions of the body.
5. Capacity to steer robots through tissues to otherwise inaccessible sites.
6. Measuring the pH of bodily fluids throughout different regions of the patient in vivo.
7. Delivery of medication directly into the gastrointestinal tract and provide feedback (Mimee et al., 2018).

Strategies for the development of millirobots consider constraints such as materials, budget, time, and the design of prototypes that can be easily reproducible in terms of cost and manufacturing. The creation of a robot prototype with three primary functions entails the development of its initial function, which is the capacity to steer and turn along two axes. This might be made possible by tendons comprised of threads that traverse through the satellite lumens of multi-lumen tubing, whereby the robot would be made to move by pulling on individual cables. Since the tubing must be flexible, thermoplastics are preferred over other materials; however, it must not be susceptible to kinking under the pull forces of the threads.

The second function of the robot prototype is heat generation, which is achieved by running power through CNT threads. The purpose of this function is to create localized hyperthermia at specific sites within patients, which could directly treat cancer and tumors when combined with procedures such as radiotherapy (Repasky et al., 2018). This application was challenged by GNPs, which can be heated remotely from outside the body, but was still implemented. The final application added to the robot was a type K thermocouple, which was connected to the CNT threads. For additional components, see Figure 5.16b.

Function Generator: Creates an AC current that runs through the CNT threads to create the heat required to induce hyperthermia.

Oscilloscope: Displays the signal that the Function Generator is creating.

DC Motors and Threads: Used to control the direction of the millirobot.

Multimeter: Employed to monitor the voltage that passes through the thermocouple to produce heat.

Two DC Power Supplies: Each used to power their own individual motors.

Thermocouple: Used to measure the heat generated by the CNT fibers.

Multi-lumen Tubing: Individual lumen along the sides, control steering, while the center lumen controls the CNT fibers and thermocouple.

CNT Threads: Black CNT thread is wound around the 3D printed component closest to the thermocouple, which is then encapsulated by a larger component. Drawings of these components can be seen in Figure 5.17.

5.7 PROTOTYPE HYDRAULIC POWERED MILLIROBOT

Our continuing work at the University of Cincinnati is focusing on building hydraulic robots with mm dimensions. Our recent work on the development of a hydraulic powered millirobot is summarized in this section. This chapter provides general suggestions for the development of robots using different propulsion and cutting methods. Each strategy would be appropriate for a particular type of robot regarding the size, torque, steering method, manufacturing technique, and cost. It was decided that a hydraulic robot would be further developed, as the hydraulics allowed for downward scaling, while sustainably providing useful torque (Vinnakota et al.,, 2024a). The hydraulic design is based on the momentum of fluid flow rather than on a pressurized hydraulic system (Figure 5.18). This was done because the robot was 3D printed (Vinnakota et al., 2023). Currently, the 3D printing of a mm size robot at a 10 μm print resolution cannot hold small enough tolerances for the fabrication of a high-pressure robot design. Note that tolerances can be added when building a device. Thus, a Peltier wheel type of robot was printed based on a design that used the momentum of water flow to operate the rotary cutting tip of the robot. This robot was described in detail by Vinnakota et al. (2024a). Initially, an Ø8 mm hydraulic robot was built and tested. The robot design was improved to increase the torque and to make the robot smaller to enable potential use in additional applications. A Ø2 mm hydraulic robot design was also developed. The different sizes of robot rotors are shown in Figure 5.19. This development involved only the working body of the robot. Figure 5.20 shows a prototype Ø4 mm millirobot with a cutting tip. The steering mechanism was not developed, as

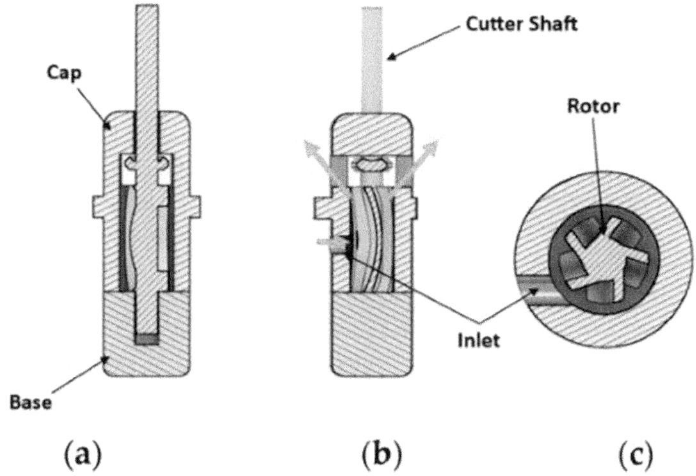

FIGURE 5.18 Drawing of a 4 mm OD single rotor millirobot with a 2 mm OD rotor: (a) Cross section showing the rotor, cap, and base; (b) the blue arrows are fluid flow lines in the inlet plane; (c) cross section through the inlet of an Ø850 μm housing. (This figure is open access from the MDPI article: Vinnakota et al. (2023).)

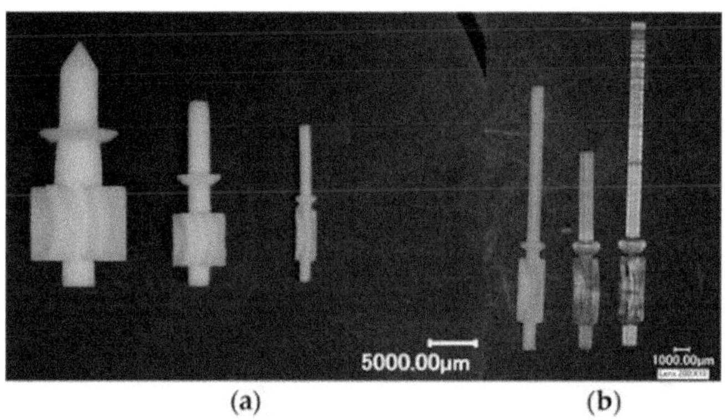

FIGURE 5.19 Millirobot rotors fabricated by 3D printing: (a) Progressively smaller rotors to demonstrate the scaling down of the robot size. The OD of the largest rotor is Ø10 mm, and the smallest is Ø2 mm. (b) The 2 mm OD rotors were printed using three different materials. (This figure is open access from the MDPI article: Vinnakota et al. (2023).)

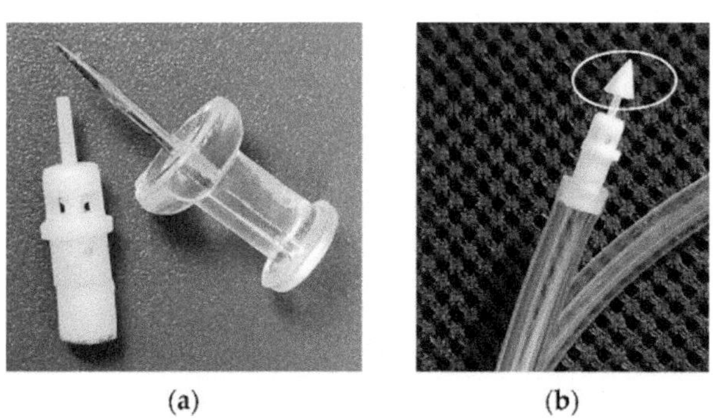

FIGURE 5.20 Prototype Ø4 mm millirobot: (a) Robot next to a pushpin for scale; (b) robot with conical cutting tip on the shaft. (This figure is open access from the MDPI article: Vinnakota et al. (2024b).)

a commercial steerable catheter was planned to be used to control the motion of the robot body. The Ø2 mm robot was tested in gelatin to simulate tissue and was shown to have the capacity to slowly cut through the gelatin. The performance of the robot based on limited experimental bench testing, including computer simulation results, is presented in Vinnakota et al. (2023, 2024a, 2024b, 2025). Finally, the Ø2 mm robot was scaled down to Ø1 mm, which was initially tested and shown to operate (Vinnakota et al., 2024a, 2024b, 2025). However, there are certain improvements required for this robot to have practical utility. The robot body was 3D printed using a polymer; thus, the shaft on the robot was not sufficiently strong to sustain the bending moment forces from the cutting of tissue. Therefore, a metal needle was employed to form the shaft, while the cutting tip was pressed onto the stainless-steel needle shaft. This approach provided adequate strength; however, the manufacturing method was difficult. The development of the Ø1 mm robot is still in progress, which may be the smallest robot available that can produce useful work in cutting tissue. The Ø1 mm robot demonstrated that 3D printing can be used to produce a hydraulic device that operates at very high RPM. Future work will include the further optimization of the performance and tolerances of the robot, attachment of its body to a steerable catheter, and validation of its functionality through a series of preclinical trials. The Ø1 mm robot has also served to verify some of the predictions made in this chapter, thus encouraging our vision that millirobots will become new and advanced tools for clinical use.

5.8 CONCLUSIONS AND GOING FORWARD

Controlling the tether movement that drives a millirobot will modernize multiple current medical devices, including robotic arms/catheters. This will also introduce a platform for the development of an extensive armamentarium of mini-surgical and universal tools to improve healthcare, while reducing costs. Newly emerging scientific and engineering knowledge coupled with AI will enable the design of many classes of miniaturized devices that are smaller, smarter, and more powerful, using hydraulics and nanotube wires, with greater manipulative freedom using flexible tethers, pull wires, and hydraulic pressure. It is hoped that this chapter will assist with advancing miniaturized robot technologies to facilitate the near-term emergence of millirobots for clinical use.

Research into new strategies for medical device development (e.g., micro and millirobotics) provides hope for the improved treatment of cancers. However, another significant hurdle is that the required basic research and development have been slow due to the lack of funding. Thus, it is difficult to know specifically which new approaches for treatments may be effective in bringing meaningful advances that should be funded. In addition, funding agencies are reluctant to fund nascent-stage projects that are high risk, even though they may ultimately bring high returns. Strategic combinations are also important (e.g., surgery, radiation, chemotherapy, with the new millirobot delivery of drugs and microsurgery) for the treatment of cancers. We propose that simple prototype millirobots should be developed and put in the hands of surgeons who can best evaluate their clinical utility, comment on their merits, and engage in furthering their development.

5.9 FUTURE WORK

Having tested the three primary millirobot functions described above, the steering application was fraught with the most challenges. It was prone to many recurring issues that originated from tendon loosening and entanglement. In future designs, a self-tightening mechanism/force applied to the threads to keep them taught (by default) would be the simplest solution for the millirobot in its current form. Existing commercial steerable catheters should also be tested for steering different robot modules. The multi-lumen tubing was preferred to the other methods of steering due to its flexible material; however, lacing the threads, as well as inserting the CNT and thermocouple, was problematic. Considering that the prototype was a scaled version of the final robot, new types of manufacturing methods will likely need to be developed when building these true scale Ø1 mm millirobots.

Potential approaches might include the extrusion of tendons in conjunction with the multi-lumen, or integration of CNT fibers within the tubing. It has been indicated that the oscillating tip motion should be restricted to the robot tip, as shearing through tissues requires significant forces to overcome friction. This suggests that a hydraulic millimeter scale cutting device, which might be 3D printed and metallized, may be preferable for traversing/cutting through tissues. Based on the current Ø1 mm prototype (Vinnakota et al., 2024a), it is roughly estimated that a fully functional steerable millirobot prototype with the capacities to cut tissue, deliver nanoparticles and drugs, and perform biopsies might be developed within five years at a cost of $2.6 million.

ACKNOWLEDGMENTS

We would like to thank the following companies for their assistance with the development of the millirobot prototype tethers and collection of data: Duke Extrusion, Qosina, Nanoscribe, Microlumen, and Vention Medical. Our special thanks are extended to the staff (Brett Heier) of Duke Extrusion for suggesting suitable materials that met the millirobot specifications; Bryan Adams, of Qosina, for suggesting the barbed fittings and other attachments for the millirobot prototypes; Alexander Legant of Nanoscribe for suggestions on the assembly of micro-lumen tubes with carbon nanotubes; Catherine of Microlumen for further discussions and recommendations of materials for the production of the micro-lumen tubes; and Thomas McClain of Vention Medical for sharing his knowledge of forces and the extension of tubes. The authors also thank Mr. Jason Bassi of Boston Microfabrication for recommendations on the use of 3D printers for the fabrication of many copies of the differently sized robots. Finally, the authors thank the following medical advisors at the University of Cincinnati who are guiding millirobot development: Rifat Wahab, Casey Faber, Dan Ionascu, Caroline Morrison, Sandra Starnes, John Perentesis, John M. Racadio, and Nicole Hilvert.

REFERENCES

3D Microprint GmbH. 2025. http://3dmicroprint.com/ (accessed January 23, 2025).

Abdelmohsen LKEA, Peng F, Tu Y, Wilson DA. Micro- and nano-motors for biomedical applications. *Journal of Materials Chemistry B*. 2014;2(17):2395–2408.

Ahn CH, Kim YJ, Allen MG. A planar variable reluctance magnetic micromotor with fully integrated stator and coils. *Journal of Microelectromechanical Systems*. 1993;2(4):165–173.

Alagarsamy KN. Carbon Nanotube Biomedical Wire [Master's capstone project]. Cincinnati (OH): University of Cincinnati; 2017. 45 p. Submitted to the Graduate School in partial fulfillment of the requirements for the degree of Master of Engineering in Biomedical Engineering, Department of Mechanical and Materials Engineering, College of Engineering and Applied Sciences.

Ali A, Sakes A, Arkenbout EA, Henselmans P, van Starkenburg R, Szili-Torok T, Breedveld P. Catheter steering in interventional cardiology: Mechanical analysis and novel solution. *The Proceedings of the Institution of Mechanical Engineers, Part H*. 2019 Dec;233(12):1207–1218.

American Cancer Society. Key statistics for lung cancer. Atlanta: American Cancer Society; 2025. https://www.cancer.org/cancer/types/lung-cancer/about/key-statistics.html (accessed August 11, 2025).

ASRM (American Society for Reconstructive Microsurgery). 2024. http://www.microsurg.org/patients/whatis/ (accessed January 23, 2025).

Bakar AA, Nakajima M, Takeuchi M, Fukuda T. Catalytic driven micro-nanorobots fabricated by direct laser writing. *Micro-NanoMechatronics and Human Science (MHS), 2014 International Symposium*, November 10–12, 2014.

Ballard JB, Dick DD, McDonnell SJ, Bischof M, Fu J, Owen JH, et al. Atomically traceable nanostructure fabrication. *J Vis Exp*. 2015 Jul 17;(101):e52900.

Barbot, A., Decanini, D., & Hwang, G. Controllable Roll-to-Swim motion transition of helical nanoswimmers. In 2014 IEEE/RSJ International Conference on Intelligent Robots and Systems (IROS) IEEE. 2014:4662–4667.

Begley S, "John McCain has died. For brain cancers like his, 'research is our only hope.'" *STAT News*, August 25, 2018 (accessed January 23, 2025). https://www.statnews.com/2018/08/25/mccain-glioblastoma-research/

Boston Microfabrication. 2025. https://bmf3d.com/.

Boston Scientific. 2025. https://www.bostonscientific.com/en-EU/medical-specialties/interventional-radiology/btg.html (accessed August 11, 2025).

BTG Vascular. 2025. https://www.btg-im.com/en-US/Interventional-Vascular (accessed January 23, 2025).

Butowski N, University of California, San Francisco (personal communication, 2019).

Büttgenbach S. Electromagnetic micromotors-design, fabrication and applications. *Micromachines*. 2014; 5(4):929–942.

CareFusion. 2025. https://www.bd.com/fr-ca/resource-and-education/documentation-landing-page/university-health-network-transforming-the-or-supply-chain-through-integration-analytics-and-standardization (accessed January 23, 2025).

Carstensen EL, Parker KJ. Physical models of tissue in shear fields. *Ultrasound in Medicine and Biology* 2014;40(4):655–674.

CDC. Heart Disease Facts. October 24, 2024a. https://www.cdc.gov/heart-disease/data-research/facts-stats/index.html#cdcreference_2 (accessed December 05, 2024).

CDC. Cancer Data and Statistics. December 03, 2024b. https://www.cdc.gov/cancer/data/index.html (accessed December 05, 2024)

Choi J, Kim J. Batch-processed carbon nanotube wall as pressure and flow sensor. *Nanotechnology*. 2010; 21(10):105502.

Galt Medical. 2025. https://galtmedical.com/ (accessed August 11, 2025).

CTCA (Cancer Treatment Centers of America). Minimally Invasive Six Pillars Approach for Tumor Resection – Brain Surgeries – The American Center for Spine and Neurosurgery – ACSN. 2021. https://www.acsneuro.com/brain-surgeries/minimally-invasive-six-pillars-approach-for-tumor-resection (accessed August 11, 2025).

Dhar J, Samanta J, Nabi Z, Aggarwal M, Conti Bellocchi MC, Facciorusso A, Frulloni L, Crinò SF. Endoscopic ultrasound-guided pancreatic tissue sampling: Lesion assessment, needles, and techniques. *Medicina (Kaunas)*. 2024 Dec 7;60(12):2021.

Dogangil G, Davies BL, Rodriguez y Baena F. A review of medical robotics for minimally invasive soft tissue surgery. *Proceedings of the Institution of Mechanical Engineers Part H*. 2010;224(5):653–679.

Domschke A, Boehm FJ. Quandary – Are Molecularly Manufactured Burgers Imbued with the Life Force? (Essay in response to the question, How Should Humanity Steer the Future?) – *FQXi Foundational Questions Institute*. April 2014. https://forums.fqxi.org/d/2085-quandary-are-molecularly-manufactured-burgers-imbued-with-the-life-force-by-frank-josef-boehm-and-angelika-domschke (accessed January 23, 2025).

Engh JA, Podnar G, Kondziolka D, Riviere CN. Toward effective needle steering in brain tissue. Conference Proceedings of the IEEE Engineering in Medicine and Biology Society (EMBC).2006;1: 559–562.

Erni S, Schürle S, Fakhrae A, Kratochvi BE, Nelson BJ. Comparison, optimization, and limitations of magnetic manipulation systems. *Journal of Micro-Bio Robotics*. 2013;8(3):107–120.

Espalin D, Muse DW, Medina F, MacDonald E, Wicker RB. 3D printing multi-functionality: Structures with electronics. *International Journal of Advanced Manufacturing Technology*. 2014;72(5–8):963–978.

Frasson L, Ferroni F, Ko SY, Dogangil G, Rodriguez Y, Baena F. Experimental evaluation of a novel steerable probe with a programmable bevel tip inspired by nature. *Journal of Robotic Surgery*. 2012;6(3): 189–197.

Freitas RA Jr. *Nanomedicine, Volume I: Basic Capabilities*. Georgetown, TX: Landes Bioscience, 1999. http://www.nanomedicine.com/NMI.htm

Freitas RA Jr. Energy Density. *IMM Report* 2019 June 25; 50. http://www.imm.org/Reports/rep050.pdf.

G.O.B.A. World of Microsurgery. 2025. https://www.goba.dk/ (accessed January 23, 2025).

Gao W, Wang J. Synthetic micro/nanomotors in drug delivery. *Nanoscale*. 2014;6(18):10486–10494.

Garcia-Gradilla V, Orozco J, Sattayasamitsathit S, Soto F, Kuralay F, Pourazary A et al. Functionalized ultrasound-propelled magnetically guided nanomotors: Toward practical biomedical applications. *ACS Nano*. 2013;7(10):9232–9240.

Garimella R, Eltorai AE. Nanotechnology in orthopedics. *Journal of Orthopaedics*. 2016;14(1):30–33.

Ghosh A, Fischer P. Controlled propulsion of artificial magnetic nanostructured propellers. *Nano Letters*. 2009;9(6):2243–2245.

Glioblastoma (Wikipedia). 2025. https://en.wikipedia.org/wiki/Glioblastoma (accessed January 23, 2025).

Goldman YE, McCray JA, Ranatunga KW. Transient tension changes initiated by laser temperature jumps in rabbit psoas muscle fibres. *Journal of Physiology*. 1987;392:71–95.

Golestanian R, Liverpool TB, Ajdari A. Propulsion of a molecular machine by asymmetric distribution of reaction products. *Physical Review Letters*. 2005;94(22):220801.

Gou P, Kraut ND, Feigel IM, Bai H, Morgan GJ, Chen Y et al. Carbon nanotube chemiresistor for wireless pH sensing. *Scientific Reports*. 2014;4:4468.

Hambley R, Hebda PA, Abell E, Cohen BA, Jegasothy BV. Wound healing of skin incisions produced by ultrasonically vibrating knife, scalpel, electrosurgery, and carbon dioxide laser. *Journal of Dermatologic Surgery and Oncology*. 1988;14(11):1213–1217.

Haworth KJ, Fowlkes JB, Carson PL, Kripfgans OD. Towards aberration correction of transcranial ultrasound using acoustic droplet vaporization. *Ultrasound in Medicine and Biology*. 2008;34(3): 435–445.

He Q, Zhu L, Lemons DE, Weinbaum S. Experimental measurements of the temperature variation along artery-vein pairs from 200 to 1000 microns diameter in rat hind limb. *Journal of Biomechanical Engineering*. 2002;124(6):656–661.

Hengsbach S, Lantada AD. Rapid prototyping of multi-scale biomedical microdevices by combining additive manufacturing technologies. *Biomed Microdevices*. 2014;16(4):617–627.

Hildebrand T, Ma Q, Heyward CA, Haugen HJ, Nogueira LP. Advanced soft tissue visualization in conjunction with bone structures using contrast-enhanced micro-CT. *Journal of Medical Imaging (Bellingham)*. 2024;11(6):066001.

Hogg T. Using surface-motions for locomotion of microscopic robots in viscous fluids. *Journal of Micro and Bio Robotics*. 2014;9(3):61–77.

Hogg T. Energy dissipation by metamorphic micro-robots in viscous fluids. *Journal of Micro-Bio Robotics*. 2016;11:85–95.

Hogg T, Sretavan DW. Controlling tiny multi-scale robots for nerve repair. In: *Proceedings of the 20th National Conference on Artificial Intelligence (AAAI-05)*; 2005 Jul 9–13; Pittsburgh, PA. Menlo Park (CA): AAAI Press; 2005:1286–1291.

Hohmann JK, Renner M, Waller EH, von Freymann G. Three-dimensional μ-printing: An enabling technology. *Advanced Optical Materials*. 2015;3:1488–1507.

Hong WK, Lee C, Nepal D, Geckeler KE, Shin K, Lee T. Radiation hardness of the electrical properties of carbon nanotube network field effect transistors under high-energy proton irradiation. *Nanotechnology*. 2006;17(22):5675–5680.

Huang TY, Sakar MS, Mao A, Petruska AJ, Qiu F, Chen XB et al. 3D printed microtransporters: Compound micromachines for spatiotemporally controlled delivery of therapeutic agents. *Advanced Materials*. 2015;27(42):6644–6650.

Ibele M, Mallouk TE, Sen A. Schooling behavior of light-powered autonomous micromotors in water. *Angewandte Chemie International Edition*. 2009;48(18):3308–3312.

IJINT Internet Journal of Interventional Medicine. 2013. http://ispub.com/IJINT (accessed January 23, 2025).

Introduction to Carbon Hybrid Materials. Proceedings of the Nanotechnology Materials and Devices Workshop. University of Cincinnati, May 22–23, 2018, Cincinnati, Ohio. https://ceas.uc.edu/research/centers-labs/nanoworld.html (accessed January 23, 2025).

Jain S, Hirst DG, O'Sullivan JM. Gold nanoparticles as novel agents for cancer therapy. *British Journal of Radiology*. 2012;85(1010):101–113.

Journal of *Micro-Bio Robot*. 2025. https://link.springer.com/journal/12213 (accessed January 23, 2025).

Kagan D, Benchimol MJ, Claussen JC, Chuluun-Erdene E, Esener S, Wang J. Acoustic droplet vaporization and propulsion of perfluorocarbon-loaded microbullets for targeted tissue penetration and deformation. *Angewandte Chemie International Edition in English*. 2012;51(30):7519–7522.

Karachaliou N, Mayo-de-las-Casas C, Molina-Vila MA, Rosell R. Real-time liquid biopsies become a reality in cancer treatment. *Annals of Translational Medicine*. 2015;3(3):36.

Kim S, Qiu F, Kim S, Ghanbari A, Moon C, Zhang L et al. Fabrication and characterization of magnetic microrobots for three-dimensional cell culture and targeted transportation. *Advanced Materials*. 2013;25(41):5863–5868.

Lee YW, Ceylan H, Yasa IC, Kilic U, Sitti M. 3D-printed multi-stimuli-responsive Mobile micromachines. *ACS Applied Materials & Interfaces*. 2021;13(11):12759–12766.

Leshansky AM, O Kenneth O, Gat O, Avron JE. A frictionless microswimmer. *New Journal of Physics*. 2007;9(5):145.

Li G, Stanford University (personal communications, 2019).

Lissandrello CA, Gillis WF, Shen J, Pearre BW, Vitale F, Pasquali M et al. A micro-scale printable nanoclip for electrical stimulation and recording in small nerves. *Journal of Neural Engineering*. 2017;14(3): 036006.

Stanford University, Lung Cancer. 2025. https://med.stanford.edu/ctsurgery/clinical-care/thoracic-surgery-services/lung-cancer.html (accessed January 23, 2025).

Ma J, Yeow JTW, Chow JCL, Barnett RB. A carbon nanotube-based radiation sensor. *International Journal of Robotics & Automation*. 2007;22(1):49.

Magdanz V, Sanchez S, Schmidt OG. Development of a sperm-flagella driven micro bio-robot. *Advanced Materials*. 2013;25(45):6581–6588.

Mahar B, Laslau C, Yip R, Sun Y. Development of carbon nanotube-based sensors—A review. *IEEE Sensors Journal*. 2007;7(2):266–284.

Manesh KM, Cardona M, Yuan R, Clark M, Kagan D, Balasubramanian S, Wang J. Template-assisted fabrication of salt-independent catalytic tubular microengines. *ACS Nano*. 2010;4(4):799–804.

Martel S. Microrobotics in the vascular network: Present status and next challenges. *Journal of Micro-Bio Robotics*. 2013;8(1):41–52.

Mathieu JB, Martel S, Yahia L, Soulez G, Beaudoin G. Preliminary investigation of the feasibility of magnetic propulsion for future microdevices in blood vessels. *Biomedical Materials and Engineering*. 2005;15 (5):367–374.

Mei Y, Huang G, Solovev AA, Ureña EB, Mönch I, Ding F et al. Versatile approach for integrative and functionalized tubes by strain engineering of nanomembranes on polymers. *Advanced Materials*. 2008;20(21):4085–4090.

MicroLumen. 2024. http://www.microlumen.com/medical-tubing/custom (accessed January 23, 2025).

Mimee M, Nadeau P, Hayward A, Carim S, Flanagan S, Jerger L et al. An ingestible bacterial-electronic system to monitor gastrointestinal health. *Science*. 2018;360(6391):915–918.

Mirkovic T, Zacharia NS, Scholes GD, Ozin GA. Fuel for thought: Chemically powered nanomotors out-swim nature's flagellated bacteria. *ACS Nano*. 2010;4(4):1782–1789.

Moreira P, Misra S. Biomechanics-based curvature estimation for ultrasound-guided flexible needle steering in biological tissues. *Annals of Biomedical Engineering*. 2015;43(8):1716–1726.

Mozaffarian D, Benjamin EJ, Go AS, Arnett DK, Blaha MJ, Cushman M et al. Heart disease and stroke statistics—2016 update: A report from the American Heart Association. *Circulation*. 2016;133(4): e38–360.

Mushtaq F, Guerrero M, Sakar MS, Hoop M, Lindo A, Sort J et al. Magnetically driven Bi_2O_3/BiOCl-based hybrid microrobots for photocatalytic water remediation. *Journal of Materials Chemistry A*. 2015;3:23670.

Nanoscribe. Microfabrication, 3D Printing on the Micrometer Scale. 2025. http://www.nanoscribe.de/en/ (accessed January 23, 2025).

Nanoworld Laboratories. 2025. https://ceas.uc.edu/research/centers-labs/nanoworld.html (accessed January 23, 2025).

Nelson BJ, Kaliakatsos IK, Abbott JJ. Microrobots for minimally invasive medicine. *Annual Review of Biomedical Engineering*. 2010;12:55–85.

Paxton WF, Kistler KC, Olmeda CC, Sen A, St Angelo SK, Cao Y, Mallouk TE, Lammert PE, Crespi VH. Catalytic nanomotors: Autonomous movement of striped nanorods. *Journal of the American Chemical Society*. 2004;126(41):13424–13431.

Peng J, Qin Z, Chen S. Design and fabrication of an electromagnetic micro-motor for intravascular ultrasound (IVUS) imaging. In: *2016 IEEE International Ultrasonics Symposium (IUS)*; 2016 Sep 18–21; Tours, France. Piscataway (NJ): IEEE; 2016. p. 1–4.

Perica K, Tu A, Richter A, Bieler JG, Edidin M, Schneck JP. Magnetic field-induced T cell receptor clustering by nanoparticles enhances t cell activation and stimulates antitumor activity. *ACS Nano*. 2014;8(3):2252–2260.

Qiagen Technology. *Using the submarine for liquid Biopsy Technology*. 2025. http://www.qiagen.com/us/?bhn_mid=29282237 (accessed January 23, 2025).

Qiu F, Fujita S, Mhanna R, Zhang L, Simona BR, Nelson BJ. Magnetic helical microswimmers functionalized with lipoplexes for targeted gene delivery. *Advanced Functional Materials*. 2015;25(11): 1666–1671.

Repasky EA, Evans SS, Dewhirst MW. Temperature matters! And why it should matter to tumor immunologists. *Cancer Immunology Research.* 2018;1(4):210–216.

Research and Markets. Catheters Market – Growth, Trends, COVID-19 Impact, and Forecasts (2021–2026). 2022. https://www.researchandmarkets.com/reports/4987202/catheters-market-growth-trends-covid-19 (accessed January 23, 2025).

Schulz M, Shanov V, YeoHeung Y. (eds.). *Nanomedicine Design of Particles, Sensors, Motors, Implants, Robots, and Devices.* Artech House, 2009b.

Schulz MJ, Li W, Ruff B, Venkatasubramanian R, Song Y, Suberu B, Cho W, Salunke P, Sowani A, Yin J, Mast D, Shanov V, Dong Z, Pixley S, Hu J, Muratore C. Chapter 11: Sensorcyte Artificial Cells for Human Diagnostics and Analytics. In *Nanomedical Device and Systems Design: Challenges, Possibilities, Visions.* Boehm FJ. (ed.), CRC Press. 2013.

Schulz MJ, Li W, Shanov V, Yun Y, Jayasinghe C, Salunke P et al. Chapter 1: A Nanotechnology Framework for Medical Innovation. In *Nanomedicine Design of Particles, Sensors, Motors, Implants, Robots & Devices.* Schulz M, Shanov V, YeoHeung Y. (eds.), Artech House. 2009a.

Seeger W. *Microsurgery of the Brain: Anatomical and Technical Principles.* Springer Science & Business Media, 2013.

Sitti M, Ceylan H, Hu W, Giltinan J, Turan M, Yim S, Diller E. Biomedical applications of untethered mobile milli/microrobots. *Proceedings of the IEEE.* 2015;103(2):205–224.

Sivasankar M, Durairaj RB. Brief review on nano robots in bio medical applications. *Journal of Advanced Robotics & Automation.* 2012;1(1):1000101.

Szabo TL. *Diagnostic Ultrasound Imaging: Inside Out.* 2nd ed. Waltham, MA, USA: Elsevier Academic Press, 2014.

Takato M, Yokozeki Y, Saito K, Uchikoba F. Development of electromagnetic MEMS motor without winding wire and application to microrobot. In: *Proceedings of the 2016 IEEE International Conference on Advanced Intelligent Mechatronics (AIM)*; 2016 Jul 12–15; Banff, AB, Canada. Piscataway (NJ): IEEE; 2016. p. 7576844.

Tang X, Zhang D, Li Z, Chen J. An omni-directional wall-climbing microrobot with magnetic wheels directly integrated with electromagnetic micromotors. *International Journal of Advanced Robotic Systems.* 2012;9(1):1–9.

Taylor RH, Anderson JH. Developing novel computing methods, interfacial technologies, and computer-integrated surgical systems to revolutionize surgical procedures in the 21st century. *NSF ERC for Computer-Integrated Surgical Systems and Technology, 1998 to 2008*, The John Hopkins University. 2000. https://www.nsf.gov/pubs/2000/nsf00137/nsf00137d.htm (accessed January 23, 2025).

Thamphiwatana S, Li J, Gao W, Zhang L, Wang J. Artificial micromotors in the Mouse's stomach: A step toward in vivo use of synthetic motors. *ACS Nano.* 2015;9(1):117–123.

Tottori S, Zhang L, Qiu F, Krawczyk KK, Franco-Obregón A, Nelson BJ. Magnetic helical micromachines: Fabrication, controlled swimming, and cargo transport. *Advanced Materials.* 2012;24(6):811–816.

Trimmer WSN, Gabriel KJ. Design considerations for a practical electrostatic micro-motor. *Sensors and Actuators.* 1987;11(2):189–206.

UCSF Health. 2025. http://www.ucsfhealth.org/education/heart_and_vascular_disease_the_facts/ (accessed January 23, 2025).

Vinnakota M, Bellur K, Starnes SL, Schulz MJ. Design and modeling of a miniature hydraulic motor for powering a cutting tool for minimally invasive procedures. *Micromachines.* 2023;14(7):7.

Vinnakota M, Bellur K, Starnes SL, Schulz MJ. U.S. Utility Patent Application for "Rotary Microhydraulic Cutter for Trans-Tissue Navigation," U.S. Serial No. 18/756,618, Filed June 27, 2024a.

Vinnakota M, Bellur K, Starnes SL, Schulz MJ. Scaling a hydraulic motor for minimally invasive medical devices. *Micromachines.* 2024b;15(1):1.

Vinnakota M, Bellur K, Starnes SL, Schulz MJ. Hydraulic micromotor powered dissector for minimally invasive therapy. *Discover Mechanical Engineering.* 2025;4(4).

Virani SS, Alonso A, Aparicio HJ, Benjamin EJ, Bittencourt MS, Callaway CW, et al. Heart disease and stroke statistics—2021 update: A report from the American Heart Association. *Circulation.* 2021;143(8):e254–e743.

Wang J, Gao W. Nano/microscale motors: Biomedical opportunities and challenges. *ACS Nano.* 2012 Jul 24;6(7):5745–51.

Weathers SP. Anderson Cancer Center (personal communication, 2019).

Wen P, Dana-Farber Cancer Institute, Society for Neuro-Oncology (personal communication, 2019).

Xiong W, Zhou YS, He XN, Gao Y, Mahjouri-Samani M, Jiang L et al. Simultaneous additive and subtractive three-dimensional nanofabrication using integrated two-photon polymerization and multiphoton ablation. *Light: Science & Applications.* 2012;1:e6.

Yu C-J, Han K, Kim E, Suh S, Lee YM, Kim JY. Physical and magnetic characteristics of carbon nanotubes radiated by proton beams. *Journal of the Korean Physical Society*. 2007;50(5):1524–1527.

Yun YH, Bhattacharya A, Watts NB, Schulz MJ. A label free electronic biosensor for detection of bone turnover markers. *Sensors (Basel)*. 2009;9(10):7957–7969.

Yun YH, Dong Z, Shanov V, Heineman WR, Halsall HB, Bhattacharya A, et al. Carbon nanotube electrodes and biosensors. *Nano Today*. 2007;2(6):30–37.

Yung A, University of Texas MD Anderson Cancer Center, National Cancer Institute, (personal communication, 2019).

Zeeshan MA, Grisch R, Pellicer E, Sivaraman KM, Peyer KE, Sort J, et al. Hybrid helical magnetic microrobots obtained by 3D template-assisted electrodeposition. *Small*. 2014;10(7):1284–1288.

Zeus. 2025. https://www.zeusinc.com/wp-content/uploads/2025/03/Zeus-Catalog-G-V1R14.pdf (accessed August 11, 2025).

Zhang S, Carberry D, Nieminen TA, Rubinsztein-Dunlop H. Hydrodynamics of micro-objects near curved surfaces. *Proc SPIE*. 2015;9548:95481C

Zhu W, Li J, Leong YJ, Rozen I, Qu X, Dong R et al. 3D-printed artificial microfish. *Advanced Materials*. 2015;27(30):4411–4417.

Zygomalas A, Kehagias I, Giokas K. Miniature surgical robots in the era of NOTES and LESS: Dream or reality. *Surgical Innovation*. 2015;22(1):97–107.

6 Shifting Paradigms in Healthcare Infrastructures

Hospitals, Clinics, and Medical Personnel

Frank J Boehm

CONTENTS

6.1 A BRIEF HISTORY HOSPITALS AND CLINICS

Some of the earliest evidence for the existence of a "hospital" per se dates back to approximately 1000 BC in ancient Mesopotamia, when royal physicians in the City of Assur (its ruins located in what is now Northern Iraq) treated ill court singers in rudimentary hospitals/nursing homes (Aitken et al., 1984; Cilliers and Retief, 2002). In the 5th century BC, the first institutionalized healthcare facilities emerged in India within and around Buddhist monasteries, as did the then fledgling nursing profession. As Buddhism extended to the East during the same time period, convalescent homes, almshouses, and elementary hospitals evolved in China and Southeast Asia (Major, 1954; Philips and Allan 1993; Chrystal, 2000).

During the Greco-Roman era (between the 8th and 6th century BC), Asclepieia (healing temples) served as the harbingers of modern-day hospitals (Thompson and Goldin, 1975), of which 300 are known to have existed in ancient Greece. Subsequently, Asclepieia were developed and thrived across the Roman Empire till 391 AD. During the 1st century AD, the Roman valetudinaria emerged, which first exemplified the formal hospital, albeit they were only available to the military. Eventually, however, similar establishments began to serve the civilian populace. It was during the Christian era that the first instance of the equivalent of a modern hospital emerged, which offered long-term care to civilians.

Extending to the 10th century it was rare to find hospitals outside monasteries; however, with the expansion of Europe, hospitals became more diversified and increasingly established by their proponents in large cities. Concurrent with the "Golden Age of Islam" was the construction of superb hospitals in Muslim countries. It was clearly evident by the 13th and 14th centuries that the advancements of Islamic hospitals surpassed those built by the Christians (Cilliers and Retief, 2002; Guthrie, 1958).

In Europe, with the dissolution of monasteries by King Henry VIII in 1540, the church withdrew its support of hospitals. Nevertheless, several continued to receive direct endowments from the crown due to petitions from the London citizenry (Waddington, 2003). In the 17th century, research into the circulatory system proceeded at St. Bartholomew Hospital, whereas during the 18th century, some of the core principles of modern surgery were established, and in the late 19th century advances in the nursing profession ensued (Robinson, 1993).

DOI: 10.1201/9781003680093-6

During the same time period with the colonization of the Americas, the first hospital (Hospital San Nicolás de Bari, Santo Domingo, Distrito Nacional Dominican Republic) was established in 1503, which was in operation until the mid-18th century, which currently sits in ruins (De Micheli, 2005). Immaculate Conception Hospital in Mexico City was the first to be established in North America in 1524 to provide care for the poor (De Micheli, 2005).

With the Age of Enlightenment (17th and 18th centuries) came the advent of the modern hospital, which began to distinguish between the provision of relief for the poor and strictly medical care. These facilities were staffed with qualified medical personnel, with the aim of applying modern techniques to treat patients. Acute cases began to be addressed in isolation, and distinct departments were created for patients with different conditions and levels of severity.

Of the cities that existed in the 13 Colonies of the United States at the time, Philadelphia was the most rapidly expanding; thus, the first hospital (Pennsylvania Hospital) was established there in 1751 (Morton and Woodbury, 1895). Hospitals primarily cared for the poor for much of the 19th century, while the financially able could afford to be treated by physicians in their own homes. However, during the mid-19th century, hospitals, as well as the practice of medicine itself, became more professionally managed. Following the devastation of the influenza epidemic (1918), there was a significant upsurge in the funding and construction of public hospitals in the United States, which incorporated the more extensive use of technologies and the development of various vaccines. However, there were multiple issues with hospitals at this time:

- segregation by race and gender, which was commonplace until the late 1960s.
- the prevalence of smoking.
- low standards for cleanliness and sterilization.
- religious groups building their own hospitals that adhered to their dietary restrictions.
- pain management was overlooked.
- patients had no say in/options for their treatments, as it was thought they would be incapable of understanding (or traumatized by detailed knowledge of) their conditions. Thus, the doctor's word was the last word.
- doctors performing surgical procedures in street clothes.
- little protection from X-rays.
- most babies were born at home.
- elderly who were sick were cared for at home.
- no nurse practitioners.
- surgeries could be viewed by the public, open air operating theaters.
- multibed wards.
- most surgeons were male.
- physician owned for profit institutions.
- minimal regulatory oversight.

(Harrington, 2019)

During the Great Depression (1932), a dedicated census revealed a shift away from privately owned to public hospitals. Of 6,562 registered hospitals, 776 government run general hospitals were at 77.1% capacity, whereas 3,529 nongovernmental general hospitals were at only 55.9% capacity. From 1909 to 1932, it was found that hospital bed numbers increased six times faster than the general population, which led the Council on Medical Education and Hospitals to state in 1933 that the US was "over hospitalized" (Figure 6.1) (Hospital Service, 1933).

To accommodate significantly increased demand following World War II, the US government strongly supported (via the 1947 Hill Burton Act) the development and expansion of community hospitals. In the 1950s/1960s, the National Institutes of Health undertook to dramatically expand medical research (for- and non-profit) (Wall, 2022). Further, in 1965 both Medicare and Medicaid were created to fund medical care for the elderly and those in poverty, respectively (Callahan and Wasunna, 2006).

FIGURE 6.1 Mayo Clinic Hospital, Saint Mary's Campus, Rochester, MN, USA. (Reproduced from WikiMedia Commons under Creative Commons Attribution-Share Alike 3.0 Unported license.)

However, hospital care costs increased drastically from 1966 to 1976 to effectively double typical patient-per-day costs. Two-thirds of these increased costs were attributed to "massive expansions in hospital payroll and non-payroll expenses—including 'profits'" (Stevens, 1999, p. 1999). In the US alone, the number of hospitals expanded from 247 (1960) (Wall, 2022) to 6,090 (2021), with operating expenditures in 2021 of $1.2 trillion (American Hospital Association, 2025). According to a 2021 report by the World Health Organization, "Global spending on health more than doubled in real terms over the past two decades, reaching US$8.5 trillion in 2019, or 9.8% of global GDP. But it was unequally distributed, with high income countries accounting for approximately 80%" (WHO, 2021). Subsequent to the COVID-19 pandemic, aggregate global health expenditures in 2022 were US$9.8 trillion, or 9.9% of global GDP (WHO, 2024).

6.2 ROLES OF HOSPITALS AND CLINICS IN THE GHCE ERA

For the most part, with the emergence and implementation of a (hypothetical, for now) mature Global Health Care Equivalency (GHCE) system over the next few decades, hospitals and clinics (with few exceptions – see below) will likely be repurposed or consigned to obsolescence. This may likely be due to the advent of molecular manufacturing (MM) within the same time frame (Domschke and Boehm, 2014), which would serve as the prerequisite enabling technology for GHCE. The authors envisage that, for the purposes of GHCE, MM might be manifest as "Factory at Home" (F@H) systems which would eventually be massively distributed on a global scale. These countertop microwave oven-sized F@H units (in addition to being able to produce a plethora of consumer items, including gourmet foods – see *Molecular Manufacturing: Emergence of the Grand Equalizer*, CRC Press (Boehm, 2025) will have the capacity to synthesize a standardized set of sophisticated autonomous nanomedical devices (ANMD).

Each class of ANMD will be endowed with a dedicated range of nanomedical capabilities (e.g., ultrahigh resolution in vivo diagnostic imaging, hyperthermia, or cavitation for the eradication of cancer cells and tumors, immune system augmentation, cellular, and even organellar resolution nanosurgical repair, etc.), which will cumulatively have the capacity to address virtually any human ailment, including aging.

Exceptions for the use of hospitals and clinics in the era of GHCE would include their serving as large capacity dedicated emergency medical facilities to treat and monitor mass traumatic injuries in cases of natural disasters, or other urban/regional scale public emergencies. Less severe forms of these injuries would be able to be rapidly and efficiently treated in one's own home (if physical access to one's home is available, and the F@H devices are functional) via specialized emergency class nanomedical devices. On a typical daily basis, hospitals and clinics would still be required for the treatment of serious physical injuries (e.g., compound fractures, broken bones, extensive serious burns, severe lacerations/bleeding, bullet wounds, radiation sickness, traumatic shock, etc.), which require immediate medical interventions. That above said, telemedical physicians and surgeons will likely have an extensive range of nanomedical devices at their disposal to assist with, expedite, and conduct follow-up tasks for any procedure.

Further, certain hospitals and clinics would likely operate as dedicated centers to facilitate childbirth. Additional specialized classes of nanomedical devices might assist with the birthing process, as they could continuously monitor all of the health parameters of the mother and child and take immediate reparative action, if required, should there be an emergency, such as childbirth distress in the mother or child, or hemorrhaging.

These significant changes will likely occur due to the advent of mature globally implemented massively distributed MM and GHCE, where the majority of the general population worldwide would have personalized access to advanced diagnostic and therapeutic ANMD that are fabricated in their own homes. Prior to the commencement of any diagnostic or therapeutic procedures, personalized compact/modular non-invasive medical scanning technologies would be engaged that have the capacity to generate comprehensive physiological reference profiles down to molecular resolution. These data would subsequently be uploaded to all prescribed ANMD to facilitate and guide their diagnostic and therapeutic activities. These profiles would encompass the full range of an individual's unique genomic, proteomic, bacteriome, virome, mycobiome, archaeome, eukaryome, and biomolecular parameters (Hopson et al., 2020).

Further, complementary scans would initially be performed for the entire vascular system (down to capillary level) and lymphatic system using a specialized Vascular Cartographic Scanning Nanodevice (VCSN) (Figure 6.2), as well as the entire gastrointestinal system with a Gastrointestinal Micro Scanning Device (GMSD) (Figure 6.3) (Boehm, 2013).

FIGURE 6.2 Conceptual Vascular Cartographic Scanning Nanodevice (VCSN) (images by Frank J. Boehm, 2013).

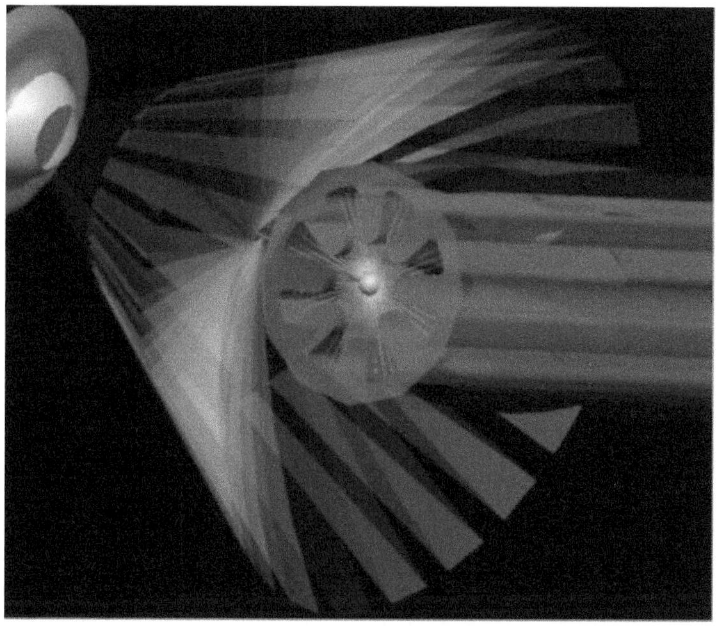

FIGURE 6.3 Conceptual Gastrointestinal Micro Scanning Device (GMSD) (image by Frank J. Boehm, 2013).

Both of these systems would have the capacity to generate micron resolution imaging, as well as ultrahigh resolution spectrographic data to facilitate outbody diagnostic AI systems. The resulting data would be conveyed via safe/secure, randomly nested/refreshed quantum encryption and stored on highly decentralized, blockchain administered, digital medical interfaces complete with personalized medical AI/AGI/ASI agents and virtual/holographic avatar consultations with medical personnel.

6.3 ROLES OF MEDICAL PERSONNEL IN THE GHCE ERA

Rapidly approaching transformations in the workplace via AI, robotics, and other advanced technologies are the precursors to the Fourth Industrial Revolution, the consequences of which we can only speculate. Cumulative technological disruptions have brought us to the advent of being able to reinvent ourselves. The current tsunami of reinvention is likely to wash over virtually every sector we can imagine. It will likely be the case that once the GHCE system begins to be consolidated into daily life, all healthcare stakeholders will see their jobs being modified or redundant, as adaptations are made to accommodate ever accelerating advances in science and technology.

With the advent and broad acceptance/implementation of mature GHCE, the roles of healthcare professionals (HCPs) spanning the medical spectrum are likely to be significantly transformed, with some undertaking completely new roles that seamlessly align with the new paradigm, while others adjust or move on from their former roles that become redundant. GHCE will undoubtedly have advantages and disadvantages for HCPs, and the required adaptations may be significant. Consequently, it may be a matter of clearly identifying and elucidating the key innovations that drive GHCE toward envisaging and implementing new healthcare workflows. From a historical perspective, every occupation that exists was created as a response to a need, which also applies to the healthcare domain. The establishment of GHCE will likely be akin to what motivated successful adaptations to new eras in the past. It will center on striking a balance/equilibrium between what we, as the global population, gain versus what is lost, and how organically this transformation takes place.

The question to be addressed prior to reinventing ourselves once again via GHCE is a remarkable and critical one. What will the needs be under this new paradigm of attending, consulting, monitoring, diagnosing, and treating individuals? (We note here that healthcare under GHCE will not be meant only for patients, but everyone.) If we discuss these aspects stepwise, telemedicine as it currently operates will advance to be replete with far more enhanced UX/UIs (user experience/user interfaces), while the monitoring of one's physiological metrics will be achieved by a dedicated monitoring class of nanorobots and ultrasensitive nanotech/AI-driven wearables.

Regular follow-up visits to a clinician will not be required, as the system itself will send notifications for self-monitoring, where nursery/physician consultation will be suggested/requested by the system if required. A patient-centered healthcare system (HCS) would be the goal and mission. Although human decision-makers will still have roles to play (albeit increasingly more decisions will be made by medical AI/AGI/ASI with initial human oversight and follow-on monitoring), this complex but agile GHCE model would auto-optimize protocols and workflows and thus associated decisions as well. The GHCE system will have as an important facet of its primary directive, constant self-improvement for the benefit of patients, toward cost-effectively optimizing their wellness and health.

A PDSA (Plan-Do-Study-Act) cycle (aka the Deming cycle) (Essraidi and Chraibi, 2024; Lu et al., 2022), which is a proven management strategy that supports constant upgrading and standardization across numerous industries, represents well what is anticipated for the GHCE era. Other needs will undoubtedly be unveiled; initially via GHCE pilot test beds that are introduced worldwide to gauge their operational efficacy across diverse countries and cultures, and subsequently when formally implemented globally to serve large populations.

Under GHCE the current acute lack of time experienced by clinicians to engage in research, recruit patients for clinical trials, support patient empowerment, or simply to spend on staying healthy (e.g., sleeping after brutal 24+ hour shifts) and being with their families would no longer be at issue. Physicians should not be in the operational field for as much time as presently. They should transfer their know-how and serve as leaders/guides toward empowering citizens to take greater responsibility for their own health, as well as being advisors at all required levels (primary/secondary education, families to carry out a healthy life, government for a profitable investment in healthcare Research and Development (R&D) and Innovation, team up in advisory and operational boards to undergo new healthcare challenges, insights, breakthroughs, and advances).

According to Dr. Bertalan Meskó, the trend trajectory looks to be that physicians become increasingly engaged with and utilize technologies such as AI and robotics in the near future (Meskó, 2024). For instance, it is estimated that by 2026, 100 million wearable devices in the United States will be connected to the Internet (Pawar and D'Souza, 2025). Since these devices increasingly integrate a health tracking feature, patients will inevitably transmit data from such devices to various health facilities via secure links. Consequently, physicians are duty bound to comprehensively understand the functionalities of these devices to accurately interpret their data.

In envisioning some of the anticipated transitions for certain HCS specialists, experts suggest that the following changes will occur over the next decade, which are framed in the context of this chapter toward mature GHCE):

a. Surgeons remotely perform robotic and AI-driven surgeries (Figure 6.4) with the aim of higher precision and expedited post-surgery recovery times (Chatterjee et al., 2024; Li et al., 2024; Moschovas et al., 2025a; Moschovas et al., 2025b)
b. Physicians include factors such as exercise, nutrition, and R&D participation (from observational and exploratory clinical studies to clinical drug validation trials) in patient diagnoses and therapeutic strategies. Healthy citizens as well as patients are made aware of

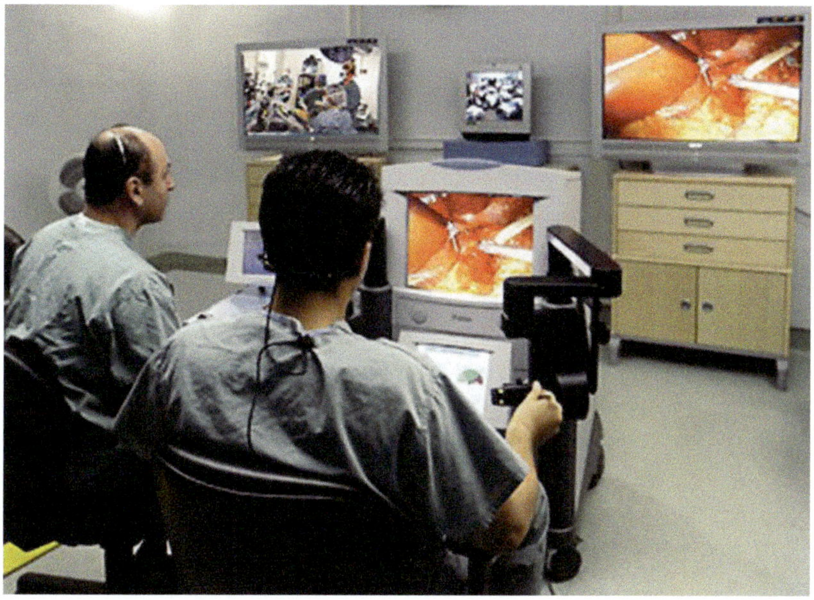

FIGURE 6.4 Telesurgery. (Reproduced from WikiMedia Commons under Creative Commons Attribution-Share Alike 4.0 International license.)

specific scientific needs and advances. Clinicians and scientists promote the use of wearables by the general public and suggest/recommend their participation in public health studies to facilitate medical and scientific advances. Virtual physician and surgeon-mediated telemedical consultations with, and recommendations for, patients become increasingly ubiquitous (Ateriya et al., 2018; Kanevsky et al., 2019; Riva and Gamberini, 2000a; Riva and Gamberini, 2000b; Singh et al., 2020; Truijen et al., 2022).

c. Nurses will provide increasingly greater psychological services to patients, well care checkup/monitoring, etc. To cover the patient's needs in a personalized way, they will work more closely with psychologists (Anne Crooks and Wingo, 2022; Morsy et al., 2022; Schuelke et al., 2019; Rouleau et al., 2023; Worlikar et al., 2023). And a digitally transformative Hospital at Home will make domestic 24/7 telemedical consultations available and reduce hospitalization length (Zychlinski, 2024).

6.4 CONCLUSION

It is conceivable that the successful establishment and operation of small globally distributed pilot GHCE test beds might transition over time to the implementation and proliferation of a more formal and standardized fully mature GHCE system. Nanomedicine would facilitate this expansion via molecular manufacturing as the engine that creates the necessary nanomedical devices that anyone might require, on demand (albeit still under prescription by physicians and surgeons) at anytime, anywhere across the globe.

REFERENCES

Aitken JT, Fuller HWC, Johnson D. *The Influence of Christians in Medicine*. Nottingham: Intervarsity Press, 1984.

American Hospital Association. *Fast Facts on U.S. Hospitals*. 2025. https://www.aha.org/statistics/fast-facts-us-hospitals (accessed March 26, 2025).

Anne Crooks E, Wingo NP. The power of presence in virtual teaching and practice environments. *Nursing Clinics of North America*. 2022;57(4):525–538.

Ateriya N, Saraf A, Meshram VP, Setia P. Telemedicine and virtual consultation: The Indian perspective. *National Medical Journal of India*. 2018;31(4):215–218.

Boehm FJ. *Nanomedical Device and Systems Design – Challenges, Possibilities, Visions*. CRC Press, 2013.

Boehm FJ. *Molecular Manufacturing: Emergence of the Grand Equalizer*. CRC Press. 2025.

Callahan D, Wasunna AA. *Medicine and the Market: Equity v. Choice*. Baltimore: Johns Hopkins University Press, 2006.

Chatterjee S, Das S, Ganguly K, Mandal D. Advancements in robotic surgery: Innovations, challenges and future prospects. *Journal of Robotic Surgery*. 2024;18(1):28.

Chrystal D. *The Cambridge Encyclopedia. Fourth Edition*. Cambridge: University Press, 2000.

Cilliers L, Retief FP. The evolution of the hospital from antiquity to the end of the Middle Ages. *Curationis*. 2002;25(4):60–66.

De Micheli A. En torno a la evolución de los hospitales. *Gaceta Médica de México*. 2005;141(1):59.

Domschke A, Boehm FJ. (Essay in response to the question, How Should Humanity Steer the Future?) – "Quandary – Are Molecularly Manufactured Burgers Imbued with the Life Force?" *FQXi Foundational Questions Institute*. April 2014. http://www.fqxi.org/community/forum/topic/2085 (accessed March 25, 2025).

Essraidi J Sr, Chraibi M. Implementing industry-inspired quality practices in healthcare: A study from the Pathology Department of the University Hospital Center of Tangier. *Cureus*. 2024;16(10):e71377.

Guthrie D. *A History of Medicine*. Edinburgh: Nelson & Sons, 1958.

Harrington J. What a Hospital Looked Like 100 Years Ago. 24/7 Wall St., LLC. October 2019. https://247wallst.com/special-report/2019/10/04/what-a-hospital-looked-like-100-years-ago/#:~:text=There%20were%20no%20nurse%20practitioners,diagnosing%20diseases%2C%20and%20starting%20treatment.&text=Hard%20to%20believe%20now%2C%20but,in%20cafeterias%20and%20waiting%20rooms (accessed September 14, 2021).

Hopson LM, Singleton SS, David JA, Basuchoudhary A, Prast-Nielsen S, Klein P et al. Bioinformatics and machine learning in gastrointestinal microbiome research and clinical application. *Progress in Molecular Biology and Translational Science*. 2020;176:141–178.

Hospital Service in the United States: Twelfth Annual Presentation of Hospital Data by the Council on Medical Education and Hospitals of the American Medical Association. NA *Journal of the American Medical Association*. 1933;100(12):887.

Kanevsky J, Safran T, Zammit D, Lin SJ, Gilardino M. Making augmented and virtual reality work for the plastic surgeon. *Annals of Plastic Surgery*. 2019;82(4):363–368.

Li Y, Raison N, Ourselin S, Mahmoodi T, Dasgupta P, Granados A. AI solutions for overcoming delays in telesurgery and telementoring to enhance surgical practice and education. *Journal of Robotic Surgery*. 2024;18(1):403.

Lu MN, Zhang BL, Dai QH, Fu XH. Application of the plan-do-check-act cycle in shortening the decision to delivery interval time. *Risk Management and Healthcare Policy*. 2022;15:1315–1323.

Major RH. *A History of Medicine*. Springfield Ill: Charles C. Thomas, 1954.

Meskó B. *A Guide to Artificial Intelligence in Healthcare*. Leanpub. 2024. https://leanpub.com/ArtificialIntelligenceinHealthcare (accessed March 26, 2025).

Morsy H, Scott C, Jairath R, Ghetti C, Chu C, Sutcliffe S, Lowder JL. Telemedicine: The new "Virtual reality" of female pelvic medicine and reconstructive surgery? *Female Pelvic Medicine and Reconstructive Surgery*. 2022;28(3):e80–e87.

Morton TG, Woodbury F. *The History of the Pennsylvania Hospital 1751-1895*. Times Printing House, Philadelphia: 1895.

Moschovas MC, Palmer K, Barani G, Corder C, Carnegie P, Dhanji T, Patel E, Paxi A, Figueiredo JC, Luís B, Matos A, Dohler M, Patel V. Expanding access to high-quality surgical care in Africa: A landmark achievement in telesurgery. *Journal of Robotic Surgery*. 2025a;19(1):124.

Moschovas MC, Saikali S, Rogers T, Dohler M, Mcdonald M, Patel E, Marquinez J, Gamal A, Magnuson J, Patel V. Exploring the teleproctoring potential of telesurgery: The first remote procedures performed simultaneously between orlando and Shanghai. *International Brazilian Journal of Urology*. 2025b;51(4):e20250083.

NanoApps Medical, Inc. 2025. https://www.nanoappsmedical.com/about/ (accessed March 26, 2025).

Pawar P, D'Souza J. Wearable Devices Statistics by Companies, Products, Shipment, Users, Demographics and Region. Coolest Gadgets. February 05, 2025. https://www.coolest-gadgets.com/wearable-devices-statistics/ (accessed March 26, 2025).

Philips E, Allan T. *The Divine Campaigns AD 1100-1200.* Amsterdam: Time Life Books, 1993.

Riva G, Gamberini L. Virtual reality as telemedicine tool: Technology, ergonomics, and actual applications. *Technol Health Care.* 2000a;8(2):113–127.

Riva G, Gamberini L. Virtual reality in telemedicine. *Telemed J E Health.* 2000b;6(3):327–340.

Robinson JO. The royal and ancient Hospital of St Bartholomew (Founded 1123). *Journal of Medical Biography.* 1993;1(1):23–30.

Rouleau G, Richard L, Côté J, Ramirez-Garcia MP. The relational virtual nursing practice model: A web-based nursing intervention. *Nursing Science Quarterly.* 2023;36(2):164–173.

Schuelke S, Aurit S, Connot N, Denney S. Virtual nursing: The new reality in quality care. *Nursing Administration Quarterly.* 2019;43(4):322–328.

Singh RP, Javaid M, Kataria R, Tyagi M, Haleem A, Suman R. Significant applications of virtual reality for COVID-19 pandemic. *Diabetes & Metabolic Syndrome.* 2020;14(4):661–664.

Stevens R. *In sickness and in wealth: American hospitals in the twentieth century.* Baltimore: Johns Hopkins University Press; 1999. p. 286–287.

Thompson JD, Goldin G. *The hospital: a social and architectural history.* New Haven (CT): Yale University Press; 1975.

Truijen S, Abdullahi A, Bijsterbosch D, van Zoest E, Conijn M, Wang Y, Struyf N, Saeys W. Effect of home-based virtual reality training and telerehabilitation on balance in individuals with Parkinson disease, multiple sclerosis, and stroke: A systematic review and meta-analysis. *Neurological Science.* 2022;43(5):2995–3006.

Waddington K. *Medical Education at St. Bartholomew's Hospital, 1123-1995.* Boydell & Brewer, 2003.

Wall BM. History of Hospitals, University of Pennsylvania School of Nursing. 2022. https://www.nursing.upenn.edu/nhhc/nurses-institutions-caring/history-of-hospitals/ (accessed January 07, 2022).

World Health Organization. *Global expenditure on health: public spending on the rise?* Geneva: World Health Organization; 2021. Licence: CC BY-NC-SA 3.0 IGO.

World Health Organization. *Global spending on health: emerging from the pandemic.* Geneva: World Health Organization; 2024. Licence: CC BY-NC-SA 3.0 IGO.

Worlikar H, Coleman S, Kelly J, O'Connor S, Murray A, McVeigh T, Doran J, McCabe I, O'Keeffe D. Mixed reality platforms in telehealth delivery: Scoping review. *JMIR Biomed Engineering.* 2023;8:e42709.

Zychlinski N, Fluss R, Goldberg Y, Zubli D, Barkai G, Zimlichman E, Segal G. Tele-medicine controlled hospital at home is associated with better outcomes than hospital stay. *PLOS ONE.* 2024;19(8):e0309077.

7 Technophysics, Smart Health Networks, and the Biocryptoeconomy

Quantized Fungible Global Health Care Equivalency Units for Health and Well-Being

Melanie Swan

CONTENTS

DOI: 10.1201/9781003680093-7

7.1 INTRODUCTION

The core theme of this book is envisaged as Global Health Care Equivalency (GHCE). This is the idea that in the coming decades, synergies between nanotechnology, nanomedicine, and AI may make it possible for any individual worldwide to have equal access to advanced, cost-effective, nanomedical diagnostic and therapeutic technologies, no matter where they live, or under what conditions. Numerous technical, social, political, and economic steps will undoubtedly be required for the realization of this vision. From a technical perspective, a greater understanding of the specific causes of disease and other health challenges, and their resolution, will be a prerequisite for the development of sophisticated diagnostic and therapeutic technologies that meet public health demands.

The conceptual arc of this chapter is as follows. Section 7.1 defines the problem space as disease causality and proposes that instantiating health as a "math problem," with a data science and smart networks approach, may lead to improved probabilities for resolution. Section 7.2 discusses a novel vector in the data science approach, technophysics, and smart networks, particularly, in Section 7.3, how this applies to the health context. Section 7.4 proposes GHCE units as an econophysics concept that can be realized with smart health networks. Section 7.5 expands on this idea by presenting smart health network applications using blockchain technology, in a biocryptoeconomy, for the secure automation of multifleet items in the body. Section 7.6 connects the proposed smart health network solutions to an initial theoretical model for disease causality and the restoration of health. The broader implications of this work are considered along with risks and limitations, and the chapter concludes.

7.2 THE PROBLEM SPACE: DISEASE CAUSALITY AND PREVENTIVE MEDICINE

Starting from first principles, one reason that more disease conditions are not yet resolved is that the causality is complex, variable, and unknown (Clendenin, 1873; Muula, 2016). Even if doctors had exhaustive medical resources and complete patient information at their disposal, it would not necessarily guarantee cures. Diagnoses and treatments can only go as far as current medical advances allow. Pathology causality and the means for treating and preventing disease are complex, particularly for widespread common conditions such as cardiovascular disease and cancer.

Cardiovascular disease is the leading cause of death, both in the United States (30 percent) and globally (22 percent) (Pagidipati and Gaziano, 2013), even though there is some claim to 50 percent or more of these deaths being preventable (Patel et al., 2015). The second biggest killer, cancer, comprises ~18 percent of U.S. deaths (National Cancer Institute, 2025). Disease tracking organizations such as the Centers for Disease Control and Prevention and the American Heart Association excel at reporting annual statistics (Benjamin et al., 2017); however, there is scant evidence of a comprehensive plan for addressing and resolving these conditions. Since much of disease causality is unknown, it is also not clear how to apportion responsibility and blame in public health systems. This situation has culminated in high health care costs, poor outcomes (Papanicolas et al., 2018), and misaligned incentives (Loewenstein et al., 2012).

Consequently, the current work proposes a potential solution to the "causality problem." The *causality problem* stems from the issues that the exact causes of disease and the mechanisms for providing preventive medical treatment for improved health outcomes are unknown. There could be many objections to this formulation; however, this is the general problem articulation that frames this chapter. The broader research question is: how can public health challenges be defined as a problem that can be solved? *Health challenges* is a term that is understood to include an extensive breadth of health and wellness issues that do not have obvious solutions at present. These might encompass disease, preventive medicine, obesity, anti-aging, addiction, and cognitive enhancement (i.e., any affliction of the flesh-based body).

7.3 THEORETICAL MODEL DEVELOPMENT: DATA SCIENCE APPROACH

One potential strategy for making progress on the problem of solving public health challenges is through the use of a data science approach, as outlined in Figure 7.1. A *data science approach* is one that employs very large data sets (the average company currently manages 160 TB of data, and scientific data sets may be much larger (IDG. 2016)) and computational methods to extract patterns and insights from these data. To the extent that health challenges can be formulated as a "math problem" (i.e., a format that is quantifiable and formal), they might be resolved via a data science approach.

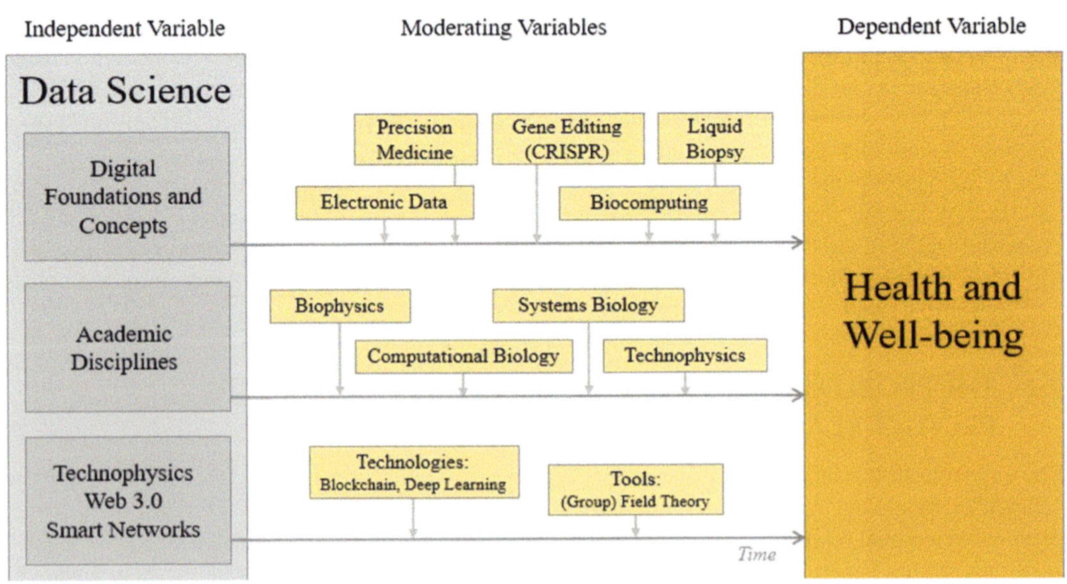

FIGURE 7.1 Theoretical model: instantiating health challenges as a data science problem. (Image created by the author.)

Clearly, many research problems are already framed mathematically and computationally. The distinction here involves the initial engagement of a systems-level approach, by casting a medical problem as not just a focused (reductionist) situation, but rather as a multifaceted health challenge. Second is the application of a wider range of tools from data science than may have been brought to bear on the problem previously, where methods are one dimension and data content is another.

A facet of the disease causality problem is that very large health data sets (scaling to millions of individuals) are not available but could serve as prerequisites for the recognition of disease patterns. By analogy to progress in other fields, the *big data argument* applies, whereby having large enough data sets is what makes the critical difference in pattern recognition and computational learning. With large enough data sets, straightforward algorithms can be run over them to obtain insights. This has proven true in many areas of machine learning and artificial intelligence (Halevy et al., 2009; Le et al., 2013) and might likewise be true for the health domain. A theoretical model of data science as a casual approach to health and well-being is outlined in Figure 7.1.

The premise is that causality is multivariable for health and well-being and that a data science approach can address this. The variable of concern, the target quantity to increase, is health and well-being (the dependent variable). The model proposes that a data science approach (the independent variable) can have an influential effect on the target, increasing health and well-being. Specifically, within the data science approach are three vectors: digital foundations and techniques, academic disciplines, and technophysics: web 3.0 smart networks. Within each of these vectors, there are detailed factors (moderating variables), which provide the next level of detail as to how precisely the target quantity (improved health and well-being) may be increased. The moderating variables are organized chronologically along a timeline as relates to their appearance and development. This model could be implemented as a logistic regression.

Some of the potential benefits of employing a causal model such as this one include using it as a roadmap for establishing an improved causal understanding of health challenges, outlining plans for their resolution, and organizing a public health system based on GHCE. A further benefit of the establishment of greater causality is then being able to instill levers of accountability in the medical system over which actors can actually have an effect.

7.4 ELEMENTS OF THE THEORETICAL MODEL

7.4.1 Digital Foundations and Techniques

Five digital foundations and techniques will now be discussed that might have significant roles to play in the advancement of cures for diseases and health maintenance. The first involves the conversion of data from manual to digital formats. In the short term, one vision is digitalized, interoperable, patient-controlled electronic medical records (EMRs). In the medium term, these data might be shared by users in privacy-protected computing models and the health data marketplace. In the longer term, there could be personal health data clouds with large data sets that include omics profiles (genomics, proteomics, RNA expression, methylation, metabolomic, and microbiomic data), daily health check data, Fitbit-type activity tracking data, sleep quality, environmental exposure, and other data streams.

Some of the potential benefits of having a personal health data cloud might involve the correlation of longitudinal data, through which it could be determined what is "normal" versus pathological for particular individuals. More relevant health focused groupings of individuals could be established for disease resolution using haplotype or underlying health conditions as opposed to demographics and other methods, which are too general and not reflective of health status.

A second and related foundation is a new paradigm for the practice of medicine that focuses on the individual (n = 1) using Quantified Self and precision medicine approaches. Systems biologist

Leroy Hood proposed the term "P4 medicine" in 2013, which refers to a future class of medicine that is predictive, preventive, personalized, and participatory (Hood, 2013). The "participatory" element is taken up in the Quantified Self movement, with individuals who track quantitative data about themselves, often for the purpose of solving health, athletic performance, or wellness issues (Swan, 2013).

The "personalized" element is taken up in the public health domain as precision medicine. *Precision medicine* is an individualized approach to disease based on the patient's unique combination of genes, environment, and lifestyle. There are several notable precision medicine initiatives underway. In the public health sector, the U.S. National Institutes of Health (NIH) One Million Volunteer Precision Medicine Initiative was announced in 2016 (Kaiser, 2016). In the private sector, there is Craig Venter's Health Nucleus program, which published study results in 2018 (n = 209) (Perkins et al., 2018).

The findings were that MRI and other screening detected early-stage cancer in 2 percent of participants, and age-related chronic diseases that required prompt medical attention in 8 percent (Cross, 2017). Leroy Hood's private venture, Arrivale, also published results in 2018 (n = 108) (Price et al., 2017). The findings claimed that, over time, measurement with personal data clouds can improve the understanding of health and disease, including early transitions to disease states. Both studies have drawn a variety of critiques, which include the validity and novelty of the results.

A third foundational advance that might have a significant role in the advancement of disease cures is *gene editing*, which is enabled through a technology called CRISPR/Cas9 (CRISPR: Clustered Regularly Interspaced Short Palindromic Repeats) (Doudna and Sternberg, 2017). CRISPR is an immune response that is found in the DNA of bacteria and archaea. The genome of a cell stores copies of DNA snippets that have previously attacked the cell, where after the cellular security system employs these exemplars to detect and delete any similar edits to its genome.

The implication for humans is that the CRISPR technique might be harnessed as a general-purpose precision technology for deleting unwanted genomic sequences related to disease. Research was published in August 2017, which reported the first successful CRISPR germline editing of human embryos (Ma et al., 2017). CRISPR/Cas9 technology was used to make a germline-specific DNA repair, correcting the MYBPC3 mutation (related to hypertrophic cardiomyopathy) in human embryos.

A fourth foundational advance is the *liquid biopsy*. Instead of the invasive extraction of a tissue sample as in a regular biopsy, a *liquid biopsy* measures the blood-based biomarkers of disease (Hauptman and Glavač, 2017). Although the idea is clear, it remains challenging, as the requisite quantities of blood-based biomarkers are difficult to detect. Irrespective, one firm, Silicon Valley-based Grail, says that it will have tests on the market in 2019 in Hong Kong, for a cancer commonly found in Southern China, nasopharyngeal carcinoma (Molteni, 2018).

The fifth foundational digital advance is *biocomputing*, which is not meant as the application of computer science to biological research, but in another sense, in terms of using biocomputing materials to perform calculations. In situ computational models might harness biology as their own computational substrate. Local computational methods within the body might be incorporated within biochips with microfluidics, nanocomputing, and biophysics methods for medical applications such as diagnosis and drug delivery.

7.4.2 New Academic Disciplines

In addition to digital foundations and techniques, there are several academic fields that are accelerating biological study and pathology resolution. The first discipline is **Biophysics**, an interdisciplinary science that applies physics approaches and methods to the study of biological phenomena. The second field is **Computational Biology**, a complement to many fields of biological investigation as the application of data analytics, mathematical modeling, and computational simulation. The third discipline is **Systems Biology**, which investigates health and biology as a complex set of networks that include gene regulatory networks, as well as cellular, chemical, and mechanical signaling networks (Davies, 2014). The fourth field is **Technophysics**, the application of physics to the study of technology (particularly networks), including biology.

7.4.3 Technophysics: Web 3.0 Smart Networks

The field of technophysics is emphasized as a novel development that could have a significant impact on cures for diseases and health maintenance. Technophysics is the study of technological systems with physics principles, particularly the smart network technologies that are emerging as part of the next phase of the Internet, Web 3.0. Concerning the eras of the Internet, Web 1.0 dealt with static information, and Web 2.0 configured social and shareable information, whereas Web 3.0 is emerging as intelligent self-operating networks, such as blockchain economic systems and deep learning neural networks, to facilitate the patterns of human interaction and control of the natural world. This chapter extends previous work in *technophysics* (the application of physics to the study of technology) and *smart networks* (intelligent self-operating global computational networks) (Swan and dos Santos, 2018) by applying these concepts to health and biology.

7.5 TECHNOPHYSICS AND SMART NETWORKS

7.5.1 Introduction to Technophysics and Smart Networks

Technophysics is the application of physics concepts and methods to technology domains, particularly for the study of smart network technologies, algorithms, and computation. The complex behaviors and fast-paced emergence of smart network technologies, such as blockchain economic networks and deep learning pattern recognition systems, suggest that rigorous methods for their study are required and that physics would be an appropriate foundation. *Smart networks* comprise autonomous intelligent networks that are a new form of global computational infrastructure in which intelligence is built directly into the software, such that an increasing degree of autonomous operation is facilitated. More formally, smart networks are state machines that make probabilistic guesses about the reality states of the world and act upon this basis.

Blockchain distributed ledgers and deep learning systems are some of the most prominent examples of emerging smart network technologies (Table 7.1). However, several other types of smart network technologies (autonomous software operations over networks) already exist and have been operating for some time. These include unmanned aerial vehicles (drones), particularly those that have autonomous strike capabilities (Singh et al., 2017), high-frequency trading (HFT) which now comprises 55 percent of U.S. equities trading volume (Miller and Shorter, 2016) the real-time bidding (RTB) market for advertising, and energy grids which engage in automated load-rebalancing. Other emerging and envisioned smart network technologies include smart city IoT (Internet of Things) sensor ecologies, automated supply chain logistics networks, cloudmind industrial robotics coordination networks (cloud-connected smart machines), personal robotic assistant networks, and space-based automation networks.

7.5.1.1 Deep Learning Chains

Certain species of smart network technologies, such as blockchain and deep learning, are in a special class in that they are both smart network technologies themselves and can serve as *control* technologies for other smart network technologies. Blockchains and deep learning nets may be employed to control other multifleet Internet-connected smart network technologies such as UAVs, autonomous driving fleets, medical nanorobots, and space-based asteroid mining machines.

TABLE 7.1
Smart Network Types and Operational Focus

	Smart Network Species	Smart Network Operations	Status
1	Unmanned aerial vehicles	UAV drones with autonomous strike capability	E
2	Programmatic trading (high-frequency trading (HFT))	Algorithmic trading (55% U.S. equities), auto-hedging	E
3	Real-time bidding (RTB) market for advertising	Automated digital advertising placement and inventory management	E
4	Energy smartgrids	Automatic electricity load-balancing & transfer on power grids	E
5	Blockchain economic networks	Transaction validation, confirmation, and recording	E/M
6	Deep learning networks	Object identification (Idtech [a]), pattern recognition, optimization	E/M
7	Smart City IoT sensor landscapes	Traffic navigation, data feeds, markets, and acquisition	M
8	Industrial robotics cloudminds	Industrial coordination (cloud-connected smart machines)	M
9	Supply chain logistics nets	Automated sourcing, ordering, shipping, receiving, payment	M
10	Personal Robotic Assistant Nets	Personalization, backup, software updates, fleet coordination	F
11	Space: Aerial Logistics Rings	In situ resource provisioning and asynchronous communication	F

Note: E: Existing, M: Emergent, F: Future Possibility.
[a] IDtech (identification technology): technological capability to recognize objects; similar to FinTech, RegTech, TradeTech, and HealthTech; technologies that digitize, standardize, and automate their respective domains.
Source: The table is created by the author.

The convergence of blockchain and deep learning is immersed in the idea of *deep learning chains*, which are a smart network control technology with properties of both underlying technologies: the secure automation, audit-log tracking, and validated transaction execution of blockchain, combined with the object and pattern recognition technology (IDtech) of deep learning.

For example, deep learning chains might be critical in autonomous driving fleets for tracking what the vehicle does (blockchain), while identifying objects in its driving field (deep learning). Deep learning chains could likewise be applied to the human body as a smart network control technology for medical nanorobots that facilitates the identification of pathogens (deep learning), while tracking and expunging them (blockchain).

7.5.2 Physics Is the Foundational Science

There is a history of different disciplines discovering that behavior arising within their domains could be explained with physics models. This is completely plausible, given that physics is perhaps the most foundational of the sciences, which attempts to describe the fundamental laws of reality and the functionality of matter, light, and energy. Wigner was one of the first to draw attention to the broad applicability of physics as an explanatory model in his 1960 paper, "The Unreasonable Effectiveness of Mathematics in the Natural Sciences" (Wigner, 1960).

Physics models have now been demonstrated in many domains, including natural systems (biology, ecology, and environmental science) and social systems (economics and political science). From these foundations, technophysics is defined by analogy to other fields such as biophysics (the application of physics to the study of biology) and econophysics (the application of physics to the study of economics). In myriad ways, technology is already "applied physics," since it is based on the principles of physics (e.g., the conductivity of semiconductors and the capacity of optical waveguides to transmit data globally); thus, the study of any of its facets is implied by physics methods.

7.5.2.1 Technophysics Examples

Technophysics is the study of technology using physics-based models, concepts, and methods, specifically, the explanation of behavior observed in smart networks with physics models. To observe technophysics in practice, it helps to have some specific examples, which are summarized in Table 7.2 and discussed below. One technophysics example is the application of complexity methods, in the form of Shannon entropy and information theory, to evaluate blockchain consensus algorithms (the automated transaction clearing mechanism).

Research has found that proof-of-stake consensus algorithms are poorly formed because they exhibit high complexity (dos Santos, 2017, 2019). High complexity in a financial market system can be destabilizing as the system may tend toward chaoticity, perhaps making flash crashes and other aberrant behavior more likely. The other main kinds of blockchain consensus algorithms are based on proof-of-work and are better formed in this context because they exhibit very low complexity. This is a technophysics example of using physics formalisms to evaluate an emerging smart technology network, and to quantitatively evaluate which kinds of design parameters may be more structurally sound.

A second technophysics application involves blockchains and quantum physics. One anticipated idea is that blockchain systems could be secured by quantum cryptography, if and when such technology becomes commercially feasible (Kiktenko et al., 2018). A related notion is that since quantum cryptography would only be adding a quantum layer to standard blockchain protocols, next-generation projects might instantiate the blockchain itself with quantum computation. One proposal is to create a blockchain using quantum particles that are entangled in time (encoding the blockchain into a temporal GHZ (Greenberger-Horne-Zeilinger) state of photons) (Rajan and Visser, 2019). Data would be encoded on a quantum particle, which would store the history of all its predecessors in a way that could not be hacked without destroying it. Similar to the first example, this case deploys physics principles in the design of smart network technology.

A third application area of technophysics involves the use of energy-entropy trade-off models. In statistical physics, systems are often evaluated such that free energy is minimized (Merhav, 2010). A deep learning network can be structured this way, as an energy function (Hamiltonian), in which the goal is to minimize the cross entropy (inefficient coding of information) (Zhang and Sabuncu, 2018). The energy function is employed in the deep learning network as an overall organizing

TABLE 7.2
Technophysics Examples: Physics Models that Describe Smart Network Behavior

1. Complexity methods using Shannon entropy find greater complexity (and possible chaoticity) in blockchain systems using proof-of-stake algorithms rather than proof-of-work algorithms (dos Santos, 2017, 2019)
2. Energy landscape minimization (via Hamiltonian) obtains deep learning system convergence on an answer (Choromanska et al., 2015)
3. UAV systems display undesirable emergent behavior: thrashing, deadlock, resource-starving, phase change (Singh et al., 2017)

Source: The table created by the author.

principle to describe the activity of the system's processing units (perceptrons). Specifically, the energy function (Hamiltonian) is defined as a combination of the parameterized summing of Gaussian random variables (processing nodes) and their weights (Choromanska et al., 2015).

7.5.2.2 Background and Genesis of Technophysics

One advance among many in physics since the 1970s has been the expansion of research into the study of disordered systems, which are those marked by the presence of asymmetry in a many-particle system (e.g., spin glasses, structural glasses, and polymer networks). The point being that methods have been generated to evaluate the collective behavior of systems that have heterogeneous components (in contrast to previous models, which focused on systems with homogeneous components). Subsequently, this has allowed physics methods to be applied to a much wider range of problems outside of its traditional domain and opened up new fields of inquiry such as biophysics, econophysics, and technophysics.

Even without the formulation of technophysics (physics methods applied to the study of technology), scientific disciplines such as statistical physics, information theory, and computer science are converging on some of the same types of problems, including those that arise in disordered systems. These problem domains include system-critical behavior and phase transitions. Statistical physics suggests that the reason for phase transitions in all of these domains could be a structural one that hinges on the existence of a glass transition (a structural change in the geometry of the set of solutions) near a constraint.

All three fields employ probability theory as one strategy for the analysis of phase transitions: statistical physics with spin glass transition formulations, information theory with error-correcting (LDPC) codes, and computer science with randomized algorithms and probabilistic combinatorics (Mézard and Montanari, 2009). Therefore, uniting these problem definitions and solutions into a global formulation, such as technophysics, might have broad applicability both within these fields and far beyond. The idea is to consolidate the problem detection, concepts, methods, approaches, and analysis from fields such as statistical physics, information theory, and computer science into one area, technophysics, which might also include economic principles and game-theoretic elements as design principles.

7.5.2.3 Theoretical Formulation: Smart Network Field Theory

The technophysics discussion and examples suggest the types of formalizations that might be required. One practical tool might be a *smart network field theory*, which is an *effective field theory* (a means of approximating) a smart network system for the purpose of characterizing, monitoring, and controlling it, particularly for criticality detection and multifleet item orchestration.

Developing effective field theories is a known technique in physics, particularly in quantum mechanics and materials science (Giorgi et al., 2004). An effective field theory is a type of approximation, or effective theory, which explains an underlying domain in simpler terms that make the system easier to investigate and operate. Specifically, effective field theories are useful because they span classical and quantum domains, and more generally, different levels in complex systems with phase transitions. The known technique of defining effective field theories is a good starting point for defining a smart network field theory.

One of the most prominent examples of an effective field theory is Quantum Field Theory (QFT). QFT extends Quantum Mechanics to descend down a system level in order to incorporate the energetic fields within which particles operate and are reinterpreted as quantized field states (Jaffe and Witten, 2000). QFT is the theoretical framework that facilitates the construction of quantum mechanical models for subatomic particles in particle physics and quasiparticles in condensed matter physics. Particles are treated as excited states or quanta of an underlying field. The implication is that these fields and units of field quanta (similar to the GHCE units that are defined later in this chapter) become the relevant locus of system investigation and manipulation through the effective field theory. Effective field theories are also employed in other areas of physics, including particle physics, statistical mechanics, condensed matter physics, general relativity, and hydrodynamics.

Effective field theories provide a means to engage a system at different scales. It is easier to measure different aspects of a given system at different levels. For example, the elliptical orbits of the planets are easier to calculate with Newtonian gravity than with general relativity. Likewise, computing the energy spectrum of a Hamiltonian in Quantum Mechanics is straightforward, but much more difficult at the QFT level, as there is too much data as pertains to all of the specific particle movements to easily calculate an energy function. Effective field theories are a means of understanding and manipulating systems at different levels.

7.5.2.3.1 Purpose and Use of Smart Network Field Theories

The purpose of having one or more smart network field theories, and related optimal control theories, is the characterization, monitoring, and control of smart network systems. One of the most practical applications is the automated coordination of multifleet items. Another is detecting and managing system criticality, such as potential phase transitions. The next section explores the concepts of technophysics, smart networks, and smart network field theories in the context of health and biology.

7.6 TECHNOPHYSICS AND SMART NETWORKS IN THE CONTEXT OF HEALTH

7.6.1 Internet-Connected Technologies with Multifleet Items

Large-scale networks are a feature of contemporary reality. They are comprised of thousands, millions, and billions of elements and thus require a smart network field theory or other similar mechanism for the automated characterization, monitoring, and control of their activity. There are different "species" of modern technology that have immense fleets of Internet-connected items. The Internet-connected aspect of technologies, their tentacular nature (Cassou-Noguès, 2016), typically allows for a variety of automated software updating and remote monitoring, while providing manufacturers with rich troves of user data. Thus, personal data privacy is an ongoing negotiation and concern.

In the consumer domain, some of the most familiar examples of multifleet items connected to the Internet across a large-scale network include tablets, smart phones, and apps. Other examples are smart TVs, Nest thermostats, home security systems, Roomba-type vacuum cleaners, Wi-Fi scales, and robotic personal assistants (such as Jibo, Lynx, Einstein, and Beam). Extending from the home, the connected car is another domain of Internet-connected technology. Modern automobiles are linked to the cloud through a vehicle gateway that provides a variety of safety, navigation, information, entertainment, electronic tolling, insurance, and maintenance applications. It is estimated that 400 million vehicles will be Internet-connected by 2025, in contrast to 237 million in 2021 (Placek, 2023). Although some countries, such as Germany, France, and the United Kingdom, were expected to reach nearly 100 percent connected car penetration by 2020 (propelled by the eCall mandate, the EU's automatic emergency call system) (Millman, 2018), this is not yet the case to date. Placek (2022) estimated that 44 percent of all operating vehicles in Europe could have basic or advanced connectivity by 2025, increasing to 50 percent by 2030. Beyond the home and car, other sophisticated smart network technologies with multifleet items include drones, supply chain logistics nets (such as automated shipment-receiving systems), blockchain peer-to-peer hosts, deep learning processing nodes, smart city IoT sensors, and autonomous driving vehicles. All of these smart network technologies require secure, automated, multifleet item coordination, monitoring, control, and data analytics.

7.6.2 Very-Large and Very-Small-Scale Smart Networks in Space and Biology

Smart networks are perhaps the most visible at the human-scale of daily operations; however, they might also be extremely important for tracking and automating operations at very large and small scales. In the very large space domain, smart network technologies might aid in many applications.

TABLE 7.3
The Human Body in Numbers (Sample Statistics)

	Biological Element	Number	Reference
1	Cells	37.2 trillion	Eveleth, 2013
2	Neurons (glial support cells)	100 billion (1 trillion)	Herculano-Houzel, 2009
3	Genes	20,000	Ponomarenko et al., 2016
4	Proteins	100,000	Ponomarenko et al., 2016
5	mRNA: different transcripts	12,000	Qiagen, 2018
6	mRNA: molecules in a single cell	360,000	Qiagen, 2018

Source: The table is created by the author.

These could include securely automating the development of asteroid mining, space civilizations, planet terraforming, and asynchronous space-based communications. Likewise, in the very small domains of biology and health, there are numerous potential smart network applications.

The most notable example of an existing smart network operation in biology is *optogenetics*, which is essentially a "a light switch for a field of neurons"; the technique of introducing a microbial opsin (a light-sensitive protein) that can be expressed in neurons to make their electrical activity controllable by light (Boyden, 2015). A field of neurons expressing the protein can thus be switched on and off. Disorders such as epilepsy and retinitis pigmentosa might be targeted with optogenetics, the latter of which is currently in human clinical trials (Reardon, 2016). Optogenetics is a *biological smart network field theory*, in that cells are controlled as a field as opposed to individually. This allows a larger class of biological phenomena to be addressed.

The body is replete with large-scale networks with multifleet items, which are coordinated by biological mechanisms (Table 7.3). For example, the human body is comprised of 37.2 trillion cells (Eveleth, 2013), whereas the brain has 100 billion neurons and 10 times as many glial cells (which surround and support neurons) (Herculano-Houzel, 2009). The sheer scale of human biology suggests that certain types of automated data collection, monitoring, and intervention with smart network technologies will be required to even begin to comprehend the underlying biological terrain, prior to progressing to more rigorous phases of pathology resolution and preventive medicine.

One useful source of biophysics statistics is Harvard Medical School's BioNumbers (bioNumbers.org). This is a database of biological numbers such as size, rates of activity, concentration, energetics, and error rates. Some example data points include cell membrane thickness (4–6 nm), the characteristic concentration of a signaling protein (10 nM), and the free energy of ATP hydrolysis (40–60 kJ/mol) (Harvard Medical School, 2025).

7.6.3 Conceptual Paradigms for Integrating the Body with Smart Networks

The body is a complex organism with myriad levels of organization. Therefore, it makes sense to have a theoretical framework for addressing how the intricacies of medical technologies are going to correspond to the complexities of the body. The question is to identify the appropriate conceptual paradigms to define the biology and the smart network technologies together for their optimal functionality.

To successfully deploy smart network technologies within the human body, a map of the terrain is required, as well as a link between the terrain and the interventional technology. Figure 7.2 outlines such a theoretical framework for connecting the body to advanced medical therapies using smart network theory. Three aspects of smart network theory as a conceptual model are proposed here, and others may develop over time. A conceptual framework is important for designing, structuring, and organizing the development of smart network technologies for health.

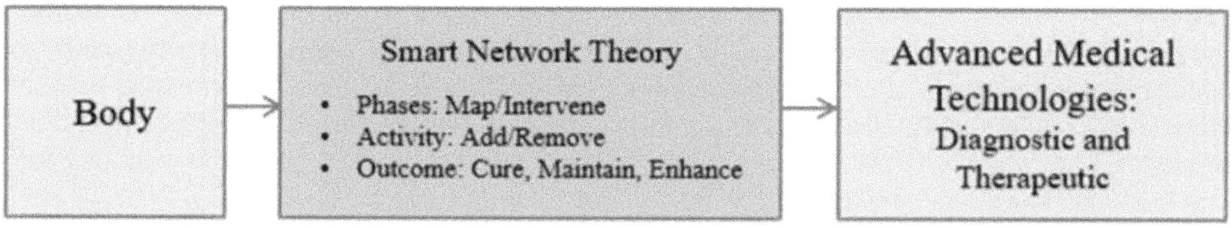

FIGURE 7.2 Theoretical framework for connecting the body to therapies. (Image created by the author.)

A smart network theory is one strategy for mapping and describing the conceptual model for how large-scale diagnostic and interventional technologies might correspond to the complex hierarchical biological operations they are attempting to survey and repair. Three elements that a conceptual model might contain are the phases, activities, and outcomes of the intended therapies as outlined below.

- **Phase: Mapping and Intervention**. An initial element in a conceptual framework for integrating human biology with smart network technologies could be the phase of activity. The two broadest categories of activity are mapping and intervention. The steps are to first map and inventory what is there (including what is "normal" for each individual) and, second, to survey and maintain the territory on an ongoing basis. This process would require that the biological mapping already completed by medical and other efforts be extended to the appropriate sequential levels of detail for specific individuals. Subsequently, every individual would possess a "body file."
- **Activity: Addition and Removal**. A second element in a conceptual framework for integrating human biology with smart network technologies might be structured by the functional operation that a smart network system, or specific nanomedical therapy, will be conducting in vivo. This would involve the simple paradigm of removing unwanted elements and adding necessary or desirable elements to the body. There could be different design standards for smart health applications that focus on exactly what needs to be removed and added. Some applications could focus on purging the body of pathogens, irreducible waste, and other unwanted elements, while others could target the addition and chaperoned uptake of nutrients, blood oxygenation elements or antioxidants. neural transmitter stimulants, and other wanted elements.
- **Outcome: Cure, Maintenance, Enhancement.** A third element in a conceptual framework for integrating human biology with smart network technologies will involve targeted outcomes. The three main categories of health outcomes are pathology resolution, health maintenance (preventive medicine), and enhancement, each of which has different application design requirements. Temporality is a dimension of outcome, in that the optimal timing for the introduction of certain therapies (particularly as relates to health maintenance and enhancement) is not yet clear. The biological issue is that most health systems function without any problem in younger organisms but degrade over time. However, when and how degradation ensues, which might guide precisely when to introduce a given intervention, is not yet known. Thus, the temporal dynamics of integrating human biology with smart network technologies will require clarification and quantification, and conceptual models will assist with outlining these processes.

7.6.3.1 Internet Connectivity: Wireless Body Area Networks

One of the initial steps that is already underway to prepare for the delivery of advanced nanomedical therapies and smart network health technologies is Wireless Body Area Networks (WBANs). Internet connectivity is a prerequisite for many kinds of advanced health devices and applications, and Internet protocol standards need to be established for how data is to be transmitted to and from the body. There is ongoing progress in developing standards for WBANs to support health care applications (Filipe et al., 2015; Sethi and Sarangi, 2017). The idea is to have an unobtrusive, secure (Kumar and Mukesh, 2013) mechanism for transmitting/receiving data to personal health data clouds through a *body gateway* (analogous to a connected car's vehicle gateway). Smart network technologies could facilitate data transfer using WBANs as the transport layer. Subsequently, WBANs could serve as the communications portal required for disease intervention, health maintenance, and enhancement, whether externally or internally onboard the patient for brief or extended durations.

7.6.4 Advanced Nanomedical Technology Applications

This section describes a slate of advanced emerging and envisioned nanomedical technology applications, many of which will require smart network technologies and a smart network field theory (or related mechanism) for their efficient operation. Human biology requires that the relevant scale for targeting cellular operations is on the order of microns, micrometers (µm or 10^{-6} m), and nanometers (10^{-9} m). For reference, a typical cell is relatively large (Ø25 µm = 25,000 nm), cellular membranes are 4–6 nm thick, and a typical DNA duplex is 2 nm wide (Harvard Medical School, 2025). Table 7.4 provides a list of advanced nanomedical technologies in various stages of development.

To date, the first four advanced nanomedical technology applications listed in Table 7.4 have already been clinically implemented in humans. These include nanoparticles and quantum dots, for dynamically controlled drug delivery, and the imaging of specific cells and tissues, respectively. Basic tissue engineering applications (supplemental stents, conduits, and bladder transplants)

TABLE 7.4
Advanced Nanomedical Technology Applications

	Medical Technology	Application	Status	References
1	Nanoparticles	Targeted drug delivery	E/C	Mirza and Siddiqui, 2014
2	Quantum Dots	Biomolecular imaging	E/C	Petryayeva et al., 2013
3	Tissue Engineering	Conduit and hollow organ replacement	E/C	Atala, 2011; Adamowicz et al., 2017
4	Brain-computer Interfaces (BCIs)	Brainwave control of a computer (example: cochlear implant)	E/C/M	Nicolas-Alonso and Gomez-Gil, 2012
5	DNA Origami (2D and 3D nanostructures)	Lock-box drug delivery system	E	Kumar et al., 2016
6	Biomolecular interface (organic-inorganic)	Adhesion, wound healing, surgery	E/M	Fattahi et al., 2014
7	3D Bioprinting	Replacement cells, personalized therapeutics	E/M	Ventola, 2014
8	DNA Nanostructures	Organic in situ building structure	M	Bujold et al., 2018
9	Synthetic Biology (Bioengineering)	Disease diagnosis, devices, parts	M/F	Brown et al., 2012; Ibrahim et al., 2015
10	Medical Nanorobots	Pathogen removal, health maintenance	F	Fahy and Wowk, 2015
11	Cryobiology, Cryonics, Cryogenomics	Low-temperature preservation of function	E/F/F	Fahy and Wowk, 2015; Wowk et al., 2018
12	Brain/Cloud Interface (B/CI)	Advanced brain diagnostics/therapeutics and cognitive enhancement	F	Martins et al., 2019

Note: E: Existing, C: Clinical (in Clinical Use), M: Emergent, F: Future Possibility.

Source: The table is created by the author.

and brain-computer interfaces (324,000 worldwide cochlear implants (ASHA, 2018)) have likewise been employed for humans.

More speculatively, considering the emerging and envisioned technologies listed in Table 7.4, a particularly promising area of ongoing development is *DNA nanotechnology*, which involves the use of nucleic acids and structures as building materials (Seeman, 1998). The significant advantage of DNA is that it is a molecular-scale building material that is already in situ, organic, and rejection-resistant, which might be exploited to create micro- and nanostructures with great precision (Boehm, 2004). It may be employed to create nanoscaffolding harnesses to support cells and provide extracellular matrix (ECM) structures, promote certain types of protein expression (including through optogenetics and synthetic biology-engineered booster cells), and generate protective polymeric nanocoatings for cellular components. Beyond these concepts, the notion of nanomedical robotics is heralded as one of the most potentially transformative technologies for the longer-term future of medicine.

7.6.4.1 Nanomedical Robots

Robert Freitas Jr. envisaged the concept of autonomous *nanomedical robots* in 1999 (Freitas, 1999; Fahy and Wowk, 2015), which are nanorobots that might be fabricated through Molecular Manufacturing (MM) (the controlled assembly of individual atoms and/or molecules) and designed to perform specific medically related tasks in the body. Similar to macroscale robots, nanomedical robots may be comprised of many thousands of nanomechanical components such as bearings and gears. Nanomedical robots could have a power supply for energy and propulsion, sensors to guide their medical procedures, and an onboard quantum computer to control their navigation, communications, and behavior.

Nanorobots would need to be very miniaturized (e.g., a bloodstream traversing nanorobot would need to pass through the smallest capillaries (~Ø3 μm) and therefore be smaller than red blood cells (Ø7 μm = 7,000 nm), perhaps on the order of ~Ø1 μm). The components that make up a nanorobot would need to be smaller still, for example, 1–10 nm in size (Freitas, 2012). Some classes of medical nanorobots that have been articulated by Freitas include respirocytes, clottocytes, vasculoids, and microbivores (Fahy and Wowk, 2015). Nanofactories that will enable MM have been outlined as a manufacturing platform to produce molecular machines such as medical nanorobots (Drexler, 1992).

Medical nanorobots have been contemplated for the treatment of any type of pathology, including cancer, aneurysms, kidney stone removal, and the elimination of DNA mutations (Saadeh and Vyas, 2014; Agrawal et al., 2024; Sivasankar and Durairaj, 2012). Medical nanorobots could be employed to maintain and protect the human body from pathogens (Domschke and Boehm, 2018) and also to aid in surgery, diagnosis and testing, gene therapy, cancer detection, and nanodentistry (Freitas, 2000). One proposal is that medical nanorobots might be continually circulating through the human body as a patrol and security system to perform a variety of beneficial health maintenance and enhancement operations.

The deployment of medical nanorobots to combat the most significant life-threatening pathologies (e.g., heart disease and cancer) (as mentioned, comprising 30 percent (Pagidipati and Gaziano, 2013) and 24 percent (National Cancer Institute, 2018) of U.S. deaths) would be a clear objective. Recyclable medical nanorobots might be used to deliver vascular support and activate drugs at specific sites in the body as nanoparticles do now. Another related application could be using medical nanorobots to provide directed electrical stimulus to the heart and brain, for example, using ultrasound to dissolve blood clots (Marosfoi et al., 2015). They could reside more permanently or for fixed durations within patients for preventive medicine and general maintenance including cellular repair and rejuvenation.

Ultimately, medical nanorobots might perform myriad biophysical maintenance and augmentation tasks in the body. An important activity for medical nanorobots could be metabolic waste removal since the body is unable to dispose of different classes of waste on its own. Inside cells, one therapy would involve the removal, or in situ degradation, of lipofuscin (a recalcitrant metabolic waste material that accumulates within cell lysosomes despite normal breakdown processes) (Nociari et al., 2017).

In the extracellular environment, the removal of long-lived proteins (advanced glycation end products (AGEs)), which accumulate from the glycation, oxidation, and crosslinking of proteins that damage collagen and elastin, could also be of significant benefit for human health (Furber, 2006; Sharma et al., 2019). A third waste removal application could facilitate the body's ability to regulate and rid misfolded proteins that accumulate into plaques and are implicated in neurodegenerative pathologies such as Alzheimer's disease (Takahashi et al., 2017). Another proposal would involve "intestinal crawlers" that perform colon maintenance by removing embedded particles and early indications of polyps, extending the trend to minimize the invasiveness of colonoscopies (Gluck et al. 2016).

7.6.4.2 Smart Network Implications of Advanced Nanomedical Technologies

The wide range of potential advanced nanomedical technologies in use and development suggests that smart network concepts (technologies and tools such as field theories) will likely be essential in the conception, design, and deployment of such future-class therapies.

7.6.5 Smart Health Networks

Health futures that feature advanced nanomedical technologies and a greater intensity of human-technology integration suggest the inception of smart health networks. As described earlier, *smart networks* are intelligent self-operating networks. By extension, *smart health networks* would comprise a class of smart networks that autonomously characterize, monitor, and intervene to facilitate optimal health, particularly in the secure automated control of multifleet therapeutic nanoscale entities that reside in the human body for brief or extended durations. A *biological smart network field theory* (bSNFT) is one tool that might be employed to coordinate the control of multifleet entities in the body. Conceptually, this is similar to having an "optogenetics for medical nanorobots," in the sense that multifleet items can be controlled as a field (and "field" might be based on energy, gradients, or other parameters).

Smart health networks might provide for the automated monitoring and control of health with smart network tools (e.g., field theories) and technologies (e.g., blockchain systems for tracking, and deep learning networks for pathology detection). Such smart health networks or *smart bionetworks* could assist in stewarding the deployment of advanced nanomedical technologies.

7.7 Smart Health Networks and Global Health Care Equivalency Units

The overall aim of this chapter is to develop a vision of health futures based on advanced nanomedical therapies and a greater intensity of human-technology integration. This section discusses how a concept such as GHCE units is the next logical step that unfolds from this vision, as a codified, standardized tracking unit for the marshalling and deployment of worldwide health services.

7.7.1 Standardization of Global Health Care Equivalency Units

The standardization of measures and service units is a critical step in the development, maturity, and professionalization of any field. Naming the concept GHCE units implies that they are standardized units of measurement that might be transferred and exchanged as any other standardized units. A definition can be stated that *GHCE units* comprises a future concept of standardized, fungible, quantized units of global health care services, particularly nanomedical diagnostic and therapeutic technologies that are extensively available worldwide. In the future, GHCE units might be traded, monitored, and aggregated as a new economic statistic.

7.7.2 GHCE Units: Econophysics Quantization and Fungibility

A good model for GHCE units comes from physics, economics, and their portmanteau, econophysics (the study of economics using physics principles). This is because these areas are based on both standardized units of measure and also units that are packaged, quantized, and fungible. These features correspond well with the notion of GHCE units. In economics, money or other forms of currency are the standardized and transferable units of measure. Likewise, in physics, quantized packets of light, energy, particle movement (spins), and electron shell orbits are standardized units.

In physics, an important form of standardization revolves around quantization, essentially the packetization or discretization of units into discrete allotments or quanta. In many physical systems, these units cannot be arbitrary, but only organized into certain quantified denominations. For example, it is not possible for wavelengths to transfer random amounts of light, but only discrete quantities of this electromagnetic energy. Economics is similarly structurally organized, wherein money has three functions: serving as a medium of exchange, a store of value, and a unit of account. The unit of account functionality indicates the fungibility and transfer of standard units (e.g., dollars, euros, or yuan) from the accounting books of one company to another. Likewise, the same GHCE units could be delivered to patients by a health service provider at the Memorial Sloan Kettering Cancer Center in New York City or at a camel mobile clinic in Samburu, Kenya (Oyaro, 2017). The assumption underlying the definition of GHCE units as a standardized unit is that they can be packaged, measured, and applied on a global basis in any location for delivery to patients, akin to light quanta or any monetary currency.

Since the future health context may be one that is characterized by the physics and economics-inspired properties of smart networks, the standardized unit's aspect of GHCE units may likewise be constituted by the fungibility of money and the quantization of statistical physics. Other associated attributes of smart network systems and smart network field theories might be employed to further manage GHCE units, with principles related to energy, entropy, and algorithmic efficiency. Theoretical models from technophysics could inform how GHCE units are packetized, quantized, instantiated, and transferred as information in a global-scale health system. One or more GHCE smart network field theories could be defined as a computational, statistical, mechanical, and information-theoretic approach. For example, using energy-entropy trade-offs to define system criticality and phase transitions from health to pathology (and vice versa) and to reduce computational complexity in the health environment. The new economic statistic measured by GHCE units could be the number of health quanta that are produced and transferred in public health and wellness systems.

7.8 APPLICATIONS: SMART HEALTH NETWORKS AND THE BIOCRYPTOECONOMY

7.8.1 Introduction to the Biocryptoeconomy

This section discusses more specifically how an exemplar smart network technology and blockchain distributed ledgers might be used to enact future-class health applications and GHCE units. Blockchain distributed ledger technology is selected as it comprises one of the most relevant, applicable, and already available smart network technologies to develop into the idea of a biocryptoeconomy and related applications. The current research extends previous work in *biocryptoeconomics* (cryptoeconomic principles applied in the biological domain) and the concept of *personal bioDACs* (packages of programs for purpose-directed activity in the biological domain) (Swan, 2016a).

7.8.2 Key Terms: Cryptoeconomics and Blockchain Technology

Cryptoeconomics is an economic paradigm based on cryptography (Swan, 2019a). More specifically, cryptoeconomics is an economic transaction system implemented within a cryptography-based software network, using cryptographic hashes (computational proof mechanisms) as a means of confirming and transferring monetary balances, assets, smart contracts, and system states. Blockchains are an institutional technology in that they displace trust from human-based institutions to software algorithms. This concept is called *trustless* trust, meaning trusting the software system, instead of human intermediaries such as a bank or a government. Somewhat counterintuitively, trustless trust removes as much human-based trust as possible, in order to make the economic system more trustworthy (Swan and de Filippi, 2017).

A *blockchain* is an immutable, cryptographic (cryptography-based), distributed (decentralized, peer-based), consensus-driven ledger (Swan, 2015a). Blockchain (distributed ledger) technology is an Internet software protocol for the instantaneous transfer of money and other forms of value (assets, contracts, public records, program states) on a global scale. Transactions are stored in an immutable cryptographically secured ledger (like a giant online spreadsheet). The transaction blocks that comprise the ledger are linked together with cryptographic hashes (short codes corresponding to large chunks of data) so that it can be constantly verified that the transaction ledger has not been modified.

Since each new transaction block calls a hash of the last block, the technology is called *blockchain*. Blockchain distributed ledger systems can be used to register and maintain digital asset inventories (such as controlled pharmaceutical substances in hospitals), as well as transfer money, assets, contracts, patient test results, or other items of value instantaneously and globally via the Internet. Blockchains (transaction blocks cryptographically linked together) are one software architecture among others in the more general class of *distributed ledger technologies*. Other topologies may include blockless blocks in the form of directed acyclic graphs, for example, IoT "tangles" as the secure transaction execution mechanism.

7.8.3 Biocryptoeconomics and Bioblockchain Health

Biocryptoeconomics is the application of cryptoeconomic principles and blockchain smart network technologies to the biological domain toward the achievement of beneficial health outcomes. Specifically, in the context of smart health networks and GHCE, biocryptoeconomics involves the notion of employing blockchains as smart network technologies for the secure automated delivery of health services, such as those provided by nanomedical multifleet robots.

The broader concept of the biocryptoeconomy is harnessing cryptographic principles and economic organizational models to automate large classes of ameliorative processes in the human body. It has been envisioned that in the farther future, fleets of medical nanorobots might be brought on board the human body for a variety of pathology resolution and enhancement activities (Fahy and Wowk, 2015) The new wrinkle is that blockchains provide a technology platform for the actual implementation of medical nanorobotic multifleets within the body, particularly inclusive of the required security and automated control features.

Just as very large scale IoT sensor networks are envisioned to be coordinated, likewise the automation of massive fleets of medical nanorobots and other nanomedical therapeutic technologies in the human body might be similarly orchestrated. Medical nanorobots could be analogous to having a fleet of IoT sensors onboard the body, coordinated by mass automation, which could be increasingly feasible and secure with blockchain smart network technology. It is not unthinkable that in several decades' time there could be one or more medical nanorobots residing within many or most types of cells throughout the human body, which are safely and securely coordinated by biocryptographic technologies. This future moonshot vision would have been incomprehensible to those who were promised only ~90 years ago, a "chicken in every pot and a car in every garage," if assured that there would be a "medical nanorobot in every cell."

7.8.4 Advanced Concept: Personal bioDACs

Different classes of medical nanorobots have been outlined, and each may require a lightweight automation and control mechanism (an "app") specific to its particular operational domain.

TABLE 7.5
Bioblockchains for Health: Personal bioDAC Apps

Health Outcome	bioDAC Applications	Biocurrency Denominations	Application Names
Pathology Resolution	Cell Monitor DAC	Cancer cells, DNA methylations, red blood cells, oxygen, amyloid beta (Aβ) plaques	PlaqueRID, MutagenBLAST
Health Maintenance	ImmuneDAC, RNAiDAC	Lymphocytes (T-cells, B-cells, NK cells), neutrophils, macrophages, cytokines, antibodies	PathogenDestroyer, ImmuneBOOST
Cognitive Enhancement	NeuroDAC, cognitive nanorobots	Ideas, potentiation, neurotransmitters, glucose	EmbeddedBCI, Cloudmind

Source: The table is created by the author.

Consequently, a biocryptoeconomic smart network technology for this might take the form of a *personal bioDAC*, which is a biological application-specific set of blockchain-based programs for automated operation that is personalized (i.e., targeted to individual-specific health parameters).

More generally, a *DAC* is a distributed autonomous corporation (Buterin, 2014), meaning a package of blockchain-based smart contracts. (Smart contracts are programs registered to a blockchain for automated execution, possibly having the four elements of legally binding contracts: two parties, consideration, and terms.) The controlled implementation of personalized medicine with medical nanorobots could be achieved by using personal bioDACs. Table 7.5 outlines some illustrative future-class personal bioDAC applications in the three main health outcome areas of pathology resolution, health maintenance, and cognitive enhancement.

7.8.4.1 Biocurrency Denominations of Personal bioDAC Apps

The various personal bioDAC applications could have different embedded biocurrencies (analogous to distinct cryptotoken money supplies in blockchain economies (Swan, 2019c)) as the denomination, measurement, action-taking mechanism (such as voting, thresholding, and quorum sensing), and resource distribution system of the app. Biocurrencies could include micronutrients, small molecules, energy (ATP), electrical charge, and ideas. The obvious biocurrencies would be those already denominated by the body and used in the applications that the bioDACs would be facilitating. Clearly nanorobotic cellular transactions would need to be tracked and monitored, including for safety, liability, remuneration, and unintended consequences (externalities) purposes.

Another example of a bioDAC is the RNAi bioDAC. This application could work with RNA transcripts: producing, blocking, and tracking siRNA and RNAi for gene silencing and translation interference in an extended application of current pharmaceutical efforts. Other bioDACs could extend the functionality of retroviral vectors and use the microbiome as a less-invasive staging site from which to source micronutrients, and provide services such as Internet connectivity, secure automated backup, energy replenishment, and drug delivery.

In the smart contract programming of bioDACs, cryptocurrency principles such as *blocktime temporality* (blockchain-based timing specifications) and *demurrage* (incentives for certain behaviors such as full resource consumption) could be specified to optimize the management and operation of the apps. For example, demurrage principles might be used for the periodic redistribution of neural biocurrencies such as ideas, with precursor neurotransmitters serotonin and dopamine (in the cognitive enhancement case), and memories with precursor neurotransmitter acetylcholine (in the dementia repair case).

This proposal has been discussed in more detail in the concept of the "Brain as a DAC" (Swan, 2015b). Principles from other blockchain-based smart network technologies, such as IPFS

(InterPlanetary File System), the file serving of content-addressable files from multiple local repositories (Benet, 2017), could likewise be configured in the biological setting. For the maintenance of optimal health, immediate on-site responses are urgently required to address stroke, aneurysm, and heart attack. Therefore, more immediate resource delivery (such as clot dissolution mechanisms and stent-like artery support) via medical nanorobots will likely be indispensable for instantaneously managing these conditions for non-fatality, or indeed efficaciously preventing them from occurring in the first place.

7.8.4.2 Brain-Computer Interfaces and Cloudminds

Neural operations are an obvious venue for bioDACs, where the idea is to provide computational support for working with the brain's ~100 billion neurons for the purposes of memory assessment, augmentation, and lifelogging (memory recording). Eventually, this could include the creation of backup copies, uploading, and coordinating *brain-computer interface (BCI)* activities, which is a technology that allows a person to control a computer using the electrical activity of the brain. In the far future, BCIs might be more internally bound to the brain, for example, in an embedded or wireless formulation.

A more invasive **embedded B/CI** (Brain/Cloud Interface) might feature the physiological integration of billions of dedicated autonomous nanorobots, which would self-migrate and directly connect with the neurons, synapses, and glial cells of the neocortex and collectively provision a stable and robust interface with the Internet edge cloud. A slightly less invasive process for a **wireless BCI** might have a constellation of autonomous nanorobots programmed to self-embed into the peripheral calvaria (neurocranium) or pericranium to form a highly organized and interconnected streaming data transfer complex that would interact with specifically targeted neuronal areas within the neocortex (Martins et al., 2019).

Advanced B/CI applications could include automated in-brain information retrieval activated by thought or subvocalization. The idea is personal voice assistants, not just externally like Alexa Echo and Google Home, but on-board interactive applications; literally "voices in your head" (to the extent permitted). Another advanced B/CI application could be participating in cloudminds. A *cloudmind* is a cloud-based collaboration of human and machine minds (again with relevant safeguards and permissions) (Swan, 2016b).

The motivations for participating in a cloudmind could be lifelogging, entertainment, exploration, intellectual discovery, and creative expression. A cloudmind is a concept in the ongoing progression of social collaboration technologies that facilitate collective endeavors. The progression includes the evolution from the corporation and the government (and earlier, the church) as the primary structures for group collaboration, to open-source and open-innovation communities, to blockchain token-based open platform economies, to cloudminds.

7.8.5 Advanced Concept: Group Field Theory: Coordinated bioDAC Action

Discussing the cloudmind concept raises the point that group coordination is an important functionality needed in biological networks and afforded by blockchain technology, particularly as controlled with a smart network field theory. Some of the proposed methods for coordinating medical nanorobotic fleets include fog computing (Bonomi et al., 2012), swarm technology (Galstyan et al., 2005), thresholding (quorum-sensing for multi-robot cooperation (Zhao et al., 2015)), chemical communication in biofluidic environments (Cavalcanti et al., 2006), and biochips based on microfluidics with nanorobotic instrumented platforms (Martel, 2008). Another method, optogenetics, was discussed earlier as an example of a biological smart network technology that is already being tested in human clinical trials (for the treatment of retinitis pigmentosa) (Reardon, 2016). A key property of optogenetics is the ability to control groups of cells as a field, rather than simply as individual cells.

The group-coordination property of biological smart network technologies such as optogenetics might be expanded into related applications. Part of the concept of blockchains is the network architecture of flat (non-hierarchical) peer-to-peer nodes, which provide services for each other. This reciprocity property can be deployed in the biological setting. Biological cells are already engaged in some degree of signaling to one another (Su et al., 2024) (although intercellular signaling is not yet fully understood), and this signaling could be enhanced and further coordinated by medical nanorobots. There could be the idea of a *nanorobotic grid*, a coordinated formation of multiple medical nanorobots organized to correspond to multiple cells or other biological units.

Medical nanorobots would have the functionality to block or distribute resources across a nanorobotic grid as a gradient or distribution function, which could be powerful in sequestering portions of diseased tissue or quickly deploying resources (such as oxygen) to various sites (Pittman, 2011). The point is that group actions, as opposed to unitary actions, could be the focus of bioDAC group fleets. These 3D "nanomedical battlefield soldiers" could be directed or self-assembled per local environmental signals into a dynamic formation or assembly of units from which to deploy requisite actions. Technophysics methods such as smart network field theories, Boltzmann distributions (concentration equalization), and Hamiltonians (energy functions) might be the most relevant.

7.8.5.1 On-demand Nanorobotic Grid Formations as a Local Bioeconomy

Blockchains are an appropriately granular tool for orchestrating the complexities of multi-agent interactions in the biological context. A more sophisticated feature is considering the coordinated resource production and delivery of ad hoc assembled nanorobotic grids as a market. The idea is that nanorobotic grids might operate as a *market machine*, monitoring their own resource production (supply), while assessing the overall demand for their resources in localized biological environments (demand). The nanorobotic grid could be a formation of medical nanorobots, assembled on demand by economic design principles, particularly to respond to an urgent situation (conceptually akin to requesting hundreds of Uber drivers to a specific location). These in-body paramedic nanorobots could support spikes in local demand, for example, for tissue oxygenation, vascular support, or clot dissolution. A nanorobotic grid is its own local bioeconomy in the sense of serving as a market mechanism for allowing supply and demand to resolve each other.

A macroscale analog is local blockchain economies emerging for the purpose of energy independence in various human communities. Individual peer nodes (home residents with solar panels, for example) produce energy, which is distributed to the local community grid to be consumed. The overall activity is objectively and unobtrusively tracked, tabulated, and settled by blockchains as the accountancy function (Chiang, 2018; Papajak, 2017).

Blockchain-coordinated *nanorobotic biogrids* could likewise coordinate activities within certain multicellular tissues (for tumor removal or plaque remediation), expanding the complexity and sophistication of their operations. Nanorobotic biogrids could participate in a community payment channel DAC (structured mechanism) to obtain (recruit) additional needed resources to the area of demand from more remote locations. This could include micronutrients, small molecules, pharmaceutical substances, electrical stimulus, chaperone systems, and enzymes. The idea is to target a more complicated class of pathology resolution, such as cancer-fighting and intransigent waste removal.

This kind of microgrid group action functionality might be useful as many disease pathologies appear to be subject to a "chicken or the egg" causality dilemma. This means that it is not clear whether the "chicken" or the "egg" comes first, in the sense that it is not known if it is the cellular environment that degrades and causes disease, or if the cells themselves initially become diseased. In cancer, heart disease, neurodegenerative disease, and other pathologies of aging, both the cellular environment and the cells are healthy in individuals without the condition, and ill-functioning in individuals with the disease. Therefore, a compound solution such as medical nanorobot grids operated with smart network field theories might provide the requisite functionality to better elucidate and attack disease causality.

7.8.6 Advanced Concept: Biosecurity, Consensus Algorithms, and Self-Mining Ecologies

As an intelligent self-operating network, a blockchain system has mechanisms built into the software which coordinate its activities. The central mechanism is the *consensus algorithm*, which is the means by which distributed nodes reach agreement about the truth state of the network and update their copy of the transaction ledger. Blockchain networks have drawn criticism for the large amount of electricity used in *mining*, the competitive peer-based process of confirming transactions, and providing cryptosecurity in decentralized networks. This raises the question of how bioblockchains (blockchain models operating in biological environments) might settle and clear transactions, while providing biosecurity.

7.8.6.1 Blockchain Biosecurity to Combat Biohacking Challenges

In the bioDAC context, there are likely to be different classes of security required by different types of activities. Mission-critical heart and brain operations would be among the most sensitive and thus require a higher level of biocryptosecurity (protection against both hacking and unintentional interference). This operationalizes into having a more robust and competitive mining effort to provide the incentives for cryptosecurity in bioblockchains.

The unequivocal requirement for biosecurity is precisely the reason why embedded medical devices might already benefit from a secure control software technology like blockchains, administered as a smart network field theory. The number of implanted devices in humans, such as cardiac monitors and insulin pumps, is only likely to increase. Although scientists opine that the risk of security attacks on implanted biodevices currently has a low probability (Abdollah and Perrone, 2017), biosecurity is an area of concern that could likely grow over time and require standardized solutions to be integrated within the underlying technology. (These kinds of biohacking problems have been considered in science fiction, for example, producing a society-wide crisis when implanted cognitive nanorobots begin to malfunction due to a virus (Bear, 1997). This raises the question of neural anti-viral software and, more philosophically, how to detect if neural implants are malfunctioning.)

At the macroscale, blockchains deployed in the physical world have different cryptosecurity requirements. Monetary transactions between humans are among the most sensitive, and IoT status chirps to the network are the least. Monetary blockchain transactions use a robust proof of work method to maintain cryptosecurity and confirm transactions. IoT transactions use a lighter weight "tangle" (unorganized) method of edge nodes, confirming a few recent transactions in order to be able to broadcast their own transaction to the network.

Mission-critical bioDACs could use similar high cryptosecurity models based on Bitcoin's proof of work algorithm (instantiating energy reserves and incentives in the biological context). For non-mission critical bioDACs, the *biomining* operation could be architected similarly to that envisioned for a smart home IoT network (Swan, 2015a). In this reciprocal model, one ecology of nodes mines (performs transaction confirmation and logging services) for another, providing independent, yet interdependent, secure transaction-logging. The game theoretic incentives and "compensation" received could be remuneration in the form of nutrients or other biocurrencies, or in the model of cooperative reciprocity.

This cooperative reciprocity is in the sense that each ecology (epithelial nanorobots are one ecology, for example) is part of a bigger biological organism and biocryptoeconomic system. One group's transactions are validated and confirmed by another mining ecology in the same bioeconomy and organism and thus are vested in the overall survival of the organism. For example, the vascular nanorobots could mine (log transactions) for the epithelial nanorobots, and vice versa, or (for greater security) in a round robin model (the epithelial nanorobots mine for the respirocyte nanorobots).

The idea is that one nanorobot ecology provides the mining operation (and possibly other peer services such as credit (future resource availability) and payment channels (ongoing resource delivery)) for another. Likewise, neural bioDACs could mine for the cardiac bioDACs (as they require the same high-grade security, validation, and anti-hacking measures), whereas digestive system bioDACs could mine for immune system bioDACs. Biomining would likely be a batched process of internal logging on lightweight clients in situ, uploaded periodically over the body gateway to personal health data clouds. This is because there would be optimized energy constraints governing the onboard processing and storage capabilities of nanorobotic bioDACs.

Interdependent bioblockchain ecologies (e.g., groups of nanorobotic biogrids) could mine for each other, in a congruent participatory decentralized manner. In situ biosecurity could be generated in that each ecology has the incentive to both maintain the network by accurately recording the transactions of other ecologies, since their own survival is also at stake, and also to have their own bona fide valid transactions recorded under the same rationale (Swan, 2015b).

7.8.7 Biocongruent Design: Biomimicry, Geoethics, and Natural Cadence

Having discussed some of the detailed advanced concepts of bioblockchains operating in the human body, this section briefly enumerates some principles for responsible design. The broader problem space of future health assumes the more intense integration of human physiology and technologies. Thus, a slate of responsible design principles might facilitate progress in a way that supports both ethical practices and technical excellence.

Since the design domain for advanced medical technologies is the human body, perhaps the first principle to promote is related to biomimicry, or more specifically biomimicry parallelism, which can be formulated as *biocongruency*. The premise is that any type of technology that ingresses the human body or is otherwise employed to orchestrate its processes would be most successful to the extent that it is biocongruent. *Biocongruent* refers to being in alignment with the structure, rhythms, size, substances, and cadence of the natural systems of the body. A biocongruent design principle would utilize a given biological system as a starting point (in contrast to a technical feature or capability, for example) and expand or limit biofunctionality from within that biological system. An example of biocongruent design would be to effectively block a cancer cell's main mode of growth, angiogenesis, attacking it on its own terms through the negation of new blood vessel formation.

In order for ubiquitous blockchain-based bioDACs to be successful in the endeavor of medically tracking, monitoring, assessing, and intervening, they would need to be in a format and operational cadence that is in exquisite alignment with that of the human body (e.g., biocongruent). Responsible design in the biological environment of the body would also do well to incorporate principles that have been outlined for geoethical nanotechnology design (Rothblatt, 2013).

7.8.7.1 Self-Regulating Homeostatic BioDACs: Self-Instantiating and Self-Retiring

This section outlines an idea for self-instantiating and self-retiring homeostatic bioDACs. Life-cycle planning is a feature of responsible design. The human body already excels at homeostatic functions (e.g., ongoing equilibrium management), and self-retirement in the form of apoptosis (e.g., programmed cell death) (Elmore, 2007) and other cellular senescence and death mechanisms (Tait et al., 2014) that are engaged when cells cease to perform their critical functions. The idea of bioDACs as market mechanisms suggests that perhaps they might sense not only the demand for what they can provide in situ but also the need to be created in the first place. One way that bioDACs could "self-create" is that they are pre-existing on a unitary basis but have the capacity to dynamically self-assemble into biogrids to tackle specific problems or diseases, akin to an on-site paramedic, and subsequently disassemble.

Another way that bioDACs might "self-create" is literally. The more sophisticated suite of activities in which bioDAC entities might engage may pertain to instantiating themselves, accessing resources, providing services to human or machine customers, supporting transaction-logging services in the biomining ecology, maintaining and repairing themselves, and eventually, retiring. Biocongruent design could include cryptoapoptosis as a biosecurity measure, and the self-senescing and self-retiring of bioDAC. The point is that medical nanorobots would have a more biocongruent design to the extent that they might be self-instantiated in the body (including automatically) with rejection-resistant DNA nanostructures and biocomputational modules, rather than being introduced from external sources. Medical nanorobots might indeed mirror the life cycle processes of biological cells.

If biocryptoeconomic nanorobots were to be truly autonomous bioDACs, they might be imbued with a means for sensing a requirement for their genesis in the tradenets of *biodemand* within the body. Subsequently, they could initiate a resource crowdfunding strategy (an economically principled means of generating itself), commence operations upon the biotoken offering's successful completion, provide required services, and then self-retire when the demand for its operation has been dissolved. This concept is akin to that of a self-owned Uber-type car that creates itself per sensing demand on a smart city tradenet grid, self-funds, self-operates, self-maintains, and self-retires (Kelion, 2015). In the human body, at the onset of a cancer or pre-cancer (e.g., via cellular threshold levels for mutational DNA copies being exceeded), there could be a trigger for a self-initiated bioDAC crowdfunding to support rapid in situ cell cancer eradication.

7.9 BIOTECHNOPHYSICS DISCUSSION AND IMPLICATIONS

7.9.1 Incorporating Biocryptoeconomics into the Data Science Model for Health

This section conveys a detailed discussion of advanced medical technologies, GHCE units, and biocryptoeconomy smart network applications developed in this chapter, to the theoretical model proposed at the beginning, for disease causality and the production of health and wellness. Figure 7.3 illustrates how the new ideas are incorporated into the model. The overall research hypothesis is that framing health challenges as a data science problem might be the most efficacious strategy for solving them.

The most significant new development within the data science approach is technophysics: Web 3.0 smart networks, which is extensively elaborated on in the chapter (biotechnophysics as technophysics in the biological context). Figure 7.3 outlines details of the biotechnophysics conceptual framework. Technophysics, which involves the application of physics concepts, principles, and

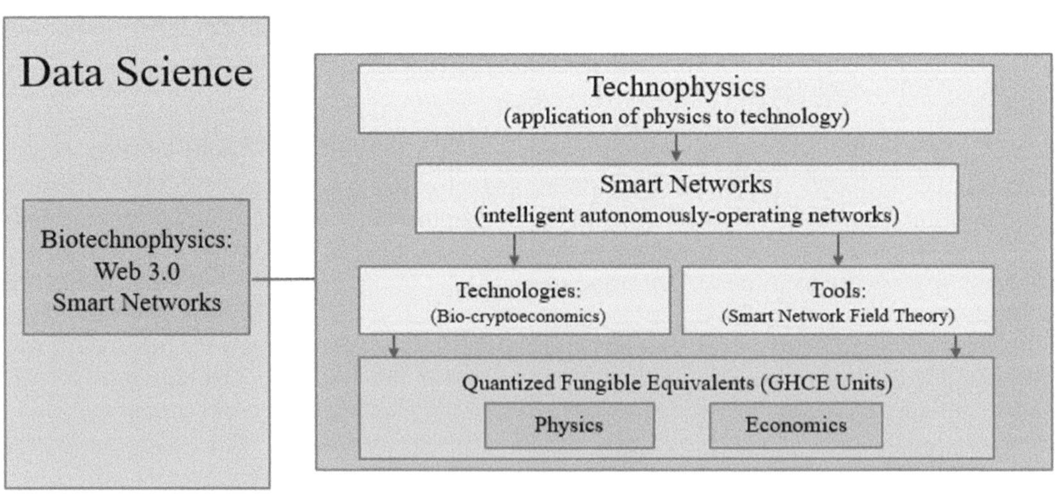

FIGURE 7.3 Smart health networks and global health care equivalency (GHCE) units. (Image created by the author.)

methods to the study of technology, is the highest-order level of organization. Within technophysics, smart networks are formulations that refer to the emergence of intelligent, autonomously operating networks as a contemporary feature of reality. Smart networks include blockchain economic systems and deep learning pattern recognition neural networks. There are at least two domains that exist within smart networks for their control: the technologies themselves, such as the biocryptoeconomic model developed in this chapter, and tools such as smart network field theories. The most relevant application of smart network technologies and tools for this work includes the establishment of quantized, fungible GHCE units, based on physics and economics principles. These GHCE units would be employed as a standardized weighted points with which to optimally deliver worldwide nanomedical diagnostic and therapeutic treatments in the future.

7.9.2 Risks and Limitations

There are certainly many critiques that could be levied against this work. One of the most obvious caveats is that this analysis likely understates the degree and complexity of the current research being carried out in biology and medicine. However, many preliminary and complex ideas are presented here, with the aim of oversimplification and unwitting omission being the development of larger-scale concepts and arguments.

A second critique might be that the current proposal is quite speculative and forward-looking and does not address technical feasibility. One issue pertains to the provision of a more specific elucidation of how the human body *biogateway* (analogous to a connected car's vehicle gateway) might operate. Another issue relates to the considerable challenges associated with in situ energy recruitment, connectivity, memory, processing, and data storage, all of which will be critical toward the realization of these concepts. The author is keenly cognizant that a much broader slate of concerns will need to be addressed if the concepts proposed here are to evolve to the pragmatic phases of engineering design and implementation.

A third issue, which has not been addressed as yet, relates to the intricately complex systemic nature of the human body. Biology is a complicated adaptive system within which the activation of a single element can have many unknown and unintended cascading consequences throughout the system. For example, many interventions that produce results in model organisms related to life extension fail to translate to the more complex domain of the human organism. Plasticity might be a factor (e.g., the recruitment of other cells to fulfill a similar function) if original sites become damaged. Any potential intervention with advanced nanomedical technologies could have significant systemic impacts. The idea would be to gradually translate current methods already in use, such as nanoparticle targeted drug delivery, to more sophisticated and directed therapies. Additional societal apparatuses may be necessary, such as more extensive bioethical review boards and real time intervention monitoring.

7.10 CONCLUSION

This chapter provides a blueprint for how a smart network technology, blockchain distributed ledgers, might be used in the body to fight disease, and produce health and well-being, and how its contribution might be measured and deployed as tradeable GHCE units. The research extends previous work in the areas of *technophysics* (the application of physics to the study of technology) and *smart networks* (intelligent autonomously operating networks) (Swan and dos Santos, 2018), as well as in the biological domain regarding *biocryptoeconomics* (cryptoeconomic principles applied in the context of biology) and *personal bioDACs* (packages of programs for purpose-directed activity in the body and the brain) (Swan, 2016a). To the extent that health futures are based on advanced medical therapies and a greater intensity of human-technology integration, a concept such as GHCE units could be a key enabling technology. The notion of GHCE units may serve as the next logical step toward implementing a standardized tracking unit for the marshaling and deployment of a democratized, massively distributed, and equalized worldwide health service, particularly as relates to nanomedical diagnostics and therapeutics.

The further potential consequences of this work are twofold. First, several novel conceptual formulations are proposed that might facilitate ongoing development in many areas. Some of these theoretical advances include the concepts of technophysics, smart networks, smart health networks, biocryptoeconomic design principles, personal bioDACs, biological smart network group field theories, and biocongruent design principles; together encapsulated as *biotechnophysics*. The aim of these formulations is to provide the theoretical foundations for future advances, with the concepts in general, and specifically as deployed in biological settings. For example, one useful advance of this work is describing a group field theory for the secure and automated coordination of medical nanorobot fleets in the body.

A second important consequence of this work is that it causes a separate point to come into sharper view. Smart networks (intelligent self-operating global computational infrastructures) are an emergent feature of reality. A concurrent manifestation is that not only are new fields emerging for the study of smart networks and other modern phenomena (disordered systems, for example) but also new *kinds* of fields may be developing. A "new kind of science" field could be evolving that is a *metascience* field, not in the usual sense of metascience (the scientific study of the conduct of science), and also not a "new kind of science" field as the computational complement to every field (Wolfram, 2002).

A new kind of metascience field, a *metascience composite*, is meant here in the sense of gathering similar concepts, methods, tools, and approaches used in multiple science fields into a consolidated formulation that might itself constitute a new class of science. The task of technophysics is to identify similar problems and solution methods arising in statistical physics, information theory, and computer science, such as system criticality, phase transition, saturation, and the group activation of field-many items.

One example of such a *metascience composite* is that probability theory is employed in the same kinds of models to define spin glasses in statistical physics, error-correcting (LDPC) codes in information theory, and randomized algorithms and probabilistic combinatorics in computer science (Mézard and Montanari, 2009). Another example of a *metascience composite* is that information theoretic operations (such as energy-entropy trade-offs) are used in many scientific disciplines in diverse ways at the surface level, but also have the same foundational structure (Verlinde, 2017). The point is that not only are new fields emerging in the scientific study of modern phenomena, but also possibly new types of fields, and that domains of high impact and complexity, such as biology, future medicine, and Global Health Care Equivalency, could spur significant advances in these areas.

GLOSSARY

Bioblockchain Software that used distributed ledger blockchain technology to operate in the biological environment, particularly the secure automation and coordination of multifleet item properties.

Biocomputing A computational model that employs biocomputing materials to perform calculations, particularly in situ computational models that might harness biology as its own computational substrate.

Biocryptoeconomics Cryptoeconomic principles and blockchain distributed ledger systems deployed in the biological domain, particularly in the convergence of blockchain technology and nanotechnology applications, using blockchain smart networks to control vast fleets of medical nanorobots in the human body.

Biocryptoeconomy Cryptoeconomic principles applied in the biological domain.

Biocurrency The unit of denomination of activity for biological resource discovery, transfer, exchange, and tracking. Biocurrencies might include cancer cells, DNA methylations, epigenetic mutations, red blood cells, oxygen, plaques, lymphocytes, neutrophils, macrophages, cytokines, antibodies, neurotransmitters, glucose, and potentiation.

Biotechnophysics Technophysics (physics applied to the study of technology) in the biological domains.

Blockchain (distributed ledger) Technology An immutable, cryptographic (cryptography-based), distributed (peer-based), consensus-driven ledger. Blockchain (distributed ledger) technology is a software protocol for the instantaneous transfer of money and other forms of value (assets, contracts, public records, program states) globally via the Internet.

Cloudmind A cloud-based collaboration of human and machine minds (with safeguards and permissions). "Mind" is generally denoting an entity with some capacity for processing, not the volitionary action and free will of a consciousness agent.

Cryogenomics A subfield that extends cryobiology (the branch of biology that studies the effects of low temperatures on living things) with DNA nanotechnology and personalized genomic profiles. For example, using DNA nanostructures to build a scaffolding or coatings around cellular components to protect them against ice crystal formation damage, or other decay and damage that may result from freezing and vitrification processes.

Cryptoeconomics An economic transaction paradigm based on cryptography; more specifically, an economic transaction system implemented in a cryptography-based software network, using cryptographic hashes (computational proof mechanisms) as a means of confirming and transferring monetary balances, assets, smart contracts, or other system states. A key concept is *trustless* trust, meaning removing as much human-based trust as possible to make the economic system trustworthy (relocating human-based trust to cryptography-based trust).

DNA Nanotechnology The design and manufacture of artificial nucleic acid structures for technological uses, particularly as a rejection-resistant in situ building material.

Deep Learning Chains A class of smart network technologies in which other technologies, blockchain, and deep learning converge as a control technology for other smart network technologies. Deep learning chains have the properties of secure automation, audit-log tracking, and validated transaction execution of blockchain, as well as the object and pattern recognition technology (IDtech) of deep learning. Deep learning chains might be employed to control other multifleet Internet-connected smart network technologies, such as UAVs, autonomous driving fleets, medical nanorobots, and space-based asteroid mining systems.

Deep Learning Neural Networks Computer programs that can identify what an object is; more technically, deep learning is a branch of machine learning based on a set of algorithms that attempts to model high-level abstractions in data by using artificial neural network architectures, based on learning multiple levels of representation or abstraction, such that predictive guesses can be made about new data.

Global Health Care Equivalency (GHCE) An envisaged conceptual future platform, wherein any individual worldwide may have the same access to advanced, cost-effective, nanomedical diagnostic and therapeutic technologies, no matter where or under what conditions they live.

Global Health Care Equivalency Units (GHCE Units) A future concept of fungible, quantized units of global health care services, particularly nanomedical diagnostic and therapeutic technologies, that are available on a worldwide basis, and possibly traded, monitored, and aggregated as a new economic statistic.

IDtech Identification technology, the technological functionality of object recognition; similar to FinTech, RegTech, TradeTech, and HealthTech; technologies that digitize, standardize, and automate operations within their respective domains.

Medical Nanorobots Molecular sized machines, fabricated through the controlled assembly of individual atoms or molecules (via molecular manufacturing (nanofactory) technologies), that are designed to perform specific medically related tasks in the human body. A blood-borne medical nanorobot might be on the order of Ø1–3 μm (1000–3000 nm), with components ranging from 1 to 10 nm in size.

P4 Medicine A health care and medical approach that is predictive, preventive, personalized, and participatory. The most feasible current application of P4 medicine envisioned is precision medicine, which primarily incorporates the personalized aspect of P4 medicine.

Personal bioDAC (Decentralized Autonomous Corporation) A package of blockchain-based smart contracts (programs registered to a blockchain for automated execution) for a purpose-driven biological activity, such as marshalling cancer detection medical nanorobots, on a personal basis (tailored to individual-specific health parameters).

Precision Medicine An individualized approach (n = 1) to disease, based on the patient's unique combination of genes, environment, and lifestyle.

Preventive Medicine Health maintenance; maintaining health such that disease conditions do not arise, or attempting to detect and address disease conditions earlier in their process, before they ever become clinically diagnosable.

Quantified Self (n = 1 medicine) An individual applying precision medicine to him/herself, engaged in the self-tracking of any type of biological, physical, behavioral, or environmental data with some kind of quantified data collection and analysis strategy, particularly toward solving a problem.

Smart Contract A software program registered to a blockchain for confirmation (time-date stamping provenance), and possibly some form of automated execution in the future. To be legally binding, smart contracts must involve the four specific elements of contracts: two parties, consideration, and terms.

Smart Health Networks (Smart Bionetworks) A class of smart networks which autonomously characterize, monitor, and intervene to facilitate health, particularly through the secure automated control of multifleet therapeutic entities residing in the body for brief or extended durations.

Smart Networks Intelligent autonomously operating networks Smart Bionetworks (see Smart Health Networks).

Smart Network Field Theory (SNFT) An effective field theory (a means of higher-level system approximation) and optimal control theory for the characterization, monitoring, and control of smart networks systems, notably criticality detection and multifleet item orchestration. Special forms of the SNFTs at the very-small and very large scale might include *Biological Smart Network Field Theories (bSNFT)* and *Space-based Smart Network Field Theories (sSNFT).*

Smart Network Technologies Exemplar smart network technologies are blockchain economic networks and deep learning pattern recognition neural networks.

Technophysics The application of physics to the study of technology; specifically, the explanation of observed behavior in smart networks with physics-based models that might include statistical physics, Brownian motion, the Schrödinger equation, Shannon entropy, information theory, and computational complexity; by analogy to biophysics and econophysics.

REFERENCES

Abdollah T, Perrone M. Hackers could remotely control defibrillator or pacemaker, U.S. warns. *The Associated Press*, 2017. https://www.cbc.ca/news/world/cybersecurity-heart-devices-implantable-1.3930997 (accessed May 23, 2024).

Adamowicz J, Pokrywczynska M, Van Breda SV, Kloskowski T, Drewa T. Concise review: Tissue engineering of urinary bladder; we still have a long way to go? *Stem Cells Translation Medicine*. 2017;6(11): 2033–2043.

Agrawal R, Deshmukh R, Garg A, Mali SN, Prajapati BG, Singh K. Nanobots: A revolutionary technology in the development of pharmaceuticals. *BioNanoScience*. 2024;15:159.

ASHA (American Speech-Language-Hearing Association). 2018. Cochlear Implants. https://www.asha.org/public/hearing/cochlear-implant/ (accessed September 6, 2018).

Atala A. Tissue engineering of human bladder. *British Medical Bulletin*. 2011;97:81–104.

Bear G. *Slant*. New York: Tor Books, 1997.

Benet J. Filecoin: A Decentralized Storage Network. *Protocol Labs*. 2017.

Benjamin EJ, Blaha MJ, Chiuve SE, Cushman M, Das SR, Deo R, et al. Heart disease and stroke statistics—2017 update: a report from the American Heart Association. *Circulation*. 2017;135(10): e146–e603.

Boehm FJ. An Investigation of Nucleic Acid/DNA-Based Manufacturing. Research contribution to the Center for Responsible Nanotechnology: Thirty Essential Nanotechnology Studies #10, August 2004. https://www.scribd.com/document/183051565/DNA-MANUFACTURING-pdf

Bonomi F, Milito R, Zhu J, Addepalli S. Fog computing and its role in the internet of things. In *Proceedings of the first edition of the MCC workshop on Mobile cloud computing (MCC '12)*. Association for Computing Machinery, New York, NY, USA, 2012:13–16.

Boyden ES. Optogenetics and the future of neuroscience. *Nature Neuroscience*. 2015;18(9):1200–1201.

Brown G, Butler PJ, Chang DW, Chien S, Clegg RM, Dewey CF et al. Cellular and molecular bioengineering: A tipping point. *Cellular and Molecular Bioengineering*. 2012;5(3):239–253.

Bujold KE, Lacroix A, Sleiman HF. DNA nanostructures at the interface with biology. *Chem*. 2018;4: 495–521.

Buterin V. DAOs, DACs, DAs and More: An Incomplete Terminology Guide. Ethereum blog. May 6, 2014. Available at https://blog.ethereum.org/2014/05/06/daos-dacs-das-and-more-an-incomplete-terminology-guide/ (accessed September 6, 2018).

Cassou-Noguès P. The syndrome of the thermometer: Two machines for reading thoughts. *SubStance*. 2016;45(2):39–60.

Cavalcanti A, Hogg T, Shirinzadeh B, Liaw HC. Nanorobot Communication Techniques: A Comprehensive Tutorial. *2006 9th International Conference on Control, Automation, Robotics and Vision*, Singapore, 2006;1–6.

Chiang D. Catalonia Seeks Energy Independence from Spain with ION Cryptocurrency. *Cryptosumer*, June 01, 2018. https://www.cashtechnews.com/2018/06/01/catalonia-seeks-energy-independence-from-spain-with-ion-cryptocurrency/ (accessed May 23, 2024)

Choromanska A, Henaff M, Mathieu M, Ben Arous G, LeCun Y. The loss surfaces of multilayer networks. In: Proceedings of the 18th International Conference on Artificial Intelligence and Statistics (AISTATS). PMLR; 2015. p. 192–204.

Clendenin W. The general causes of disease. *Public Health Papers and Reports*. 1873;1:46–53.

Cross R. This $25,000 physical has found some 'serious' health problems. Others say it has serious problems. *Science*, May 12, 2017. https://www.science.org/content/article/25000-physical-has-found-some-serious-health-problems-others-say-it-has-serious (accessed May 23, 2024).

Davies JA. *Life Unfolding: How the Human Body Creates Itself*. Oxford: Oxford University Press, 2014.

Domschke A, Boehm FJ. Chapter 19: The "Sentinel": A Conceptual Nanomedical Strategy for the Enhancement of the Human Immune System. In *Immune Aspects of Biopharmaceuticals and Nanomedicines*. Bawa R et al. (eds.), Singapore: Pan Stanford. 2018.

dos Santos RP. On the philosophy of Bitcoin/Blockchain technology: Is it a chaotic, complex system? *Metaphilosophy*. 2017;48(5):620–633.

dos Santos RP. Consensus Algorithms: A Matter of Complexity? In *Blockchain Economics*. Swan M, Potts J, Takagi S, Witte F, Tasca P. (eds.), London: World Scientific. 2019.

Doudna JA, Sternberg SH. *A Crack in Creation: Gene Editing and the Unthinkable Power to Control Evolution*. New York: Houghton Mifflin Harcourt, 2017.

Drexler KE. *Nanosystems: Molecular Machinery, Manufacturing, and Computation*. New York: Wiley-Interscience; 1992.

Elmore S. Apoptosis: A review of programmed cell death. *Toxicologic Pathology*. 2007;35(4):495–516.

Eveleth R. There are 37.2 Trillion Cells in Your Body. *Smithsonian Magazine*. 2013. https://www.smithsonianmag.com/smart-news/there-are-372-trillion-cells-in-your-body-4941473/ (accessed May 23, 2024).

Fahy GM, Wowk B. Principles of cryopreservation by vitrification. *Methods in Molecular Biology*. 2015;1257:21–82.

Fattahi P, Yang G, Kim G, Abidian MR. A review of organic and inorganic biomaterials for neural interfaces. *Advanced Materials*. 2014;26(12):1846–1885.

Filipe L, Fdez-Riverola F, Costa N, Pereira A. Wireless body area networks for healthcare applications: protocol stack review. *Int J Distrib Sens Netw*. 2015;2015:213705.

Freitas RA Jr. *Nanomedicine, Vol. I: Basic Capabilities*. Austin: Landes Bioscience, 1999.
Freitas RA Jr. Nanodentistry. *Journal of the American Dental Association*. 2000;131(11):1559–1565.
Freitas RA Jr. Welcome to the Future of Medicine. Transhumanism and Personal Identity. In *The Transhumanist Reader*. More M, Vita-More N. (eds.), Oxford: Wiley-Blackwell. 2012;67–72.
Furber JD. Extracellular glycation crosslinks: Prospects for removal. *Rejuvenation Research*. 2006;9(2):274–278.
Galstyan A, Hogg T, Lerman T. Modeling and mathematical analysis of swarms of microscopic robots. *Proceedings 2005 IEEE Swarm Intelligence Symposium, 2005. SIS 2005.*, Pasadena, CA, USA. 2005; 201–208.
Giorgi GA, Guerraggio A, Thierfelder J. *Mathematics of Optimization: Smooth and Nonsmooth Case*. London: Elsevier, 2004.
Gluck N, Shpak B, Brun R, Rösch T, Arber N, Moshkowitz M. A novel prepless X-ray imaging capsule for colon cancer screening. *Gut*. 2016;65(3):371–373.
Halevy A, Norvig P, Pereira F. The unreasonable effectiveness of data. *IEEE Intelligent Systems*. 2009;24(2):8–12.
Harvard Medical School. Key Numbers for Cell Biologists. *BioNumbers* database. 2025. http://bionumbers.hms.harvard.edu/KeyNumbers.aspx (accessed August 12, 2025).
Hauptman N, Glavač D. Colorectal cancer blood-based biomarkers. *Gastroenterology Research and Practice*. 2017;2017:2195361.
Herculano-Houzel S. The human brain in numbers: A linearly scaled-up primate brain. *Frontiers in Human Neuroscience*. 2009;3:31.
Hood L. Systems biology and p4 medicine: Past, present, and future. *Rambam Maimonides Medical Journal*. 2013;4(2):e0012.
Ibrahim F, Thio TH, Faisal T, Neuman M. The application of biomedical engineering techniques to the diagnosis and management of tropical diseases: A review. *Sensors (Basel)*. 2015;15(3):6947–6995.
IDG. 2016. *Data & Analytics Survey*. https://cdn2.hubspot.net/hubfs/1624046/IDGE_Data_Analysis_2016_final.pdf (accessed September 6, 2018).
Jaffe A, Witten E. Quantum Yang–Mills theory. 2000. https://arthurjaffe.com/Assets/pdf/QuantumYangMillsWebRevised.pdf (accessed May 23, 2024).
Kaiser J. NIH's 1-million-volunteer precision medicine study announces first pilot projects. *Science*, February 25, 2016. https://www.science.org/content/article/nih-s-1-million-volunteer-precision-medicine-study-announces-first-pilot-projects (accessed May 23, 2024).
Kelion L. Could driverless cars own themselves? *BBC News*, 2015. https://www.bbc.com/news/technology-30998361 (accessed May 23, 2024).
Kiktenko EO, Pozhar NO, Anufriev MN, Trushechkin S, Yunusov RR, Kurochkin YV et al. Quantum-secured blockchain. 2018. https://arxiv.org/pdf/1705.09258.pdf (accessed September 6, 2018).
Kumar R, Mukesh R. State of the art: Security in wireless body area networks. *International Journal of Computer Science & Engineering Technology (IJCSET)*. 2013;4(5):622–630.
Kumar V, Palazzolo S, Bayda S, Corona G, Toffoli G, Rizzolio F. DNA nanotechnology for cancer therapy. *Theranostics*. 2016;6(5):710–725.
Le QV. Building high-level features using large scale unsupervised learning. *2013 IEEE International Conference on Acoustics, Speech and Signal Processing*, Vancouver, BC, Canada. 2013;8595–8598.
Loewenstein G, Volpp KG, Asch DA. Incentives in health: Different prescriptions for physicians and patients. *JAMA*. 2012;307(13):1375–1376.
Ma H, Marti-Gutierrez N, Park SW, Wu J, Lee Y, Suzuki K et al. Correction of a pathogenic gene mutation in human embryos. *Nature*. 2017;548(7668):413–419.
Marosfoi MG, Korin N, Gounis MJ, Uzun O, Vedantham S, Langan ET et al. Shear-activated nanoparticle aggregates combined with temporary endovascular bypass to treat large vessel occlusion. *Stroke*. 2015;46(12):3507–3513.
Martel S. Nanorobots for microfactories to operations in the human body and robots propelled by bacteria. *Facta Universitatis Series Automatic Control and Robotics*. 2008;7(1):1–8.
Martins NRB, Angelica A, Chakravarthy K, Svidinenko Y, Boehm FJ, Opris I, Lebedev MA, Swan M, Garan SA, Rosenfeld JV, Hogg T, Freitas RA Jr. Human brain/cloud interface. *Frontiers in Neuroscience*. 2019;13:112.
Merhav N. Statistical physics and information theory. *Foundations and Trends in Communications and Information Theory*. 2010;6(1–2):1–212.
Mézard M, Montanari A. *Information, Physics, and Computation*. Oxford UK: Oxford University Press, 2009.

Miller RS, Shorter G. *High frequency trading: overview of recent developments.* Washington, DC: Congressional Research Service; 2016. Report No.: R44443.

Millman R. Connected cars report: 125 million vehicles by 2022, 5G coming. *Internet of Business.* 2018. https://internetofbusiness.com/connected-cars-report-125-million-vehicles-by-2022-5g-coming (accessed August 12, 2025).

Mirza AZ, Siddiqui FA. Nanomedicine and drug delivery: A mini review. *International Nano Letters.* 2014;4(94):1–7.

Molteni M. A Blood-based Cancer Test gets its first Results. *Wired*, 2018. https://www.wired.com/story/a-blood-based-cancer-test-gets-its-first-results/ (accessed May 23, 2024).

Muula AS. What causes diseases? Why the answer is complicated, and contentious. *The Conversation*, 2016. https://theconversation.com/what-causes-diseases-why-the-answer-is-complicated-and-contentious-55955 (accessed May 23, 2024).

National Cancer Institute. Cancer Statistics. 2025. https://www.cancer.gov/about-cancer/understanding/statistics (accessed August 12, 2025).

Nicolas-Alonso LF, Gomez-Gil J. Brain computer interfaces, a review. *Sensors (Basel).* 2012;12(2):1211–1279.

Nociari MM, Kiss S, Rodriguez-Boulan E. Lipofuscin Accumulation into and Clearance from Retinal Pigment Epithelium Lysosomes: Physiopathology and Emerging Therapeutics. In *Lysosomes: Associated Diseases and Methods to Study Their Function.* Dhiman P. (ed.), IntechOpen. 2017;1–29.

Oyaro K. Taking health services to remote areas. *Africa Renewal.* 2017. https://www.un.org/africarenewal/magazine/december-2016-march-2017/taking-health-services-remote-areas (accessed May 23, 2024).

Pagidipati NJ, Gaziano TA. Estimating deaths from cardiovascular disease: A review of global methodologies of mortality measurement. *Circulation.* 2013;127(6):749–756.

Papajak U. Can the Brooklyn Microgrid project revolutionise the energy market? *Medium*, 2017. https://medium.com/thebeammagazine/can-the-brooklyn-microgrid-project-revolutionise-the-energy-market-ae2c13ec0341 (accessed May 23, 2024).

Papanicolas I, Woskie LR, Jha AK. Health care spending in the United States and other high-income countries. *JAMA.* 2018;319(10):1024–1039.

Patel SA, Winkel M, Ali MK, Narayan KM, Mehta NK. Cardiovascular mortality associated with 5 leading risk factors: National and state preventable fractions estimated from survey data. *Annals of Internal Medicine.* 2015;163(4):245–253.

Perkins BA, Caskey CT, Brar P, Dec E, Karow DS, Kahn AM et al. Precision medicine screening using whole-genome sequencing and advanced imaging to identify disease risk in adults. *Proceedings of the National Academy of Sciences of the United States of America.* 2018;115(14):3686–3691.

Petryayeva E, Algar WR, Medintz IL. Quantum dots in bioanalysis: A review of applications across various platforms for fluorescence spectroscopy and imaging. *Applied Spectroscopy.* 2013;67(3):215–252.

Pittman RN. The Circulatory System and Oxygen Transport. In *Regulation of Tissue Oxygenation.* San Rafael CA: Morgan & Claypool Life Sciences, 2011.

Placek M. Connected Car European market penetration 2025–2030. *Statista*, 2022. https://www.statista.com/statistics/1237208/connected-car-market-penetration-europe/ (accessed May 23, 2024).

Placek M. Connected cars worldwide – statistics & facts. *Statista*, 2023. https://www.statista.com/topics/1918/connected-cars/ (accessed May 23, 2024).

Ponomarenko EA, Poverennaya EV, Ilgisonis EV, Pyatnitskiy MA, Kopylov AT, Zgoda VG, Lisitsa AV, Archakov AI. The size of the human proteome: The width and depth. *International Journal of Analytical Chemistry.* 2016;2016:7436849.

Price ND, Magis AT, Earls JC, Glusman G, Levy R, Lausted C et al. A wellness study of 108 individuals using personal, dense, dynamic data clouds. *Nature Biotechnology.* 2017;35(8):747–756.

Qiagen. How much RNA does a typical mammalian cell contain? FAQ ID-2946, 2018. https://www.qiagen.com/us/resources/faq?id=06a192c2-e72d-42e8-9b40-3171e1eb4cb8&lang=en (accessed September 6, 2018).

Rajan D, Visser M. Quantum blockchain using entanglement in time. *Quantum Rep.* 2019;1(1):3–11.

Reardon S. Light-controlled genes and neurons poised for clinical trials. *Nature News.* 2016.

Rothblatt M. Geoethical rules for nanotechnological advances. *J Geoethical Nanotechnol* 2013;8(1):1–7.

Saadeh Y, Vyas D. Nanorobotic applications in medicine: Current proposals and designs. *American Journal of Robotic Surgery.* 2014;1(1):4–11.

Saadeh Y, Vyas D. Nanorobotic applications in medicine: current proposals and designs. *Am J Robot Surg.* 2014;1(1):4–11.

Seeman NC. DNA nanotechnology: Novel DNA constructions. *Annual Review of Biophysics and Biomolecular Structure.* 1998;27:225–248.

Sethi P, Sarangi SR. Internet of Things: Architectures, Protocols, and Applications. *J Electr Comput Eng.* 2017;2017:9324035.

Sharma AK, Sharma VR, Gupta GK, Ashraf GM, Kamal MA. Advanced glycation end products (AGEs), glutathione and breast cancer: Factors, mechanism and therapeutic interventions. *Current Drug Metabolism.* 2019;20(1):65–71.

Singh S, Lu S, Kokar MM, Kogut PA, Martin L. Detection and classification of emergent behaviors using multi-agent simulation framework (WIP). *In Proceedings of the Symposium on Modeling and Simulation of Complexity in Intelligent, Adaptive and Autonomous Systems (MSCIAAS '17).* Society for Computer Simulation International, San Diego, CA, USA. 2017;3:1–8.

Sivasankar M, Durairaj RB. Brief review on nano robots in bio medical applications. *Advances in Robotics and Automation.* 2012;1:101.

Su J, Song Y, Zhu Z, Huang X, Fan J, Qiao J, Mao F. Cell-cell communication: New insights and clinical implications. *Signal Transduction and Targeted Therapy.* 2024;9(1):196.

Swan M. The quantified self: Fundamental disruption in big data science and biological discovery. *Big Data.* 2013;1(2):85–99.

Swan M. *Blockchain: Blueprint for a New Economy.* Sebastopol CA: O'Reilly Media, 2015a.

Swan M. Blockchain thinking: The brain as a DAC (decentralized autonomous corporation). *Technology and Society Magazine.* 2015b;34(4):41–52.

Swan M. Bio-cryptoeconomy: Nanorobotic DACs for cell repair and enhancement. *Institute for Ethics and Emerging Technologies.* 2016a.

Swan M. Bio-cryptoeconomy: nanorobotic DACs for cell repair and enhancement. Institute for Ethics and Emerging Technologies [Internet]. 2016 Aug 29

Swan M. The future of brain-computer interfaces: Blockchaining your way into a cloudmind. *Journal of Evolution and Technology.* 2016b;26(2):60–81.

Swan M. Introduction to Blockchain Economics. In *Blockchain Economics.* Swan M, Potts J, Takagi S, Witte F, Tasca P. (eds.), London: World Scientific. 2019a.

Swan M. Blockchain Economic Theory: Digital Asset Contracting Reduces Debt and Risk. In *Blockchain Economics.* Swan M, Potts J, Takagi S, Witte F, Tasca P. (eds.), London: World Scientific. 2019c.

Swan M, de Filippi P. Toward a Philosophy of Blockchain. In *Metaphilosophy*, 2017, 48. hal-01676883

Swan M, dos Santos RP. Smart Network Field Theory: The Technophysics of Blockchain and Deep Learning. October 8, 2018. https://ssrn.com/abstract=3262945 (accessed May 23. 2024).

Tait SW, Ichim G, Green DR. Die another way–non-apoptotic mechanisms of cell death. *Journal of Cell Science.* 2014;127(Pt 10):2135–2144.

Takahashi RH, Nagao T, Gouras GK. Plaque formation and the intraneuronal accumulation of β-amyloid in Alzheimer's disease. *Pathology International.* 2017;67(4):185–193.

Ventola CL. Medical Applications for 3D Printing: Current and Projected Uses. *P & T : a Peer-reviewed Journal for Formulary Management.* 2014 Oct;39(10):704–711.

Verlinde EP. Emergent gravity and the dark universe. *SciPost Physics.* 2017;2(3).

Wigner EP. The unreasonable effectiveness of mathematics in the natural sciences. *Communications in Pure and Applied Mathematics.* 1960;13(1):1–14.

Wolfram S. *A New Kind of Science.* Champaign IL: Wolfram Media, 2002.

Wowk B, Fahy GM, Ahmedyar S, Taylor MJ, Rabin Y. Vitrification tendency and stability of DP6-based vitrification solutions for complex tissue cryopreservation. *Cryobiology.* 2018;82:70–77.

Zhang Z, Sabuncu MR. Generalized cross entropy loss for training deep neural networks with noisy labels [Internet]. arXiv preprint arXiv:1805.07836. 2018

Zhao Q, Li M, Wang Z, Li J, Luo J. A Quorum Sensing algorithm to control nanorobot population and drug concentration in cancer area. *2015 IEEE International Conference on Robotics and Biomimetics (ROBIO)*, Zhuhai, China. 2015;42–47.

8 Synergies between Nanotechnology and AI toward GHCE

Godfrey Gandawa

CONTENTS

DOI: 10.1201/9781003680093-8

8.1 INTRODUCTION TO GHCE AND ITS SIGNIFICANCE

Global Health Care Equivalency (GHCE) represents a visionary concept that transcends geographical boundaries, socioeconomic disparities, and technological limitations. At its core, GHCE envisions a world where every individual, regardless of their location or social status, has equitable and timely access to advanced healthcare technologies. This transformative idea challenges the status quo, urging us to rethink healthcare delivery and prioritize inclusivity. This chapter delves into the current advancements, ethical considerations, and prospects of this interdisciplinary alliance, examines case studies where the integration of nanotechnology and artificial intelligence (AI) has successfully addressed complex health challenges, and outlines strategies for overcoming barriers toward its potentially ubiquitous adoption. Ultimately, GHCE envisions a world where the synergies between nanotechnology, nanomedicine, and AI culminate in universal healthcare equity, under which every individual has equal access to the pinnacle of medical care (Figure 8.1).

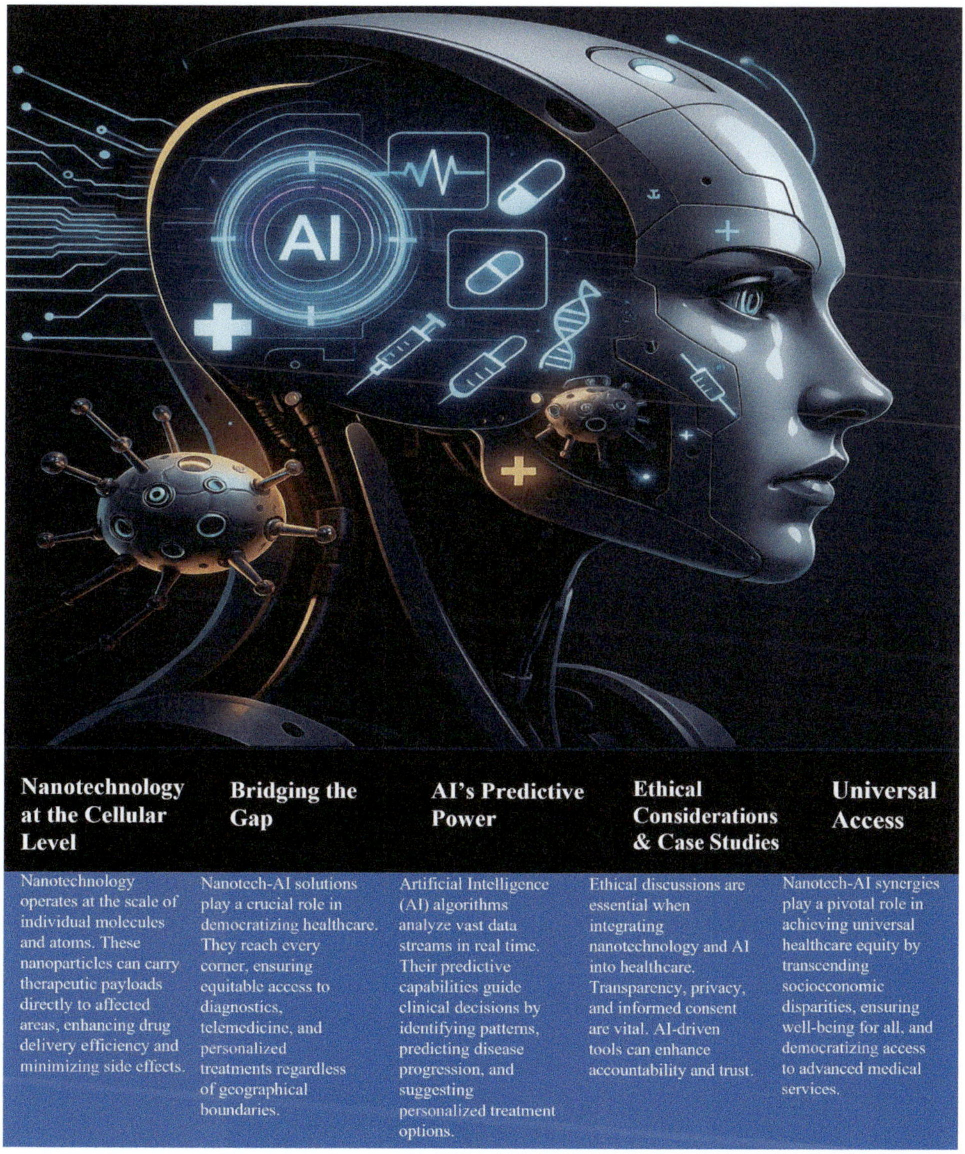

FIGURE 8.1 Nanotechnology: AI synergies for Global Health Care Equivalency (GHCE). (Image created by the author.)

8.2 TRANSFORMATIVE POTENTIAL

8.2.1 Bridging Healthcare Divides

It is unfortunately the case that profound health disparities persist globally. While some regions in the Global North benefit from leading-edge medical advancements, many in the Global South struggle with gaining access to even basic health services. Fortunately, synergies between nanotechnology, AI (AGI/ASI in the relatively near future), blockchain, and quantum computation (QC) are emerging as powerful allies toward the bridging of these divides. By harnessing these synergies, healthcare delivery might be revolutionized to ensure that even marginalized populations receive timely, effective, and dignified care. Kaushik and Suhag (2023) suggest that AI and nanotechnology are merging in certain areas, where their combination may translate to new communication technologies and insights, while AI-powered nanobots operating at the molecular level reshape healthcare and material science. In support of this, Sahota (2023) posits that AI complements nanotechnology by analyzing data, optimizing material design, and predicting behavior, which reshapes the materials that we encounter daily. Nanotechnology is being harnessed for diagnostics, therapeutics, and the prevention of diseases with successful use cases that include regenerative medicine, gene therapy, dentistry, oncology, aesthetics, and drug delivery (Malik et al., 2023).

8.2.2 Conflict Reduction and Diplomacy

GHCE is envisaged to extend far beyond clinical settings. For example, in conflict zones where access to healthcare is often disrupted, nanomedical devices imbued with AI may play pivotal roles. Remote diagnostics, telemedicine, and supportive AI-driven decision infrastructures could assist with mitigating the impacts of conflicts on civilian health. Moreover, collaborative efforts in health technologies may facilitate diplomatic relations by emphasizing our shared humanity over political disparities.

8.3 EQUITABLE ACCESS: A MORAL IMPERATIVE

8.3.1 The Urgency of Equitable Healthcare

GHCE underscores the moral imperative of equitable access to healthcare. As we witness rapid breakthroughs in nanomedicine, wearable health devices, and predictive analytics, we must ensure that these innovations extend to every corner of the globe. Health is an inalienable fundamental human right; thus, denying access based on geography or socioeconomic status perpetuates injustice. The World Health Organization (2021) posited that health equity refers to the absence of unfair, avoidable, or remediable differences between populations. Further, the WHO believes that health equity is achieved when everyone can attain their full potential for optimal health and well-being.

As the world embarks on this journey, there is a need to recognize that GHCE is not some distant dream but indeed an urgent necessity. By embracing the convergence of nanotechnology, AI, and QC, we can elevate healthcare to unprecedented levels of equity, compassion, and global well-being.

8.3.2 Nanotechnology: Enabling Precision Personalized Medicine

Operating at atomic and molecular scales, nanotechnology (asyanadhikary, 2015) and nanomedicine have emerged as potent transformative forces in medicine. Their applications extend far beyond traditional macro-level solutions, thereby enabling us to elucidate the intricacies of cellular, organellar, and molecular resolution processes with the aim of manipulating them to optimize health. In this section, we explore how nanotechnology enables precision personalized medicine, a paradigm that tailors treatments to individual patients based on their distinctive characteristics. Targeted drug delivery, diagnostics, and personalized therapies will be explored to highlight breakthroughs in nanomedicine. Further, case studies are presented that exemplify successful applications of nanomedicine in diverse healthcare settings. In broadening the discussion on advances in nanotechnology, data intensive research reveals that AI has enabled significant progress in this domain, including the development of new materials and devices (Agrawal et al., 2024). Improved characterization and analysis techniques, along with more efficient manufacturing processes, have been achieved through the integration of AI.

8.3.3 Targeted Drug Delivery

8.3.3.1 Nanoparticles as Precision Couriers

In the complex realm of medical science, where molecules dance and cells whisper secrets, nanoparticles emerge as remarkable players. Nanoparticles, ranging from 1 to 100 nanometers (nm) in size, are meticulously engineered to serve as targeted drug carriers (a role akin to precision couriers) that navigate the bloodstream (Figure 8.2). At their core, these diminutive structures are comprised of nanometric architectures that wield the power of scale to manipulate microscopic matter with great finesse. We might imagine them as molecular vessels, each designed to ferry therapeutic agents to specific destinations within the body with the mission of delivering healing payloads with pinpoint accuracy. Their advantages include high precision, enhanced efficacy, and bioavailability. In support of this articulation, Agrawal et al. (2024) posited that nanoparticles and nanocarriers enhance the efficacy of drug delivery, while AI algorithms target specific tissues and optimize drug release profiles.

8.3.3.1.1 Precision

Nanoparticles can be endowed with the uncanny ability to play favorites when they are decorated with molecules that have the capacity to selectively target specific tissues, cells, and even organelles, akin to microscopic matchmakers. When a disease lurks, they home in on their targets and bypass healthy tissues and cells. This level of precision minimizes collateral damage, which is a critical feat in the delicate realm of healing.

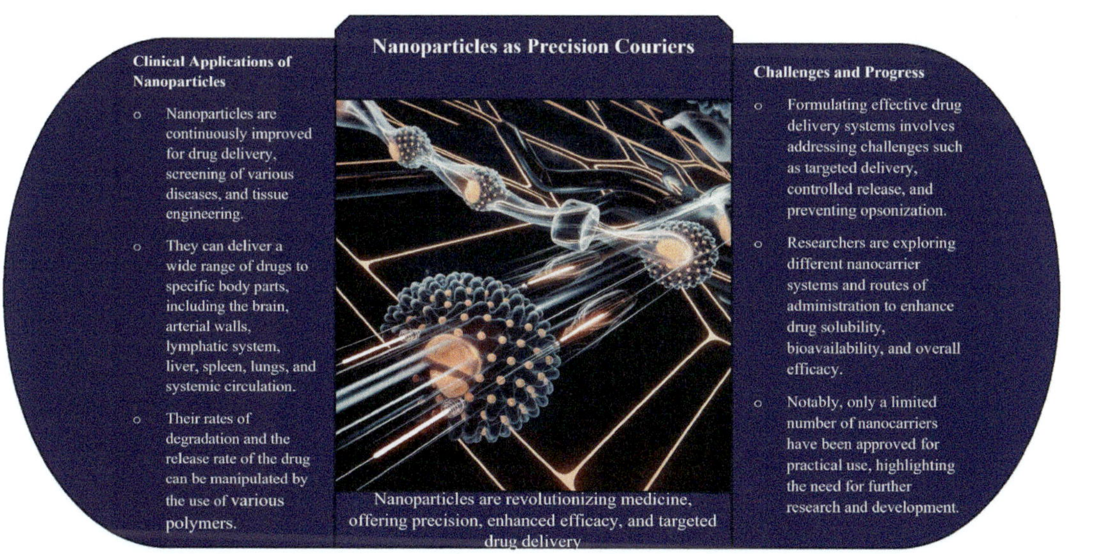

FIGURE 8.2 Nanoparticles as precision couriers. (Image created by the author.)

8.3.3.1.2 *Enhanced Efficacy*

Let us picture this—a nanoparticle equipped with a drug payload navigates through the bloodstream. It weaves past red blood cells, glides along capillaries, and reaches its designated target site—a tumor, perhaps, where it unloads its cargo. By precisely delivering drugs to disease sites, nanoparticles significantly amplify treatment outcomes; thus, the battles against illness gain an edge. Alternatively, exquisitely engineered nanoparticles such as gold nanoshells (Dheyab et al., 2023; Liang et al., 2014; Liu et al., 2008; Liu et al., 2018; Park et al., 2015) or Superpraramagnetic Iron Oxide Nanoparticles (SPIONs) (Ansari et al., 2024a; Ansari et al., 2024b; Carlton et al., 2024; Ferreira-Filho et al., 2024; Molaei, 2024) may be precisely targeted to cancerous tumors or individual cancer cells. Once they have reached and attached to their targets, they can be induced to selectively eradicate them via hyperthermia, where near-infrared laser light (to activate gold nanoshells) or a suitable magnetic field (to activate SPIONs) is employed to heat them to the point where cancer cells do not survive. Fortunately, healthy cells and tissues in proximity are unaffected. This is because the thermal threshold for cancer cells/tumors is ~42°C due to their limited cooling capacity, while healthy cells can withstand temperatures of up to 48°C or higher (Boehm and Chen, 2009; Zhu, 2006).

8.3.3.1.3 *Bioavailability*

Drug encapsulation is the secret weapon of nanocarriers, which cloak drugs and shield them from premature degradation. Imagine fragile molecules wrapped within a protective shell—a nano-sized suit of armor. This shield ensures optimal bioavailability by allowing drugs to linger longer to apply their healing influence where required. In the healthcare domain, nanoparticles play a harmonious tune—a delicate balance of precision, efficacy, and protection. As we peer into their world, we glimpse a future where healing becomes not just a science, but an art; a well-choreographed dance of particles orchestrated for our well-being.

8.4 CASE STUDY 1: LIPOSOMAL DOXORUBICIN

A nanoparticle-based liposomal doxorubicin formulation has emerged as a powerful tool in the fight against cancer. Its unique design (hydrophilic core/lipophilic bilayer shell) (Figure 8.3) enables precisely targeted drug delivery to specific types of cancer cells, or even organelles (e.g., mitochondria

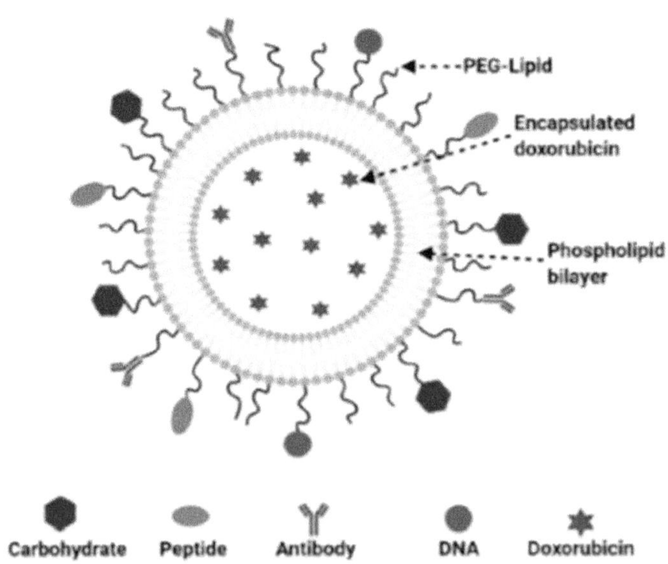

FIGURE 8.3 Examples of liposomal doxorubicin surface functionalization using various targeting ligands. (Reproduced from Makwana et al. (2021) with permission.)

and lysosomes), with minimal side effects and maximized therapeutic impacts (Makwana et al., 2021). Doxorubicin in Liposome (DOXIL®) required more than 10 years to develop due to various non-trivial challenges related to the stable encapsulation and retention of doxorubicin and the reduction of particle dimensions to <Ø100–120 nm. Doxorubicin release profiles are being further optimized using recently developed techniques such as automated nanoparticle exclusion chromatography (nPEC), which employs a monolithic polyvinylpyrrolidone coated silica column to distinguish the encapsulated liposomal drug from what is released. Further, the impacts of pH, temperature, and other parameters may be evaluated for their effects on the drug release rate (Niyonshuti et al., 2024).

8.4.1 Mechanism

At the heart of liposomal doxorubicin lies the liposome—a tiny lipid-based vesicle. We can visualize this as a microscopic bubble, its wall comprised of the same material as our cell membranes with its surface studded with receptors. Within this diminutive bubble, the potent chemotherapy drug (doxorubicin) finds refuge. The liposome serves as a protective nanocarrier, which shields doxorubicin from premature degradation, while ensuring its safe passage through the bloodstream.

But here is the magic: liposomes can be induced to have strong affinities for tumor cells when they are decorated with ligands. Consequently, they can recognize the irregularities (the overexpressed receptors) on the surfaces of cancer cells. Akin to precision couriers, they home in on these sites. Upon arrival, they may release their doxorubicin cargoes directly into the heart of the tumor. This bull's-eye hit spares healthy tissues/cells while delivering a potent blow to cancer cells. This translates to three benefits, namely, reduced cardiotoxicity, enhanced drug accumulation in tumors, and improved patient tolerance.

8.4.2 Reduced Cardiotoxicity

Since conventionally administered doxorubicin can potentially harm the heart (Sheibani et al., 2022), cardiotoxicity (a dreaded side effect) limits its use. However, liposomal doxorubicin changes the game as it can bypass the heart to selectively target tumors, thereby significantly reducing (and perhaps in the future, completely negating) cardiac damage. Patients can breathe easier knowing that their hearts are being spared (Leonard et al., 2009).

8.4.3 Enhanced Drug Accumulation in Tumors

The journeys of liposomes are strategic as they can be molecularly "programmed" to engage in passive and/or active targeting (Torchilin, 2010); thus, they can linger in proximity to tumors to cumulatively increase drug payloads. This prolonged residence ensures that doxorubicin levels within the tumor increase significantly. Cancer cells feel the heat, and their growth is arrested by this stealthy invasion.

8.4.4 Improved Patient Tolerance

Chemotherapy takes its toll in the form of nausea, fatigue, hair loss, etc.; however, liposomal doxorubicin eases these burdens. Patients experience fewer side effects (Ansari et al., 2017; Moles et al., 2023), and their quality of life is preserved. It is indeed a delicate balance: fighting cancer without sacrificing fundamental well-being.

8.5 DIAGNOSTICS: NANOSENSORS AND EARLY DISEASE DETECTION

8.5.1 Nanosensors: Detecting Disease at the Molecular Level

Nanosensors are marvels of miniaturization. These nanoscale devices—smaller than a grain of sand—possess an extraordinary talent: detecting specific molecules or biomarkers (Del Bonis-O'Donnell

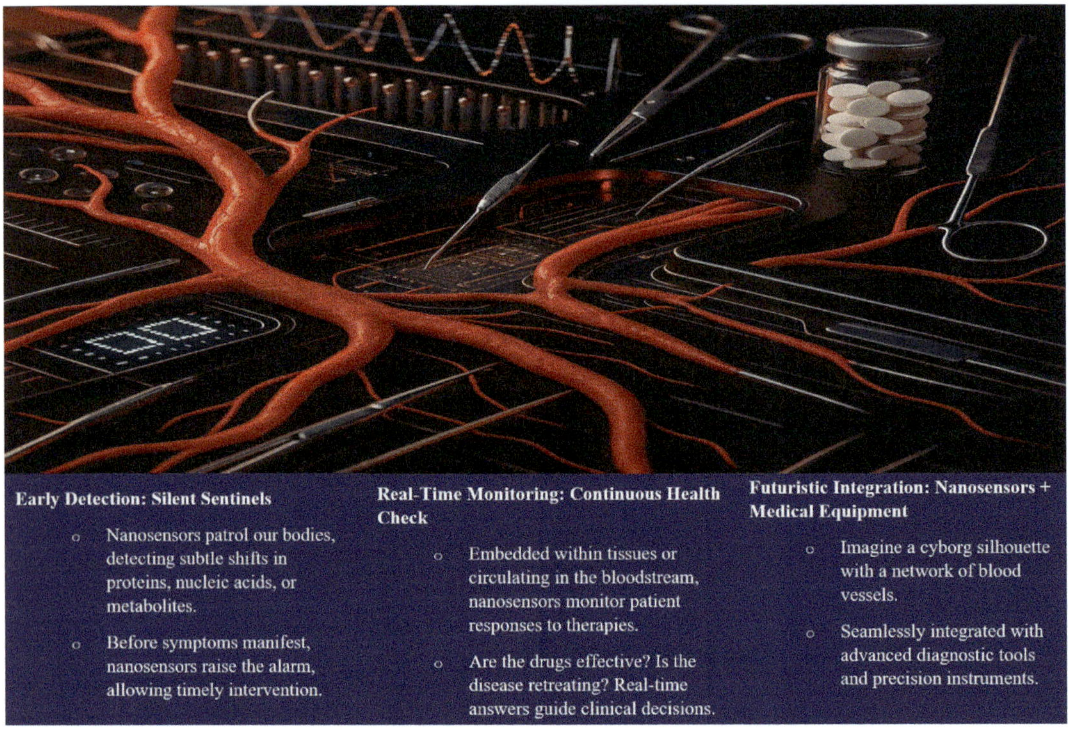

FIGURE 8.4 Nanosensors: Guardians of health. (Image created by the author.).

et al., 2021; Deng et al., 2021; Hang et al., 2021; Jin et al., 2022; Velusamy et al., 2023), with two important applications, namely, early disease detection and real-time monitoring (Figure 8.4).

8.5.2 Early Disease Detection

Nanosensors serve as silent sentinels that patrol our bodies, seeking whispers of trouble. Disease markers, including subtle shifts in proteins, nucleic acids, or metabolites, do not escape their notice (Guo et al., 2018; Huang et al., 2022; Huber et al., 2015; Leong et al., 2022; Tang et al., 2022; Yaari et al., 2021). Consequently, prior to symptoms becoming manifest, nanosensors can raise the alarm.

8.5.3 Real-Time Monitoring

Imagine continuous health checks using nanosensors embedded within tissues or circulating through the bloodstream to monitor the responses of patients to therapies. Are the drugs working? Is the disease retreating? The regular real-time quantification of biomolecular parameters will be vital toward guiding clinical decisions (Bano et al., 2023; Cheng et al., 2023; Kausar et al., 2021; Zhang et al., 2023). According to the American Hospital Association (2023), AI plays a promising role in point of patient care clinical decision support, as algorithms can analyze vast quantities of patient data. It outperformed traditional tools such as the Modified Early Warning Score (MEWS) for the assessment of clinical deterioration risks. Secinaro et al. (2021) asserted that AI analyzed patient data (e.g., electronic health records and medical images) for accurate diagnoses, while diagnostic algorithms aided in early disease detection. Further, the American Hospital Association (2023) suggested that AI's most significant application in diagnosis was medical imaging (e.g., radiologists use AI to augment clinical decisions related to lung nodule detection via CT scans). This was supported by Kumar et al. (2023), who asserted that AI leveraged medical datasets (e.g., ultrasound, MRI, mammography, CT scans) for precise disease diagnosis and feature extraction, while classification processes played a decisive role in disease predictions.

8.6 CASE STUDY 2: GOLD NANOPARTICLES FOR CANCER ANTIGEN DETECTION

Picture gold nanoparticles—tiny spheres, shimmering like sun-kissed dewdrops. Now, coat them with antibodies—molecular detectives trained to recognize specific cancer antigens. When these functionalized gold nanoparticles encounter their targets—say, rogue proteins indicative of early-stage cancer—they bind to them. The result is a color change that is visible even to the naked eye. This is diagnostics at its finest—a molecular ballet revealing hidden truths (Fan et al., 2020; Liu and Lämmerhofer, 2019; Lorenzana-Vázquez et al., 2023; Singh et al., 2018b).

8.6.1 Personalized Therapies: Nanorobots

8.6.1.1 Nanorobots: Miniature Surgeons

Nanorobots are envisaged as microscale wonders—our miniaturized surgeons. Programmed for precision, they navigate the intricate highways of our bodies. Their applications will be felt in two domains, namely, drug delivery and tissue repair (Figure 8.5).

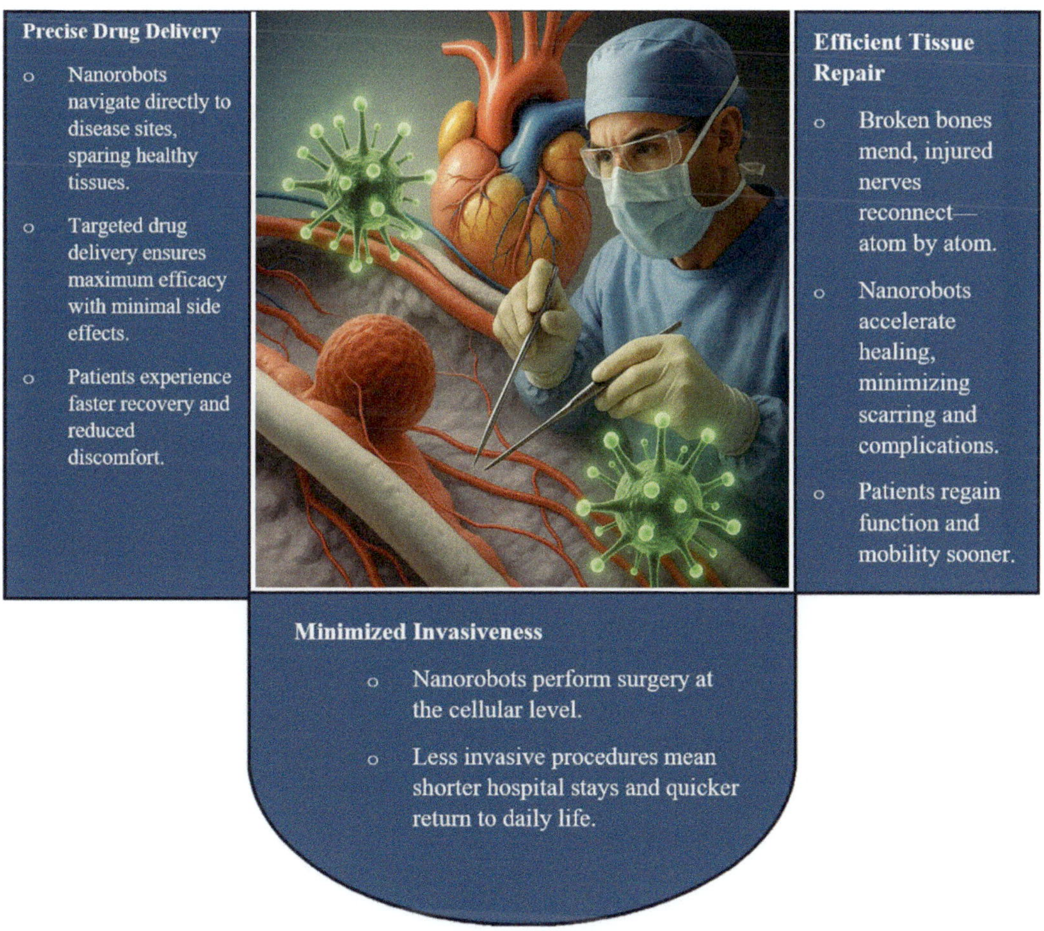

FIGURE 8.5 Nanorobots: Patient benefits. (Image created by the author.).

Drug Delivery: Imagine teams of nanorobots coursing through blood vessels, with the programmed mission of targeted drug delivery. They swim, crawl, or glide to seek out disease sites. When they arrive at their targets, be they tumors or inflamed tissues, they precisely release their therapeutic payloads. This is surgery at the cellular level, which spares neighboring healthy cells and tissues.

Tissue Repair: Damaged tissues require repair and nanorobots may heed the call (Abaszadeh et al., 2023; Altyar et al., 2023; Liu et al., 2023; Yan et al., 2024) as they will possess capacities to regenerate broken bones (Cao et al., 2022; Perán et al., 2013), reconnect injured nerves (Lou and Jonckheere, 2007; Leary, 2010; John et al., 2015; Perán et al., 2013), rebuild muscle tissues (Colapicchioni et al., 2022), etc. by way of what is in essence atom by atom healing.

AI-driven nanorobots will perform precise tasks at the nanoscale, such as drug delivery, tissue repair, or environmental monitoring. Galand (2022) proposed that nanotechnology may be leveraged for ultraprecise disease treatment, while acknowledging that although nanobots show promise as diagnostic tools, their primary application thus far has been as drug delivery agents. Since misdiagnosis is a significant issue that impacts millions of patients annually, nanotechnologies may advantageously offer less expensive and more convenient diagnostic strategies. Smart pills equipped with nanoscale sensors can detect diseases long before symptoms become apparent to patients (e.g., FDA-approved PillCam monitors the gastrointestinal tract to detect Crohn's disease, internal bleeding, or polyps). Further downstream, we may well see the emergence of passive semi-autonomous microbots (e.g., Gastrointestinal Micro Scanning Device (GMSD)) to scan the entire GIT at micron scale resolution. These systems might evolve as fully autonomous nanorobots (e.g., Vascular Cartographic Scanning Nanodevice (VCSN)), which would scan the entire human vascular system (down to capillary level) and lymphatic system to precisely detect any anomalies that may be precursors to disease (Boehm, 2013).

8.7 CASE STUDY 3: DNA-BASED NANOROBOTS IN CANCER THERAPY

The merging of nanotechnology and precision medicine holds immense promise. Our nanosurgeons can don DNA cloaks with strands of genetic code that guide their actions (Aye and Sato, 2022; Johnston and Caruso, 2008; Li et al., 2018; Ma et al., 2019; Patil et al., 2015). We might imagine them as tiny locksmiths that unlock drug compartments only in the presence of specific cancer-related signals. They infiltrate tumors and release chemotherapies precisely where required (Peng et al., 2018; Singh et al., 2018a). As we continue to unravel the mysteries of cellular processes, nanoscale innovations will revolutionize healthcare. By leveraging nanoparticles, nanosensors, and nanorobots, we inch closer to highly personalized treatments that optimize efficacy while minimizing harm. The future of medicine lies in these diminutive marvels, where science meets precision.

8.7.1 Artificial Intelligence in Healthcare: From Diagnostics to Decision-Making

AI has emerged as a transformative force in healthcare, which has already revolutionized disease detection, risk assessment, and treatment planning. In this section, we delve into the pivotal roles of AI algorithms and their impacts on patient care. From medical imaging analysis to drug discovery, AI-powered tools are reshaping healthcare delivery. However, alongside these advancements, critical challenges must be addressed as they relate to data privacy, bias, and ethical considerations. Researchers have developed AI models to predict material properties at nanometric scales. These models facilitate the design of novel materials with specific properties, such as strength, conductivity, or reactivity. According to Agrawal et al. (2024), understanding the environmental impacts of nanotechnology is essential, where AI aids in the modeling of exposure and ecological pathways.

8.8 DISEASE DETECTION AND DIAGNOSIS

8.8.1 Artificial Intelligence Algorithms for Precise Diagnoses

8.8.1.1 Medical Imaging Analysis

Within the dim glow of X-rays, the intricate dance of MRIs, and the cross-sectional mysteries of CT scans, AI algorithms find their purpose. Deep learning models—like tireless detectives—scrutinize these complex medical images. They can detect subtle abnormalities—the faint shadows and irregular contours that may often elude human eyes. Through the synergistic fusion of human expertise and machine precision, radiologists armed with AI insights are making increasingly accurate diagnoses (Hosny et al., 2018; Le et al., 2019; Lee et al., 2024b; Mikhael et al, 2023).

8.8.1.2 Early Detection

Picture an AI sentinel—an invisible watcher—analyzing patient data. Tasked with the mission of identifying whispers of trouble, it sifts through electronic health records, genetic profiles, and lifestyle details. Early signs of diseases emerge—the cryptic markers, the statistical tremors. For instance, within the quiet chambers of a CT scan, AI can spot lung nodules—tiny seeds of potential danger (Binczyk et al., 2021; Tunali et al., 2021). Armed with this knowledge, clinicians can swiftly intervene to alter the course of fate.

8.8.2 Risk Assessment and Personalized Treatment

8.8.2.1 Tailoring Treatment Paths

8.8.2.1.1 Predictive Medicine

The crystal ball of AI does not predict lottery numbers; it foresees disease progression. Patient histories unfold—a multiplicity of symptoms, treatments, and outcomes. Genetic codes divulge secrets and strands of DNA reveal vulnerabilities. Lifestyle choices—the echoes of exercise, nutrition, and stress—paint the canvas. AI synthesizes these threads to weave personalized narratives. It assesses individual risk factors—the ticking clocks, the hidden cliffs. Then, it makes recommendations: "This path, not that." Predictive medicine—a compass for clinicians—guides them toward tailored interventions for human well-being.

8.8.2.1.2 Drug Discovery

In the labyrinthine sphere of drug development, AI pulls on its digital lab coat. It pores over vast datasets—the chemical symphonies, the molecular ballets. Potential drug candidates emerge—like undiscovered constellations. AI predicts their efficacy—their dance with disease targets. It optimizes treatment regimens—dosing schedules, combinations, side effect profiles. The result? Accelerated drug discovery—a sprint toward healing (Moingeon, 2024; Paul et al., 2021). In this dance of algorithms and stethoscopes, healthcare transforms. AI becomes a silent partner to clinicians and physicians—a whisperer in the clinic, a nudge in the laboratory. As health professionals navigate this brave new world, let compassion guide their steps. For within the code and pixels lies a promise: precision, empathy, and a healthier tomorrow.

8.9 CHALLENGES AND CONSIDERATIONS FOR ARTIFICIAL INTELLIGENCE IN HEALTHCARE

8.9.1 Data Privacy and Bias

8.9.1.1 Data Security

AI thrives on data—the lifeblood of its algorithms, wherein patient records, medical histories, and diagnostic images all contribute to its knowledge base. However, within this treasure trove lies a challenge: privacy. Ensuring that patient data remains confidential is paramount. Compliance

TABLE 8.1
Key Points Related to Data Security and Bias Mitigation in the Context of AI and Healthcare

Aspect	Data Security	Bias Mitigation
Challenge	Privacy: Ensuring that patient data remains confidential is paramount	Biases: AI models may carry biases from training data, favor certain groups, and discriminate against others
Regulations	Compliance with regulations (e.g., HIPAA in the United States) is crucial	Efforts to reduce bias are essential. Professionals recalibrate and fine-tune algorithms
Secure Storage	Secure data storage (encrypted and guarded against breaches) is the professionals' duty	Fairness: Every patient deserves an unbiased diagnosis. Algorithms should serve humanity equitably

Source: Table created by the author.

with regulations (such as HIPAA in the United States) becomes a fortress. Secure data storage—encrypted, guarded against breaches—becomes the professionals' duty. For every byte of data, professionals stand guard, protecting not just data, but the trust of patients.

8.9.1.2 Bias Mitigation

AI models, akin to impressionable apprentices, learn from their training data. But what if that data serves as a vector for viral biases—those subtle hints of prejudice that can seep into algorithms? They favor certain groups while discriminating against others. Thus, efforts to reduce bias become essential through which professionals recalibrate, fine-tune, and scrub away prejudice; fairness becomes their lodestar. For within the realm of healthcare, every patient deserves an unbiased diagnosis—a level playing field where algorithms serve humanity, not perpetuate inequities (Gichoya et al., 2023; Leung et al., 2023; Mittermaier et al., 2023; Nazer et al., 2023) (Table 8.1).

8.9.2 Ethical Dilemmas

8.9.2.1 Transparency

Envisage an AI decision—a recommendation for treatment, a prediction of disease progression. Clinicians, like curious detectives, seek understanding. How did AI arrive at this conclusion? Transparency becomes the mandate. Data scientists and bioinformacists must lift the veil to reveal the inner workings—the layers of neural networks, the weights and biases (Hah and Goldin, 2022; Jabbour et al., 2023; Komorowski et al., 2018). Although clinicians agree, rest assured, they need not blindly follow; they comprehend. In transparency, trust blooms, which must be the overarching aim at the end of the day.

8.9.2.2 Human Oversight of AI

Brilliant as AI is as a sophisticated tool and assistant, it is not an infallible oracle. It aids decision-making; however, reasoned human expertise will remain irreplaceable (Schwendicke et al., 2020; Steyvers and Kumar, 2024; Vearrier et al., 2022). We can envisage a delicate surgery, where a surgeon's hands or the proxy instrumentation of a surgical robot (e.g., da Vinci robotic surgical system) is guided by AI; yet her/his judgment, experience, intuition, and empathy will ultimately steer the surgical course (Hashimoto et al., 2018; Moglia et al., 2021; Wilson, 2014). Balancing automation with hard won human clinical wisdom becomes the tightrope walk. Professionals may embrace AI, albeit they will always be mindful of the healer's touch—the warmth of human hands. In the midst of this complex interplay of algorithms and ethics, healthcare evolves. Professionals navigate the labyrinth, guided by compassion, science, and a commitment to healing. For every line of code and neural connection, it must be remembered that beyond the data points are lives—fragile, resilient, and waiting for answers. AI's impact on healthcare is/will be profound; however, responsible implementation is key. By harnessing the capabilities of AI and robotics while addressing challenges, medical professionals can achieve better patient outcomes and transform the future of medicine.

8.10 GLOBAL EQUITY AND ACCESSIBILITY: NANOTECH/AI SYNERGIES IN HEALTHCARE

The convergence of nanotechnology, nanomedicine, and AI holds the promise of democratizing healthcare. By overcoming geographical barriers and reaching underserved populations, this synergy aims to achieve equitable access to advanced medical technologies. In this section, we delve into how nanotech/nanomed/AI applications bridge gaps, enhance global health equity, and transform healthcare delivery (El-Warrak et al., 2023; Gholap et al., 2024; Weerarathna et al., 2023).

8.10.1 Nanotech/AI Synergies for Equitable Access

8.10.1.1 Remote Diagnostics

Remote diagnostics harness the power of AI algorithms to analyze patient data from distant locations. Imagine a digital bridge that connects patients and experts across vast geographical divides. Its current applications are manifest as telemedicine and point-of-care devices.

Telemedicine**—**Picture patients in a remote village, many miles away from specialized medical centers, where, through video calls, they consult with specialists. AI, serving as a silent companion, assists in diagnosing conditions. It deciphers symptoms, interprets test results, and recommends treatments. Consequently, the village becomes a virtual clinic—a hub of healing (Bellini et al., 2022; Liu et al., 2022; Sahashi and Sugimura, 2021; Teo and Ting, 2023).

Point-of-Care Devices**—**Now envision portable devices—the size of the palm of one's hand—embedded with nanosensors. These tiny sentinels detect biomarkers, monitor vital signs, and whisper health secrets. Subsequently, AI steps in to interpret real-time data. Is it an infection? A cardiac anomaly? Immediate feedback guides decisions; thus, patients, even in far flung corners of the globe, receive timely insights toward therapeutic interventions (Angehrn et al., 2020; Gasparin et al., 2022; Manickam et al., 2022; Osterwalder et al., 2023).

8.10.1.2 Virtual Telemedical Consultations

Telemedicine is a digital lifeline that leverages AI-powered chatbots and virtual consultations. It bridges gaps, erases miles, and shrinks waiting rooms. Its benefits are twofold, namely, geographical reach and reduced travel burdens (Ateriya et al., 2018; Campbell et al., 2023; Greenhalgh et al., 2016; Stephens and Greenberg, 2022).

8.10.1.2.1 Geographical Reach

Patients in rural villages, underserved regions, or areas with limited medical infrastructures gain access to expert advice. The village elder with a smartphone consults a neurologist, while a farmer with a tablet discusses arthritis; thus, geographical barriers crumble (Bhatia and Sharma, 2006; Burke, 2015; Kapur et al., 2024; Sreelatha and Ramesh, 2016; Tchero et al., 2018).

8.10.1.2.2 Reduced Travel Burden

Imagine an elderly couple—their joints creaking, their eyes weary—who can now avoid arduous journeys to urban centers; instead, they tap into virtual clinics. AI chatbots greet them, listen to their symptoms, and guide the next steps. The burden of travel lightens, as timely care arrives at their doorstep (Bator et al., 2015; Hahn et al., 2023; Mars, 2013; Webster et al., 2021).

8.10.1.3 AI-Powered Chatbots

AI chatbots don the white coats of knowledge while residing in apps, websites, and messaging platforms. Their mission includes the provision of medical information, answering health-related queries, and offering basic health advice. AI-powered chatbots have three primary advantages, namely, 24/7 availability, language adaptability, and health education. (Ghorashi et al., 2023; Kurniawan et al., 2024; Lee et al., 2024a; Lukkahatai and Han, 2024; Maniaci et al., 2024).

8.10.1.3.1 Availability 24/7

Since health concerns do not respect office hours, chatbots operate around the clock. The insomniac, worried parent, and curious teenager will all find answers. "Is this rash serious?" "What's the best diet for diabetes?" Chatbots will respond tirelessly.

8.10.1.3.2 Language Adaptability

The global village speaks myriad tongues. Multilingual chatbots—fluent in English, Spanish, Mandarin, and more—cater to diverse populations. The refugee, the traveler, the expatriate—they all seek guidance. Language barriers are dissolved by utilizing chatbots with language adaptability.

8.10.1.4 Education

Prevention is the silent hero. Chatbots disseminate wisdom regarding vaccines, hygiene, nutrition, and more. They nudge smokers toward quitting, joggers toward stretching; thus, community health blooms, one informed message at a time. In the world of bits and bytes, healthcare transcends borders. Nanotech-AI synergies become the compass—a guide toward equity, compassion, and well-being. For every chatbot response, every nanosensor whisper, the global population inches closer to a world where optimal health abounds.

8.11 CASE STUDY: EQUITABLE ACCESS IN ACTION

8.11.1 India's ASHA Workers

In India, Accredited Social Health Activists (ASHAs) play a critical role as community health workers. Their tireless efforts reach the hearts of rural villages, where healthcare infrastructures are often scarce.

8.11.1.1 Role of ASHAs

ASHAs wield mobile apps armed with AI chatbots. These digital companions serve a dual purpose, namely, health education and connecting communities.

8.11.1.1.1 Health Education

ASHAs educate villagers about health and hygiene, as well as disseminating information on preventive health measures, nutrition, maternal care, and child health.

8.11.1.1.2 Connecting Communities

ASHAs bridge the gap between villagers and health services, as they use AI chatbots to provide immediate answers to health-related queries. Whether it is a worried mother seeking advice or an elderly person with concerning symptoms, ASHAs connect them to the appropriate resources.

8.11.1.2 ASHA Impacts

ASHAs become the lifelines of health awareness, as they empower villagers with knowledge, dispel myths, and promote well-being. Timely interventions—whether for vaccinations, antenatal care, or disease management—become possible. In the quiet corners of rural India, ASHAs weave a safety net—a web of health literacy and compassion (Mohan and Gupta, 2021).

8.11.2 Telemedicine in Sub-Saharan Africa

Sub-Saharan Africa is grappling with healthcare disparities due to limited infrastructures, vast distances, and resource constraints, which create barriers to timely medical attention.

8.11.2.1 Solution

Telemedicine emerges as a beacon of hope. Powered by AI, telemedicine platforms connect patients with specialists across virtual bridges. Here is how it works.

8.11.2.1.1 Virtual Consultations

Imagine a patient in a remote village, an expanse of savannah, or dense jungle who seeks expert medical advice. Through telemedicine, they can connect with a specialist hundreds of miles away. Further, AI algorithms assist with diagnoses, treatment planning, and follow-up care (Figure 8.6).

FIGURE 8.6 Telemedicine and virtual consultations. (Image created by the author.)

8.11.2.1.2 Nanosensors Aid Diagnostics

Tiny nanosensors embedded in portable devices, like diminutive agents of technology—detect health markers. Is it malaria? Tuberculosis? Nanosensors relay data to AI systems that interpret patterns; thus, diagnoses become beneficial digital symphonies.

8.11.2.1.3 Telemedicine Impact

Telemedicine transcends borders; far-flung communities acquire access to expertise that was previously out of reach. A grandmother with a chronic condition receives personalized guidance. A child with a fever finds reassurance. Sub-Saharan Africa, once distant from medical hubs, now stands connected, which is a testament to technology's healing touch (Mupela et al., 2011).

In these narratives, entire populations witness the fusion of compassion and innovation. ASHAs and telemedicine pioneers become the unsung heroes—bridging gaps, saving lives, and ensuring that healthcare knows no boundaries. Nanotech-AI synergies transcend physical boundaries, which makes high quality healthcare accessible to all. Whether in bustling cities or remote villages, equitable access becomes a reality. As communities increasingly embrace these innovations, they will prioritize the notion of global health equity and work toward the massively distributed propagation and implementation of mature GHCE.

8.12 SYNERGIES AND CHALLENGES: INTEGRATION OF NANOTECHNOLOGY AND AI

The convergence of nanotechnology and AI represents a powerful alliance with the potential to significantly shift paradigms and transform myriad domains. As we explore their synergies, we uncover novel pathways for scientific discovery, medical breakthroughs, and technological advancements. However, this integration also presents many profound challenges that require deep and thoughtful consideration. In this section, we delve into the complementary nature of collaborative research initiatives between nanotechnology, nanomedicine, and AI, as well as the hurdles faced in this interdisciplinary enterprise.

8.12.1 Materials Discovery and Optimization

8.12.1.1 Nanomaterials

Nanotechnology enables the manipulation of materials at atomic or molecular resolution. Envision having the capacity to craft ultraprecise materials and nanodevices with integrated functionality atom by atom. Nanomaterials emerge—tiny building blocks with unique properties. But how do we even begin to design and fabricate them? Here is where AI steps in.

8.12.1.2 AI Algorithms Predict Material Properties

AI algorithms analyze vast datasets—material compositions, geometric structures, and performance metrics—as well as learn patterns and decipher correlations. When encountering a new material candidate, they predict its behavior. Will it be strong, conductive, or biocompatible? With unprecedented speed and accuracy, AI will guide us toward novel nanomaterials that are tailored for specific functionalities.

8.12.2 Precision Medicine

8.12.2.1 Nanosensors and Diagnostics

Imagine nanoscale sensors—whispers of technology—embedded within our bodies. They detect biomarkers—the subtle cues of health and disease. Is it a cancer marker? A cardiac enzyme? Early detection becomes possible. But how do we orchestrate and interpret this symphony of data?

8.12.2.2 AI Interprets Sensor Data

AI algorithms analyze sensor readings by deciphering the spikes, the dips, and the language of molecules. Are these false alarms, or critical signals that translate to the emergence of accurate diagnoses, where precision medicine takes shape? Consequently, treatment recommendations align with individual needs, wherein the patient armed with knowledge confidently faces the path ahead.

8.12.2.3 Drug Delivery

Nanoparticles—nanoscale messengers—navigate the bloodstream to carry and deliver drugs with high accuracy, akin to guided missiles. Cancer cells and inflamed tissues receive targeted payloads, while AI algorithms optimize drug release kinetics. Side effects diminish or are completely negated, and therapeutic efficacy soars.

8.12.3 Advanced Imaging and Characterization

8.12.3.1 Atomic Force Microscopy

In the imaging realm, nanotechnology can enhance Atomic Force Microscopy (AFM) techniques, providing high-resolution views of biological structures—the delicate folds of proteins, the dense convolutions of mitochondria, the intricate dances of cells. AFM probes, like gentle, unseeable fingertips, trace atomic contours. But how do we extract meaning from these images? (Table 8.2).

8.12.3.2 AI Assists in Image Analysis

AI agents have the capacity to scrutinize AFM scans to reveal subtle features—the hidden irregularities, the binding sites. Is it a virus? An aggregate of proteins? Subsequently, diagnostics improve, clinicians gain insights, and disease mechanisms unravel. AFM becomes an exquisite lens that reveals life's secrets (Table 8.2).

TABLE 8.2
Features of Advanced Imaging and Characterization

Feature	Description	Real-World Applications
Atomic Force Microscopy (AFM)	High-resolution imaging technique that enhances the elucidation of biological structures (e.g., proteins and cells). AFM probes trace contours with precision	Used in materials science, biology, and medicine to study cell mechanics, morphology, and the surface characteristics of materials
AI Assists in Image Analysis	AI algorithms analyze AFM scans to reveal subtle features such as hidden irregularities and binding sites	Improves diagnostic accuracy for diseases, supports research in molecular biology, and aids in the development of new pharmaceuticals
Simulations	AI-driven simulations predict how nanodevices will respond to various conditions like stress, temperature, or chemical interactions	Guides the design of nanomaterials and devices for use in electronics, aerospace, and medical implants

Source: Table created by the author.

8.12.3.3 Simulations

AI-driven simulations predict material behaviors. Nanodevices—tiny complex constructs—come to life in silico. How will they respond to stress, temperature, or chemical interactions? These simulations guide design choices, where efficiency, safety, and performance converge. In this interplay between atoms and algorithms, nanotechnology, nanomedicine, and AI become partners that unlock new materials, heal bodies, and peer deep into the invisible. As researchers navigate this frontier, they should let curiosity be their compass—a journey toward unprecedented innovation, compassion, and a healthier world (Table 8.2).

8.13 COLLABORATIVE RESEARCH INITIATIVES

8.13.1 Interdisciplinary Teams

Successful nanotechnology-nanomedicine-AI collaborations will involve experts from diverse disciplines—materials science (Bensaude-Vincent, 2001; Schummer, 2004), computer science, biology, physics, medicine, electronics, engineering, mechanical design, etc. These teams foster creativity and the diverse cross-pollination of ideas.

8.13.1.1 Data Sharing

Open-access datasets facilitate collaborative research. Initiatives like The Materials Project (2024) and The Nanomaterial Registry (Guzan et al., 2013) encourage data sharing and serve to accelerate discoveries.

8.14 ETHICAL CONSIDERATIONS AND RESPONSIBLE INNOVATION IN NANOMEDICINE AND AI

With continued advances in nanotechnology, nanomedicine, and AI (and likely AGI and ASI in the near future), ethical considerations become paramount. Responsible innovation ensures that these transformative technologies will benefit humanity and the planet while minimizing (or ideally, completely negating) their risks. In this section, we explore some core ethical dilemmas related to nanomedicine and AI with an emphasis on transparency, informed consent, and equitable distribution. Further, guidelines for responsible innovation toward achieving GHCE are proposed.

8.14.1 Ethical Dilemmas in Nanomedicine and AI

8.14.1.1 Transparency and Intelligibility

The challenge is that AI decisions are not always transparent or even intelligible to humans. The solution to this issue is to strive for transparency through the development of interpretable AI models. In essence, there is a need to explain how decisions are arrived at, particularly in the context of critical healthcare.

8.14.1.2 Informed Consent

As relates to informed consent, obtaining such for nanomedical research remains a challenge. One solution to this challenge is to prioritize clear and concise communication with patients. It is paramount to explain to patients the benefits, risks, and potential impacts of nanotechnology-based treatments to ensure that they clearly understand these aspects prior to voluntarily participating in related research.

8.14.1.3 Equitable Distribution

Ensuring equal access to nanomedical technologies (e.g., screening, diagnostics, therapeutics, monitoring) remains a challenge, which beckons the development of policies that prevent disparities.

These policies should encompass and address affordability, geographical barriers, marginalized populations, the promotion of inclusivity in research, and the timely delivery of treatments.

8.15 RESPONSIBLE INNOVATION GUIDELINES

8.15.1 Anticipatory Governance

8.15.1.1 Early Engagement

Early engagement should involve all stakeholders (e.g., researchers, policymakers, patients) in comprehensive discussions as relates to nanotechnology, nanomedicine, AI/AGI/ASI, and QC. Perceived and anticipated challenges should be proactively addressed.

8.15.1.2 Foresight

Foresight techniques should be employed to identify the potential risks and benefits of nanotechnology, nanomedical, and AI-driven nanomaterials, nanodevices, and nanosystem platforms. Further, the long-term downstream consequences of these innovations for humanity, the environment, and the planet will need to be considered.

8.15.2 Stewardship and Trust

8.15.2.1 Cultivate Trust

Technology-trust should be fostered across public and private sectors, where transparency, accountability, and responsible practices build confidence.

8.15.2.2 Data Privacy

Safeguard personal cognitive data and other health-related information, as well as respect privacy rights.

8.16 GLOBAL STANDARDS

8.16.1 International Cooperation

It is essential for individuals to collaborate globally toward the establishment of regulatory standards for nanotechnology, nanomedicine, and AI/AGI/ASI while harmonizing regulations and governance frameworks. The Organisation for Economic Co-operation and Development (OECD) working paper, *Responsible innovation in neurotechnology enterprises*, appears to indicate that responsible innovation in nanotechnology and AI requires a holistic approach (Garden et al., 2019). By addressing ethical dilemmas, promoting transparency, and ensuring equitable access, we can harness these technologies for the betterment of global health. There is a need to navigate this frontier with wisdom, compassion, and commitment toward the emergence of envisaged significantly paradigm shifting GHCE.

8.17 CONCLUSION: NANOTECHNOLOGY, NANOMEDICINE, AND AI IN HEALTHCARE

In the dynamic landscape of healthcare, synergies between nanotechnology, nanomedicine, and AI emerge as a transformative force. Embedded within the dynamic interactions between science and compassion, where nanoscale wonders meet AI, we uncover the seeds of transformation. Deep within the intricate dance of molecules and algorithms, healthcare transcends its traditional boundaries. As we approach the conclusion of this chapter, let us delve into the heart of these themes.

8.17.1 Personalized Precision Medicine Revolution

At the forefront of this revolution lies personalized precision medicine, a paradigm shift that embraces the notion that each of us is inherently unique. Nanotechnology, with its capacity to manipulate matter at atomic and molecular scales, becomes our ally. Imagine diminutive nanoparticles akin to molecular architects that self-assemble to construct personalized drug delivery systems. These nanocarriers navigate the bloodstream and seek out diseased cells with exquisite precision. Coupled with AI, the power to customize treatments based on genetic profiles is unlocked. No longer must we rely on one-size-fits-all approaches. Alternatively, we customize and craft therapies that harmonize with the biological symphonies of individual patients. Adverse side-effects diminish, efficacy soars, and hopes for optimal health blossom anew.

8.17.2 Early Disease Detection and Prevention

Picture a future where healthcare transcends crisis management. Nanosensors, akin to miniaturized sentinels, continuously patrol our bodies to detect cellular anomalies long before the symptoms of disease manifest. They divulge secrets—subtle shifts in biomarkers, cellular hints of trouble brewing. AI steps in to analyze this data tapestry. Predictive models emerge, replete with insights that guide their forecasts of disease trajectories, where the focus shifts from reactive treatments to proactive prevention. Imagine a world where even the hints of diseases surrender long before they can become manifest—a canvas of health painted with foresight and care.

8.17.3 Data-Driven Decision-Making

In the corridors of hospitals and research labs, data flows like a river. Nanoscale devices generate myriad streams of information—heartbeats, glucose levels, genetic snippets. AI algorithms/agents wade through this torrent, discerning patterns that remain invisible to human eyes, as they guide clinicians, optimize resource allocation, and streamline healthcare workflows. The result translates to improved patient outcomes, cost-effective care, and an opus of efficiency. Humanity stands at the crossroads of data-driven destiny, where bytes and molecules (and in the future, even atoms) converge to heal.

8.17.4 Global Equity and Accessibility

Healthcare knows no borders. Yet disparities persist—geographical chasms that separate the privileged from the marginalized. Here, nanotech-AI synergies become our bridge (Spatharou et al., 2020). Remote diagnostics leap across continents, telemedicine transcends mountains, and AI-powered chatbots prescribe healing words in every language. Whether in rural villages or bustling metropolises, equitable access becomes a reality. The promise of high-quality healthcare for all, like a shared melody, reaches every ear.

8.17.5 Ethical Considerations

As we ascend this technological summit, we must tread carefully. Privacy, like a fragile butterfly, flutters in our grasp. Informed consent—the cornerstone of trust—must remain unwavering, while data security, a fortress against digital storms, demands vigilance. AI-driven decisions, like constellations, must reveal their paths. Bias, that silent specter, weaves itself into algorithms, which must be unraveled. As we embrace the convergence of nanotech-AI, let ethics be our compass, guiding us toward a future where healing knows no prejudice.

In summary, the marriage of nanotechnology, nanomedicine, and AI elevates healthcare to unprecedented levels of equity, efficacy, and scope. As pioneers in this frontier, we remain vigilant, guided by compassion, integrity, innovation, and a commitment to global well-being. Our symphony of healing awaits its crescendo—a harmonious blend of science, empathy, and hope.

REFERENCES

Abaszadeh F, Ashoub MH, Khajouie G, Amiri M. Nanotechnology development in surgical applications: Recent trends and developments. *European Journal of Medical Research*. 2023;28(1):537.

Agrawal R, Tilak P, Devand A, Bhatnagar A, Gupta P et al. Chapter 18: Artificial Intelligence in Nanotechnology. In *Data Science and Big Data Analytics*. Mishra D. (ed.), Springer Nature Singapore Pte Ltd. 2024;239–248.

Altyar AE, El-Sayed A, Abdeen A, Piscopo M, Mousa SA, Najda A, Abdel-Daim MM. Future regenerative medicine developments and their therapeutic applications. *Biomedicine & Pharmacotherapy*. 2023;158:114131.

American Hospital Association. *How AI Is Improving Diagnostics, Decision-Making and Care*. 2023. https://www.aha.org/aha-center-health-innovation-market-scan/2023-05-09-how-ai-improving-diagnostics-decision-making-and-care (accessed June 18, 2024).

Angehrn Z, Haldna L, Zandvliet AS, Gil Berglund E, Zeeuw J, Amzal B et al. Artificial intelligence and machine learning applied at the point of care. *Frontiers in Pharmacology*. 2020;11:759.

Ansari L, Shiehzadeh F, Taherzadeh Z, Nikoofal-Sahlabadi S, Momtazi-Borojeni AA, Sahebkar A, Eslami S. The most prevalent side effects of pegylated liposomal doxorubicin monotherapy in women with metastatic breast cancer: A systematic review of clinical trials. *Cancer Gene Therapy*. 2017;24(5):189–193.

Ansari SR, Mahajan J, Teleki A. Iron oxide nanoparticles for treatment and diagnosis of chronic inflammatory diseases: A systematic review. *Wiley Interdisciplinary Reviews: Nanomedicine and Nanobiotechnology*. 2024a;16(3):e1963.

Ansari SR, Suárez-López YDC, Thersleff T, Häggström L, Ericsson T, Katsaros I et al. Pharmaceutical quality by design approach to develop high-performance nanoparticles for magnetic hyperthermia. *ACS Nano*. 2024b;18(23):15284–15302.

asyanadhikary Nanotechnology in Healthcare: The Tiny Marvels That Transform Medicine. 2015. https://eduhealth-news.com/nanotechnology-in-medicine-tiny-marvels-with-huge-impacts/ (accessed June 18, 2024).

Ateriya N, Saraf A, Meshram VP, Setia P. Telemedicine and virtual consultation: The Indian perspective. *National Medical Journal of India*. 2018;31(4):215–218.

Aye SL, Sato Y. Therapeutic applications of programmable DNA nanostructures. *Micromachines (Basel)*. 2022;13(2):315.

Bano R, Soleja N, Mohsin M. Genetically encoded FRET-based nanosensor for real-time monitoring of A549 exosomes: Early diagnosis of cancer. *Analytical Chemistry*. 2023;95(13):5738–5746.

Bator EX, Gleason JM, Lorenzo AJ, Kanaroglou N, Farhat WA, Bägli DJ, Koyle MA. The burden of attending a pediatric surgical clinic and family preferences toward telemedicine. *Journal of Pediatric Surgery*. 2015;50(10):1776–1782.

Bellini V, Valente M, Gaddi AV, Pelosi P, Bignami E. Artificial intelligence and telemedicine in anesthesia: Potential and problems. *Minerva Anestesiologica*. 2022;88(9):729–734.

Bensaude-Vincent B. The construction of a discipline: Materials science in the United States. *Historical Studies in the Physical and Biological Sciences*. 2001;31(2):223–248.

Bhatia JS, Sharma S. Telemedicine odyssey customised telemedicine solution for rural and remote areas in India. *Studies in Health Technology and Informatics*. 2006;121:22–35.

Binczyk F, Prazuch W, Bozek P, Polanska J. Radiomics and artificial intelligence in lung cancer screening. *Translational Lung Cancer Research*. 2021;10(2):1186–1199.

Boehm FJ. Chapter 1 – Exemplar Nanomedical Vascular Cartographic Scanning Nanodevice. *Nanomedical Device and Systems Design – Challenges, Possibilities, Visions*. Boca Raton, FL, USA: CRC Press, 2013.

Boehm FJ, Chen A. Medical applications of hyperthermia based on magnetic nanoparticles. *Recent Patents on Biomedical Engineering*. 2009;2:110–120. https://www.eurekaselect.com/article/40238

Burke BL Jr, Hall RW. Section on telehealth care. Telemedicine: Pediatric applications. *Pediatrics*. 2015;136(1):e293–308.

Campbell K, Greenfield G, Li E, O'Brien N, Hayhoe B, Beaney T et al. The impact of virtual consultations on the quality of primary care: Systematic review. *Journal of Medical Internet Research*. 2023;25:e48920.

Cao D, Martinez JG, Hara ES, Jager EWH. Biohybrid variable-stiffness soft actuators that self-create bone. *Advanced Materials*. 2022;34(8):e2107345.

Carlton H, Arepally N, Healy S, Sharma A, Ptashnik S, Schickel M et al. Magnetic particle imaging-guided thermal simulations for magnetic particle hyperthermia. *Nanomaterials (Basel)*. 2024;14(12):1059.

Cheng S, Zhang S, Liu R, Zeng H, Yin Y, Zhang M. Potentiometric nanosensor for real-time measurement of hydrogen sulfide in single cell. *Chemical Communications (Cambridge)*. 2023;59(14):1959–1962.

Colapicchioni V, Millozzi F, Parolini O, Palacios D. Nanomedicine, a valuable tool for skeletal muscle disorders: Challenges, promises, and limitations. *Wiley Interdisciplinary Reviews: Nanomedicine and Nanobiotechnology*. 2022;14(3):e1777.

Del Bonis-O'Donnell JT, Mun J, Delevich K, Landry MP. Synthetic nanosensors for imaging neuromodulators. *Journal of Neuroscience Methods*. 2021;363:109326.

Deng J, Zhao S, Liu Y, Liu C, Sun J. Nanosensors for diagnosis of infectious diseases. *ACS Applied Bio Materials*. 2021;4(5):3863–3879.

Dheyab MA, Aziz AA, Khaniabadi PM, Jameel MS, Oladzadabbasabadi N, Rahman AA et al. Gold nanoparticles-based photothermal therapy for breast cancer. *Photodiagnosis and Photodynamic Therapy*. 2023;42:103312.

El-Warrak L, Nunes M, Luna G, Barbosa CE, Lyra A, Argôlo M et al. Towards the future of public health: Roadmapping trends and scenarios in the post-COVID healthcare era. *Healthcare (Basel)*. 2023;11(24):3118.

Fan J, Cheng Y, Sun M. Functionalized gold nanoparticles: Synthesis, properties and biomedical applications. *Chem Record*. 2020;20(12):1474–1504.

Ferreira-Filho VC, Morais B, Vieira BJC, Waerenborgh JC, Carmezim MJ, Tóth CN et al. Influence of SPION surface coating on magnetic properties and theranostic profile. *Molecules*. 2024;29(8):1824.

Galand A. Three Ways Nanotechnology Is Changing the Healthcare Industry. *Forbes*. 2022. https://www.forbes.com/sites/forbestechcouncil/2022/07/20/three-ways-nanotechnology-is-changing-the-health-care-industry/ (accessed June 18, 2024).

Garden H, Winickoff DE, Frahm NM, Pfotenhauer S. Responsible innovation in neurotechnology enterprises. OECD Science, Technology and Industry Working Papers. 2019;(05). https://www.oecd.org/en/publications/2019/10/responsible-innovation-in-neurotechnology-enterprises_2d346c46.html (accessed October 18, 2024).

Gasparin AT, Araujo CIF, Schmitt P, Cardoso MR, Perussolo MC, de Jesus TCS et al. Hilab system, a new point-of-care hematology analyzer supported by the internet of things and artificial intelligence. *Scientific Reports*. 2022;12(1):10409.

Gholap AD, Uddin MJ, Faiyazuddin M, Omri A, Gowri S, Khalid M. Advances in artificial intelligence for drug delivery and development: A comprehensive review. *Computers in Biology and Medicine*. 2024;178:108702.

Ghorashi N, Ismail A, Ghosh P, Sidawy A, Javan R. AI-powered chatbots in medical education: Potential applications and implications. *Cureus*. 2023;15(8):e43271.

Gichoya JW, Thomas K, Celi LA, Safdar N, Banerjee I, Banja JD et al. AI pitfalls and what not to do: Mitigating bias in AI. *British Journal of Radiology*. 2023;96(1150):20230023.

Greenhalgh T, Vijayaraghavan S, Wherton J, Shaw S, Byrne E, Campbell-Richards D et al. Virtual online consultations: Advantages and limitations (VOCAL) study. *BMJ Open*. 2016;6(1):e009388.

Guo R, Yin F, Sun Y, Mi L, Shi L, Tian Z, Li T. Ultrasensitive simultaneous detection of multiplex disease-related nucleic acids using double-enhanced surface-enhanced Raman scattering nanosensors. *ACS Applied Materials & Interfaces*. 2018;10(30):25770–25778.

Guzan KA, Mills KC, Gupta V, Murry D, Scheier CN, Willis DA, Ostraat ML. Integration of data: The Nanomaterial Registry project and data curation. *Computational Science & Discovery*. 2013;6(1):014007.

Hah H, Goldin D. Moving toward AI-assisted decision-making: Observation on clinicians' management of multimedia patient information in synchronous and asynchronous telehealth contexts. *Health Informatics Journal*. 2022;28(1):14604582221077049.

Hahn Z, Hotchkiss J, Atwood C, Smith C, Totten A, Boudreau E et al. Travel burden as a measure of healthcare access and the impact of telehealth within the veterans health administration. *Journal of General Internal Medicine*. 2023;38(Suppl 3):805–813.

Hang X, He S, Dong Z, Minnick G, Rosenbohm J, Chen Z, Yang R, Chang L. Nanosensors for single cell mechanical interrogation. *Biosensors and Bioelectronics*. 2021;179:113086.

Hashimoto DA, Rosman G, Rus D, Meireles OR. Artificial intelligence in surgery: Promises and perils. *Annals of Surgery*. 2018;268(1):70–76.

Hosny A, Parmar C, Quackenbush J, Schwartz LH, Aerts HJWL. Artificial intelligence in radiology. *Nature Reviews Cancer*. 2018;18(8):500–510.

Huang J, Chen X, Jiang Y, Zhang C, He S, Wang H, Pu K. Renal clearable polyfluorophore nanosensors for early diagnosis of cancer and allograft rejection. *Nature Materials*. 2022;21(5):598–607.

Huber F, Lang HP, Zhang J, Rimoldi D, Gerber C. Nanosensors for cancer detection. *Swiss Medical Weekly*. 2015;145:w14092.

Jabbour S, Fouhey D, Shepard S, Valley TS, Kazerooni EA, Banovic N, Wiens J, Sjoding MW. Measuring the impact of AI in the diagnosis of hospitalized patients: A randomized clinical vignette survey study. *JAMA*. 2023;330(23):2275–2284.

Jin C, Wu Z, Molinski JH, Zhou J, Ren Y, Zhang JXJ. Plasmonic nanosensors for point-of-care biomarker detection. *Materials Today Bio*. 2022;14:100263.

John AA, Subramanian AP, Vellayappan MV, Balaji A, Mohandas H, Jaganathan SK. Carbon nanotubes and graphene as emerging candidates in neuroregeneration and neurodrug delivery. *International Journal of Nanomedicine*. 2015;10:4267–77.

Johnston AP, Caruso F. Stabilization of DNA multilayer films through oligonucleotide crosslinking. *Small*. 2008;4(5):612–8.

Kapur N, Sabherwal S, Sharma P, Nayab J, Koh Pei Chen P, Srivastava S, Majumdar A. Assessing the reliability of tele-refraction for real time consultation with a remote optometrist. *PLoS One*. 2024;19(6): e0299491.

Kausar H, Ambrin G, Okla MK, Alamri SA, Soufan WH, Ibrahim EI, Abdel-Maksoud MA, Ahmad A. FRET-based genetically encoded nanosensor for real-time monitoring of the flux of α-tocopherol in living cells. *ACS Omega*. 2021;6(13):9020–9027.

Kaushik S, Suhag D. Chapter 14: Nanotechnology for AI in Healthcare. In *Integrated Nanomaterials and Their Applications*. Suhag D, Thakur A, Thakur P. (eds.), Springer. 2023.

Komorowski M, Celi LA, Badawi O, Gordon AC, Faisal AA. The artificial intelligence clinician learns optimal treatment strategies for sepsis in intensive care. *Nature Medicine*. 2018;24(11):1716–1720.

Kumar Y, Koul A, Singla R, Ijaz MF. Artificial intelligence in disease diagnosis: A systematic literature review, synthesizing framework and future research agenda. *Journal of Ambient Intelligence and Humanized Computing*. 2023;14(7):8459–8486.

Kurniawan MH, Handiyani H, Nuraini T, Hariyati RTS, Sutrisno S. A systematic review of artificial intelligence-powered (AI-powered) chatbot intervention for managing chronic illness. *Annals of Medicine*. 2024;56(1):2302980.

Le EPV, Wang Y, Huang Y, Hickman S, Gilbert FJ. Artificial intelligence in breast imaging. *Clinical Radiology*. 2019;74(5):357–366.

Leary JF. Nanotechnology: What is it and why is small so big? *Canadian Journal of Ophthalmology*. 2010;45(5):449–456.

Lee J, Park J, Han HS. Using ChatGPT for kidney transplantation: Perceived information quality by race and education levels. *Clinical Transplantation*. 2024a;38(7):e15378.

Lee S, Jung JY, Mahatthanatrakul A, Kim JS. Artificial intelligence in spinal imaging and patient care: A review of recent advances. *Neurospine*. 2024b;21(2):474–486.

Leonard RC, Williams S, Tulpule A, Levine AM, Oliveros S. Improving the therapeutic index of anthracycline chemotherapy: Focus on liposomal doxorubicin (Myocet). *Breast*. 2009;18(4):218–24.

Leong SX, Leong YX, Koh CSL, Tan EX, Nguyen LBT, Chen JRT et al. Emerging nanosensor platforms and machine learning strategies toward rapid, point-of-need small-molecule metabolite detection and monitoring. *Chemical Science*. 2022;13(37):11009–11029.

Leung TI, Sagar A, Shroff S, Henry TL. Can AI mitigate bias in writing letters of recommendation? *JMIR Medical Education*. 2023;9:e51494.

Li S, Jiang Q, Liu S, Zhang Y, Tian Y, Song C et al. A DNA nanorobot functions as a cancer therapeutic in response to a molecular trigger in vivo. *Nature Biotechnology*. 2018;36(3):258–264.

Liang Z, Li X, Xie Y, Liu S. 'Smart' gold nanoshells for combined cancer chemotherapy and hyperthermia. *Biomedical Materials*. 2014;9(2):025012.

Liu C, Mi CC, Li BQ. Energy absorption of gold nanoshells in hyperthermia therapy. *IEEE Transactions in NanoBioscience*. 2008;7(3):206–214.

Liu S, Lämmerhofer M. Functionalized gold nanoparticles for sample preparation: A review. *Electrophoresis*. 2019;40(18–19):2438–2461.

Liu X, Jing Y, Xu C, Wang X, Xie X, Zhu Y et al. Medical imaging technology for micro/nanorobots. *Nanomaterials (Basel)*. 2023;13(21):2872.

Liu X, Zhao C, Wang L, Wang G, Lv B, Lv C, Xie G, Wang F. Evaluation of an OCT-AI-based telemedicine platform for retinal disease screening and referral in a primary care setting. *Translational Vision Science and Technology*. 2022;11(3):4.

Liu Y, Crawford BM, Vo-Dinh T. Gold nanoparticles-mediated photothermal therapy and immunotherapy. *Immunotherapy*. 2018;10(13):1175–1188.

Lorenzana-Vázquez G, Pavel I, Meléndez E. Gold nanoparticles functionalized with 2-thiouracil for antiproliferative and photothermal therapies in breast cancer cells. *Molecules*. 2023;28(11):4453.

Lou M, Jonckheere E. Magnetically levitated nano-robots: An application to visualization of nerve cells injuries. *Studies in Health Technology and Informatics*. 2007;125:310–312.

Lukkahatai N, Han G. Perspectives on artificial intelligence in nursing in Asia. *Asian Pacific Island Nursing Journal*. 2024 Jun 19;8:e55321.

Ma W, Zhan Y, Zhang Y, Shao X, Xie X, Mao C et al. An intelligent DNA nanorobot with *in vitro* enhanced protein lysosomal degradation of HER2. *Nano Letters*. 2019;19(7):4505–4517.

Makwana V, Karanjia J, Haselhorst T, Anoopkumar-Dukie S, Rudrawar S. Liposomal doxorubicin as targeted delivery platform: Current trends in surface functionalization. *International Journal of Pharmacetics.* 2021;593:120117.

Malik S, Muhammad K, Waheed Y. Emerging applications of nanotechnology in healthcare and medicine. *Molecules.* 2023;28(18):6624.

Maniaci A, Lazzeroni M, Cozzi A, Fraccaroli F, Gaffuri M, Chiesa-Estomba C, Capaccio P. Can chatbots enhance the management of pediatric sialadenitis in clinical practice? *European Archives of Oto-Rhino-Laryngology.* 2024;281(7):6133–6140.

Manickam P, Mariappan SA, Murugesan SM, Hansda S, Kaushik A, Shinde R, Thipperudraswamy SP. Artificial intelligence (AI) and internet of medical things (IoMT) assisted biomedical systems for intelligent healthcare. *Biosensors (Basel).* 2022;12(8):562.

Mars M. Telemedicine and advances in urban and rural healthcare delivery in Africa. *Progress in Cardiovascular Diseases.* 2013;56(3):326–335.

Mikhael PG, Wohlwend J, Yala A, Karstens L, Xiang J, Takigami AK et al. Sybil: A validated deep learning model to predict future lung cancer risk from a single low-dose chest computed tomography. *Journal of Clinical Oncology.* 2023;41(12):2191–2200.

Mittermaier M, Raza MM, Kvedar JC. Bias in AI-based models for medical applications: Challenges and mitigation strategies. *npj Digital Medicine.* 2023;6(1):113.

Moglia A, Georgiou K, Georgiou E, Satava RM, Cuschieri A. A systematic review on artificial intelligence in robot-assisted surgery. *International Journal of Surgery.* 2021;95:106151.

Mohan D, Gupta S. Role of accredited social health activists (ASHAs) in COVID-19 pandemic in India. *Journal of Social Health and Diabetes.* 2021;9(1):1–4.

Moingeon P. Harnessing the power of AI-based models to accelerate drug discovery against immune diseases. *Expert Rev Clin Immunol.* 2024;20(10):1135–1138.

Molaei MJ. Magnetic hyperthermia in cancer therapy, mechanisms, and recent advances: A review. *Journal of Biomaterials Applications.* 2024;39(1):3–23.

Moles E, Howard CB, Huda P, Karsa M, McCalmont H, Kimpton K et al. Delivery of PEGylated liposomal doxorubicin by bispecific antibodies improves treatment in models of high-risk childhood leukemia. *Science Translational Medicine.* 2023;15(696):eabm1262.

Mupela E, Mustarde P, Jones H. Telemedicine in sub-Saharan Africa: The case of Zambia. *International Journal of Telemedicine and Applications.* 2011:1–7.

Nazer LH, Zatarah R, Waldrip S, Ke JXC, Moukheiber M, Khanna AK et al. Bias in artificial intelligence algorithms and recommendations for mitigation. *PLOS Digital Health.* 2023;2(6):e0000278.

Niyonshuti II, Jayaraj S, Jiang W, Mudalige T. A robust chromatographic method for drug release profiling of liposomal doxorubicin HCl. *Journal of Pharmaceutical Sciences.* 2024;113(9):2837–2842.

Osterwalder J, Polyzogopoulou E, Hoffmann B. Point-of-care ultrasound-history, current and evolving clinical concepts in emergency medicine. *Medicina (Kaunas).* 2023;59(12):2179.

Park SE, Lee J, Lee T, Bae SB, Kang B, Huh YM, Lee SW, Haam S. Comparative hyperthermia effects of silica-gold nanoshells with different surface coverage of gold clusters on epithelial tumor cells. *International Journal of Nanomedicine.* 2015;10(Special Issue):261–271.

Patil S, Rao RS, Majumdar B. Nanorobots: Changing trends in cancer therapy. *Journal of Contemporary Dental Practice.* 2015;16(9):i–ii.

Paul D, Sanap G, Shenoy S, Kalyane D, Kalia K, Tekade RK. Artificial intelligence in drug discovery and development. *Drug Discovery Today.* 2021;26(1):80–93.

Peng R, Zheng X, Lyu Y, Xu L, Zhang X, Ke G, Liu Q, You C, Huan S, Tan W. Engineering a 3D DNA-logic gate nanomachine for bispecific recognition and computing on target cell surfaces. *Journal of the American Chemical Society.* 2018;140(31):9793–9796.

Perán M, García MA, Lopez-Ruiz E, Jiménez G, Marchal JA. How can nanotechnology help to repair the body? Advances in cardiac, skin, bone, cartilage and nerve tissue regeneration. *Materials (Basel).* 2013;6(4):1333–1359.

Sahashi S, Sugimura H. Lecture No. 10 AI and telemedicine: How is technology transforming the horizons for global health? *Japanese Journal of Clinical Oncology.* 2021;51(12 Suppl 2):i41–i44.

Sahota N. AI and Nanotechnology: Pioneering Tomorrow's Tech Evolution. November 29, 2023 https://www.neilsahota.com/ai-and-nanotechnology-pioneering-tomorrows-tech-evolution/ (accessed June 19, 2024).

Schummer J. Multidisciplinarity, interdisciplinarity, and patterns of research collaboration in nanoscience and nanotechnology. *Scientometrics.* 2004;59(3):425–465.

Schwendicke F, Samek W, Krois J. Artificial intelligence in dentistry: Chances and challenges. *Journal of Dental Research.* 2020;99(7):769–774.

Secinaro S, Calandra D, Secinaro A, Muthurangu V, Biancone P. The role of artificial intelligence in healthcare: A structured literature review. *BMC Medical Informatics and Decision Making*. 2021;21(1):125.

Sheibani M, Azizi Y, Shayan M, Nezamoleslami S, Eslami F, Farjoo MH, Dehpour AR. Doxorubicin-induced cardiotoxicity: An overview on pre-clinical therapeutic approaches. *Cardiovascular Toxicology*. 2022;22(4):292–310.

Singh HR, Kopperger E, Simmel FC. A DNA nanorobot uprises against cancer. *Trends in Molecular Medicine*. 2018a;24(7):591–593.

Singh P, Pandit S, Mokkapati VRSS, Garg A, Ravikumar V, Mijakovic I. Gold nanoparticles in diagnostics and therapeutics for human cancer. *International Journal of Molecular Sciences*. 2018b;19(7):1979.

Spatharou A, Hieronimus S, Jenkins J. Transforming healthcare with AI: The impact on the workforce and organizations. McKinsey & Company. March 10, 2020. https://www.mckinsey.com/industries/healthcare/our-insights/transforming-healthcare-with-ai#/ (accessed June 19, 2024).

Sreelatha OK, Ramesh SV. Teleophthalmology: Improving patient outcomes? *Clinical Ophthalmology*. 2016;10:285–295.

Stephens J, Greenberg GM. Asynchronous telehealth. *Primary Care*. 2022;49(4):531–541.

Steyvers M, Kumar A. Three challenges for AI-assisted decision-making. *Perspectives on Psychological Science*. 2024;19(5):722–734.

Tang X, Zhu Y, Guan W, Zhou W, Wei P. Advances in nanosensors for cardiovascular disease detection. *Life Sciences*. 2022;305:120733.

Tchero H, Tabue Teguo M, Lannuzel A, Rusch E. Telerehabilitation for stroke survivors: Systematic review and meta-analysis. *Journal of Medical Internet Research*. 2018;20(10):e10867.

Teo ZL, Ting DSW. AI telemedicine screening in ophthalmology: Health economic considerations. *Lancet Global Health*. 2023;11(3):e318–e320.

The Materials Project. 2024. https://legacy.materialsproject.org/ (accessed October 13, 2024).

Torchilin VP. . Passive and active drug targeting: Drug delivery to tumors as an example. *Handbook of Experimental Pharmacology*. 2010;197:3–53.

Tunali I, Gillies RJ, Schabath MB. Application of radiomics and artificial intelligence for lung cancer precision medicine. *Cold Spring Harbor Perspectives in Medicine*. 2021;11(8):a039537.

United Nations General Assembly. Transforming our world: the 2030 Agenda for Sustainable Development. A/RES/70/1. 2015 Oct 21. https://www.refworld.org/legal/resolution/unga/2015/en/111816 (accessed August 12, 2025)

Vearrier L, Derse AR, Basford JB, Larkin GL, Moskop JC. Artificial intelligence in emergency medicine: Benefits, risks, and recommendations. *Journal of Emergency Medicine*. 2022;62(4):492–499.

Velusamy P, Su CH, Ramasamy P, Arun V, Rajnish N, Raman P et al. Volatile organic compounds as potential biomarkers for noninvasive disease detection by nanosensors: A comprehensive review. *Critical Reviews in Analytical Chemistry*. 2023;53(8):1828–1839.

Webster J, Young P, Kiecker J. Telerehabilitation for amputee care. *Physical Medicine and Rehabilitation Clinics of North America*. 2021;32(2):253–262.

Weerarathna IN, Kamble AR, Luharia A. Artificial intelligence applications for biomedical cancer research: A review. *Cureus*. 2023;15(11):e48307.

Wilson TG. Advancement of technology and its impact on urologists: Release of the daVinci Xi, a new surgical robot. *European Urology*. 2014;66(5):793–794.

World Health Organization. Health Equity Campaign 2021. https://www.who.int/campaigns/world-health-day/2021/health-equity-campaign-2021#:~:text=On%20World%20Health%20Day%202021,have%20led%20to%20health%20inequities (accessed June 19, 2024).

Yaari Z, Yang Y, Apfelbaum E, Cupo C, Settle AH, Cullen Q et al. A perception-based nanosensor platform to detect cancer biomarkers. *Science Advances*. 2021;7(47):eabj0852.

Yan C, Feng K, Bao B, Chen J, Xu X, Jiang G, et al. Biohybrid nanorobots carrying glycoengineered extracellular vesicles promote diabetic wound repair through dual-enhanced cell and tissue penetration. *Advanced Science (Weinh)*. 2024;11(31):e2404456.

Zhang K, Wang Y, Xue J, Liang N, Wei Z. Real-time monitoring ATP variation in human cancer organoids for a long term by DNA-based nanosensor. *Analytica Chimica Acta*. 2023;1275:341608.

Zhu H. European patent EP1952825 (2006). https://europepmc.org/article/PAT/EP1952825 (accessed August 12, 2025).

9 Ethical Implications of Global Health Care Equivalency

Ingrid Vasiliu-Feltes

CONTENTS

9.1 INTRODUCTION

Frontier technology deployments, including artificial intelligence (AI), prospective artificial general intelligence (AGI), and nanoscience, have profound disruptive potential for humanity and economics. They also serve as powerful catalysts for a new era where the boundaries between physical, biological, and digital worlds become increasingly blurred. Their impacts will likely trigger a complex cascade of adaptive changes in how we live, work, and educate future generations. When examining recent advances in nanotechnology, it is evident that many scientific discoveries have triggered the emergence of novel medical devices and systems, diagnostic sensors/imaging, and therapeutic systems. Some experts believe that nanotechnology and nanomedicine (Kim et al., 2010) deployments, coupled with AI, could have the highest ever commercial and quality of care impacts on healthcare quality and safety. The potential versatility of nanomedical platforms will make them uniquely suited across myriad domains within the healthcare ecosystem (e.g., prevention, diagnosis, treatment, rehabilitation, monitoring, wellness, nutrition, longevity, and others).

Nanotechnology has the capacity to revolutionize oncological diagnostics, therapy, nuclear medicine, and imaging. Advances in nanomaterial sciences offer novel solutions that encompass targeted cancer medications, early-detection devices, robust imaging modalities, and enhanced radiotherapy adjuvants. However, these significantly innovative strides have also highlighted the need for a more

 DOI: 10.1201/9781003680093-9

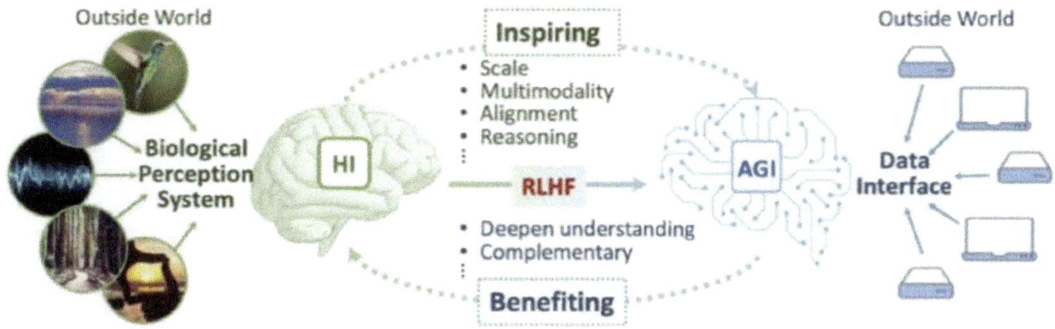

FIGURE 9.1 Evolution of AGI inspired by human intelligence. (Reproduced from Zhao et al. (2013) with permission.)

prudent approach to informed consent, given the immense nanotechnology and AI literacy gap. They also acutely emphasize the need for a more refined anthropogenic/nanotechnology/nanomedicine feedback loop that can provide oversight and continuous improvements in quality.

Similarly, AI deployments have progressed from beta pilots to large-scale healthcare and life science adoption. We are witnessing a global effort to increase awareness, education, and enhanced governance for the implementation of AI (and perhaps the near-term emergence of AGI) (Figure 9.1) across myriad domains. While the ethical deployment of AI is industry-agnostic and requires a proactive ethics-by-design approach for long-term success, most experts agree that healthcare and the life sciences require customized strategies. Although ethical principles and moral values have existed for centuries, the digital era and rapid large-scale adoption of novel technologies pose serious digital ethics challenges that need to be addressed from philosophical, legal, and self-sovereignty perspectives.

This chapter highlights how we might design initiative-taking digital ethics programs for an envisaged novel global healthcare ecosystem (Global Health Care Equivalency (GHCE)) to mitigate any potential negative consequences of technology deployments. Further, design methodologies combined with ethical principles may assist in building a human-centric global healthcare ecosystem that will be equitable, inclusive, and protect human rights. Specific digital ethics nuances associated with AI and nanotechnology applications will also be covered, including those related to various domains such as wellness, disease prevention, diagnosis, treatment, chronic disease management and rehabilitation, and longevity.

Cultural influences or socioeconomic differences that may impact ethical technology design frameworks will also be a focus of this chapter, as well as the future research that will be required to shed light on highly controversial digital ethics topics (e.g., brain-machine interfaces (BMI), brain-computer interfaces (BCI), brain/cloud interfaces (B/CI), bionic humans, human cloning, digital health twins, robotics, avatars, etc.).

Ethics has been a core anthropogenic discipline and deliberation over millennia, dating back to ancient Greece (Price, 1985). Following decades of relegation, ethics are currently experiencing a resurgence in scientific and business communities. This is due to the myriad of complex ethical issues we now encounter due to the high levels of augmented intelligence and automation that permeate our cultures and daily lives (Peterson, 2018). Scientific and business communities, as well as numerous not-for-profit and government agencies, are rightly concerned regarding ethical issues that will likely have impacts across all industries, with topics spanning algorithmic bias, social determinant discrimination, data privacy, data ownership, AI transparency, and trust, which make the headlines daily.

Industry leaders who aim to engage in the next industrial revolution and successfully manage the impacts of emerging technologies, such as (most notably) distributed ledger technologies across all life-science ecosystem domains, will be required to orchestrate a complex armamentarium of novel skills. This expertise will encompass technology and AI literacy, environmental, social, and governance (ESG) consciousness, and a mastery of digital and applied ethics (Vasiliu-Feltes and Thomason, 2021). Further, it has become evident that novel technologies such as blockchain will demand versatility in foundational ethical concepts (Gami et al., 2023). Reactive or mitigation approaches will not be an optimal solution. Ethical leadership in this digital era will be defined by upholding moral values and complemented by state-of-the-art strategic planning, a significant revision/updating of our education system, coupled with a profound transformation of current healthcare delivery systems.

9.2 EXPLORING THE GLOBAL ETHICAL LANDSCAPE

The digital era has demanded a significant revision of business, social, and individual ethics concepts, while revealing opportunities related to the optimal design of sustainable digital ethics programs in this rapidly emerging hyper-connected, hyper-automated, hyper-virtualized world. The complex issues involved in this revised global digital ethics model may be analyzed from three perspectives (ethical leadership, business, and societal ethics). The latest research publications reveal an urgent need for the responsible development and application of AI systems in healthcare, including several reports that shed light on the multifaceted ethical considerations in AI and related domains. One key focus is "Ethical Considerations in AI Development," which underlines the paramount importance of ethics in the AI development process. This emphasizes the need for a conscientious and ethical approach in designing and implementing AI systems (Wilson, 2022). Another critical dimension is "AI Ethics and the Law," where scholars delve into the legal aspects of AI ethics, scrutinizing issues such as liability and the creation of regulatory frameworks (Wilson, 2021). The dynamic interface between "AI and Healthcare Ethics" is gathering increasing attention. Researchers in this area are diligently exploring current challenges and charting a course for future ethical practices where AI intersects with healthcare, seeking to balance the advantages of AI in medicine with potential ethical dilemmas (Davis, 2021; Smith and Brown, 2022). The paper "Ethical Implications of AI in Biomedical Research" emphasizes the ethical quandaries associated with AI's role in biomedical research and advocates for the establishment of ethical guidelines and robust oversight within this domain (Taylor, 2022).

"Ethical Considerations in Nanotechnology" amplifies the overarching discussions on ethics in emerging technologies (Lewis, 2021). Additionally, "AI Ethics: Challenges and Opportunities" is a research endeavor that underscores the challenges and opportunities surrounding ethical AI development. The authors address concerns such as algorithmic bias mitigation and the promotion of fairness in AI systems (LS Johnson, 2021; WE Johnson, 2021; Johnson and Smith, 2021). "Ethical Issues in AI and Machine Learning" extends the discourse to the wider arena of AI ethics. This research area elucidates the ethical concerns that are central to AI and machine learning, encompassing issues of transparency, fairness, and accountability (LS Johnson, 2021; WE Johnson, 2021; Johnson and Smith, 2021).

9.2.1 Ethical Leadership

We are currently in the midst of the Fourth Industrial Revolution and likely entering the fifth. Blending boundaries between physical, digital, and biological domains will likely continue, where AI (prospectively AGI/ASI), nanotechnology, and other emerging frontier technologies can potentially have intensely disruptive global impacts. Many experts predict that healthcare will be deeply impacted in this new era. Leaders who endeavor to be prepared for the Fifth Industrial Revolution and successfully manage its impacts on healthcare will have a profound cognizance of technology literacy, design thinking, and a mastery of applied ethics (Vasiliu-Feltes and Thomason, 2021).

It was once adequate to be a "tech-savvy" leader; however, with the recent exponential advances in a variety of modern technologies in healthcare (e.g., the full AI portfolio, nanotechnology, nanomedicine, blockchain, IoT, IoB, BMIs, BCIs, quantum and edge computing, robotics, implants, wearables, telemedicine, VR, XR, next-generation sequencing, panomics, multiomics), healthcare leaders must have a higher degree of technical acumen to remain competitive. Further, with a focus on personalized medicine powered by genomics and the importance of designing a human-centric healthcare ecosystem, future healthcare leaders would be well advised to garner a profound understanding of these disciplines. More importantly, it has become evident that exponential medicine and healthcare will also demand broad versatility in foundational ethical concepts.

9.2.2 Business and Social Ethics

As a society, we must reflect on redefining humanity's role in the new digital era and how emerging technologies impact our society, economy, culture, education, and work. The endorsement and promotion of large-scale digital health adoption also implies and demands a moral obligation to ensure privacy, greater transparency, enhanced security, improved traceability, increased efficiency and equity, and broad exploitability. Scientific and business communities and numerous government and not-for-profit agencies are appropriately concerned about the unique ethical issues associated with a digital healthcare ecosystem.

There is a deep and complex relationship between digital business ethics and other aspects of society. Business leaders of the digital era will be expected to embody a hybrid business/digital ethics acumen. By adopting a design thinking approach for business ethics, all benefits offered by emerging technologies and scientific advances may be leveraged while maintaining a human-centric stance. In a recent publication, the author concluded that *Developing a Universal Code of Conduct for Digital Business Ethics and a Global Digital Ethics Framework* should be a moral imperative (Vasiliu-Feltes, 2022). As digital ethics advocates, we must hold ourselves accountable to establish a robust digital ethics legacy for future generations.

The modern world is undergoing rapid and profound changes in healthcare and business ecosystems due to the exponential pace of digital innovation and transformation. Although these transitions offer exciting opportunities and likely have profound disruptive paradigm shifting impacts, they will also exacerbate current digital ethics challenges and create new ones never envisioned. Business ethics in the digital era pose more complex and nuanced challenges. Several publications have highlighted an extensive range of ethical questions that remain to be answered, in conjunction with a new set of ethical challenges that will emerge as we leap into the next Industrial Revolution. The blurring of mechanical, electronic, digital, and biological domains worldwide will likely proceed at an exponential pace, encompassing technologies such as AI, AGI, ASI (Goertzel, 2017; Kejriwal, 2021; Simon, 2019; Simon, 2022), IoT, 6G, or next-generation computing, which have the potential to significantly reshape, recalibrate, or disrupt human society and the global economy.

Leaders who endeavor to embrace Industry 5.0 and succeed in managing a new digitalized hyperconnected global workforce must possess technological fluency, embrace design thinking strategies, and demonstrate a mastery of applied digital ethics, while confidently using their emotional intelligence. They should lead with purpose and embrace ethics to remain competitive while ensuring long-term sustainability for their companies. Further, given the greater consumer-centric reality and a broader acceptance of technology-*as-a-service* business models, recommendations include that current leaders be prepared to accept, manage, and adapt to significantly rapid change, and be willing to disrupt themselves to avoid being so by others. Digital data governance, proactive ethics and compliance programs, ESG-consciousness, and a focus on United Nations Sustainable

Development Goals are emerging as expectations for the new generation of leaders. Versatility in foundational ethical concepts and an inclination to adopt complex ethics frameworks for every step of technological deployments are imperative in this new business environment, which is defined by a high degree of digitalization, virtualization, and the intelligent automation of most business procedures.

An ongoing debate ensues regarding the responsibilities of business communities and the incentives or mandates required to enforce them. Moral theories must be balanced with practical corporate social responsibility. It is incumbent upon society to rethink how these moral constructs might be mindfully and efficiently applied in this digital era. A digital ethics framework must encompass ethical and digital risk analyses (a set of tools that may be deployed at various stages to assess different business strata), as well as establish and monitor digital ethics implementations via Key Performance Indicators (KPIs) and build a culture of digital ethics, while embracing an ongoing quality improvement mindset. Business leaders typically require assistance with deploying these frameworks due to a lack of digital ethics literacy and a failure to harmonize with other vital organizational units, such as Compliance, Human Resources, and Marketing, to mention a few.

Many ethics-related myths are perpetuated and exacerbated when efforts are made to address digital ethics. Furthermore, profound misunderstandings relating to the impacts of digital ethics are continually encountered in all aspects of organizational ecosystems. Last but not least, an ongoing resistance against embracing proactive approaches has been observed for some business leaders, who preferentially select reactive crisis mitigation when it comes to ethics breaches. A best-in-class cyberethics program represents one of the major potential drivers of success when operationalizing proactive digital ethics frameworks. More often than not, cybersecurity and ethics are managed in silos, thus presenting opportunities for cyberattacks or privacy breaches. Comprehensive state-of-the-art digital ethics programs must always integrate a robust cyberethics component and be fully integrated/harmonized with organizational cybersecurity systems.

Advances in applied nanotechnology, nanomedicine, and AI are at the epicenter of myriad ethical, philosophical, legal, cultural, and economic dilemmas. Defining meaningful roles for humanity amid the expanding global tech ecosystem is not a straightforward task for ethics experts; however, there appears to be a consensus worldwide that emerging technologies will significantly impact the fabric of our lives, work, entertainment, and evolution. Trustworthy and responsible technology deployments must positively define the future if we intend to avoid negative consequences, uphold core ethical principles, and abide by our established societal codes of conduct. We must understand and harmonize the complex relationships between digital ethics and human rights, while ensuring inclusivity by bridging digital divides. Researchers/authors call our attention to the multiple moral dilemmas we encounter when we attempt to balance human rights while struggling to contain a global health crisis and respect digital equity for vulnerable populations.

A recent article published in the *American Journal of Psychiatry and the Law* emphasized the risk of bias, maleficence, and how easily justice can be violated when AI technologies are not responsibly applied (Hogan et al., 2021). A recent book titled *Future Law* eloquently highlights the challenges encountered when formulating legislation for emerging technologies, and some of the core regulatory and ethical intricacies lawmakers need to consider (Edwards et al., 2020). The authors also emphasize that arts and culture mediate between technology and law. Another book, *The Ethics of Biomedical Big Data*, identifies critical societal issues by approaching the debate on the ethical deployment of new and emerging technologies, while calling for international collaboration to formulate data governance policies (Mittelstadt and Floridi, 2018) (Figure 9.2). The authors caution against exceptionalism, parochialism, and adventitious ethics in life sciences research.

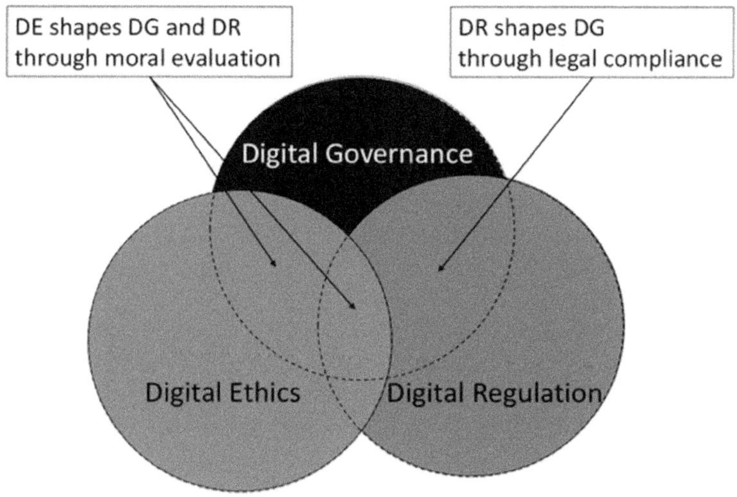

FIGURE 9.2 Relationships between digital ethics, digital regulation, and digital governance (mage derived from Floridi (2018)).

9.2.3 Navigating Digital Ethics in Healthcare

A heightened focus on genomics, neuroscience, synthetic biology, and nanoscience, combined with rapid developments in AI, computation, and data analytics in modern healthcare, has also spurred a similar increase in ethical challenges. The extensive use of digital data requires ongoing education, a shared commitment to addressing issues, and cooperative approaches to protect ethical principles. Individual needs should be carefully balanced with those of society and business when developing comprehensive digital ethics programs.

A recently published book chapter, *Ethical and Regulatory Challenges in Emerging Health Technologies*, considers the advances in biotechnology sciences (e.g., genomics, neuroscience, synthetic biology, and nanoscience) combined with rapid advances in computer and data analytics (Rizk, 2022). Inherent challenges that accompany the increasing complexity of ethical issues associated with modern medical technologies are rapidly evolving and transforming the healthcare system. In particular, data-driven systems are altering traditional healthcare delivery. Knowledge and expertise gaps for providers and users between these innovative interventions and existing healthcare paradigms create multiple ethical, regulatory, and economic challenges that may destabilize trust in the safety, fairness, and effectiveness of healthcare systems. Ritz advocates for the requirement of specialized expertise, a broader scope of analysis, and a new subfield of ethics referred to as technology ethics, or "techno-ethics," which deals with framing principles and strategies to guide the implementation and use of technologies.

The preservation of digital identity, privacy, and self-sovereignty is paramount in the digital age. When merging nanotechnology, nanomedicine, and AI to create novel solutions that have a strong potential to redefine the healthcare ecosystem, there is a moral duty to ensure the security, confidentiality, and integrity of highly sensitive data. Digital identity is a foundational element of successful ethical AI and nanotechnology deployments in any industry, which is critical for healthcare. Digital identity may be represented by an individual, organization, application, or device and includes electronic signatures, seals, website authentication, and registered delivery. From an ethical perspective, all expressions must be reflected on toward understanding the impacts of digital identity categorizations when implementing blockchain across the life science spectrum. In Cameron's landmark "The Laws of Identity," he outlines the principles of identity, where blockchain is conducive to attaining all of them (user control and consent, minimal disclosure for a constrained use, justifiable parties, directed identity, pluralism of operators and technologies, human integration, and consistent experience across contexts) (Cameron, 2005).

Another essential article by Allen and Fry (2023) describes four online identity models, each of which requires distinct digital ethics guardrails (centralized, federated, user-centric, and self-sovereign identity). He also drafted novel principles of self-sovereign identity, which should be foundational for those who develop digital ethics blockchain playbooks (user-centricity, control, access, transparency, longevity, portability, interoperability, consent, minimized data disclosure, and protection).

A more recent article by Bouras et al. (2020) provides a comprehensive review and overview of the impacts of identity management and its importance in e-healthcare. The seven criteria of identity management they outline fully apply to life sciences research: autonomy, authority, availability, approval, confidentiality, tenacity, and interoperability. The authors also provide a helpful comparison of identity management models and how each type impacts the seven criteria. Their findings suggest that decentralized models are the only ones that offer autonomy, as well as the highest authority, availability, and confidentiality. This publication also highlights challenges when using centralized, federated, or user-centric identity models, such as a lack of autonomy and interoperability, or the approval feature in both centralized and federated identity models.

9.3 KEY ETHICAL CONSIDERATIONS

The critical ethical considerations that we must consider for applied ethics for all emerging technologies are broad and encompass core ethical principles such as Autonomy, Beneficence, Non-Maleficence, Justice, and core behavioral norms including Veracity, Privacy, Confidentiality, and Fidelity.

The deployment of digital healthcare and life sciences requires a dual focus on critical concepts such as self-sovereignty and informed consent. Business leaders and ethics experts must collaborate closely to harmonize the operation of these principles while creating a sustainable program for all stakeholders. The establishment of trust and a robust data governance program is essential for long-term success. For nanotechnology, nanomedicine, and AI, which have already had profound transformative impacts on healthcare and are poised to redefine healthcare for future generations, these two elements are not only a business requirement but also a moral imperative. Further, a digital ethics playbook for AI (future AGI/lcc), nanotechnology, and nanomedicine must establish ethics guardrails for all domains within the healthcare ecosystem (wellness, longevity, diagnosis, treatment, rehabilitation, disease prevention, and surveillance).

9.4 APPROACHES FOR ETHICAL DEPLOYMENTS OF NANOTECHNOLOGY, NANOMEDICINE, AND AI

9.4.1 Ethics by Design

The exponential growth of technologies is transforming our daily lives, generating vast amounts of data, and accelerating the velocity of data flows globally. Thus, the public is often confronted with changes that move faster than they can comprehend. There is an urgent need for the analysis and communication of ethical issues associated with these advances, to scrutinize ethical questions raised by modern science, technological progress, and the Fourth Industrial Revolution. Further, to explore how to harness the speed, accuracy, and power of emerging technologies in policy research and public engagement to assist leaders, policymakers, and the public in understanding the impacts of these technologies on the economy and society at large.

A review of the digital ethics publications from global technological and ethics experts in academia, industry, law, and various not-for-profit agencies validates that digital ethics should be proactively and mindfully implemented. This, to ensure the development of technologies that embed ethics by design within organizational strata, services, or products. The establishment of global ethical governance needs to be explored, as well as the application of design thinking for the ethical implementation of emerging technologies, the complex relationships between political power and ethics, as well as future research issues for ethical technologies (Vasiliu-Feltes and Thomason, 2021).

9.4.2 Proactive Digital Ethics Programs

When evaluating the key elements that comprise proactive state-of-the-art digital ethics programs, new digital ethics codes of conduct will need to be identified. These, as well as a new digital bill of rights and data governance in conjunction with the traditional components of ethics and compliance programs. In a recent white paper, Kristin Gloria of the Aspen Institute forecasted a different future for digital rights. There are unique digital ethics considerations for all types of research and stages within the research life cycle. There are numerous domains where ethical governance must be embedded in AI, nanotechnology, and nanomedicine solutions for any research enterprise, encompassing study design, IRB review, audits, compliance, patient recruitment, consent, data privacy, addressing conflicts of interest, patient autonomy, and data ownership. Enhancing research quality and safety is paramount for upholding the principles of beneficence and non-maleficence.

Companies should understand the importance of a realistic digital ethics statement and adopt a digital ethics code of conduct for all organizational roles. Often, these statements are aspirational and inspirational in content, however difficult or impossible they may be to operationalize. An ethics statement is foundational and must align with the organization's overall mission, as well as the resources that are realistically required to uphold the ethical principles and moral values outlined therein.

Furthermore, the consequences of non-adherence to digital ethics codes of conduct must be enforced by human resource operations to ensure their effectiveness and sustainability. To successfully deploy ethical AI, nanotechnology, and nanomedicine, technical efforts must be integrated with a robust data governance program and a rigorous auditing process, while being harmonized with an established ethics playbook that incorporates legal, financial, and reputational risk management programs. A state-of-the-art digital ethics program for combined AI, nanotechnology, and nanomedicine implementation would place equal emphasis across all domains (people, processes, and product life cycles). Equally important will be alignments with cybersecurity policies and procedures, as cyberethics is often overlooked due to traditionally siloed healthcare approaches. Beyond the creation of a customized ethics statement and strategic roadmap for nanotechnology, nanomedicine, and AI deployments, it will be vital to design and implement KPIs and develop robust training and talent.

In terms of ethical governance, enterprises that aim to deploy nanotechnology, nanomedicine, or AI should initially clarify their ethical approach and decide if they will have an enterprise-wide standard for ethical governance, or organize/allocate it by department or unique role assignment. Next, key stakeholders should evaluate their current digital ethics risk, conduct a thorough failure mode analysis, and design their digital ethics strategy to allow for dynamic adaptation to all future risks, as well as unforeseen and novel emerging technologies that might enter and disrupt the market. Subsequently, important decisions will need to be made regarding the desired level of automation for the ethical governance process, and where to supplement human oversight in the human-computer interaction ecosystem without causing bottlenecks or data breaches. When operationalizing a digital ethics strategy, it is essential to align operational procedures with all other facets of the enterprise to achieve optimal adherence and a robust digital ethics culture.

9.4.3 Ethical Challenges in Converging Technology Deployment

As highlighted in the preceding sections, the implementation of any new technology poses significant ethical challenges. When novel solutions, products, or services are deployed that leverage the capacities of several frontier technologies, the inherent challenges increase exponentially, ranging from technical (e.g., interoperability and standards development) to the most complex applied digital ethics realms.

9.4.3.1 Applied Ethics in AI Deployments

Currently, applied ethics for AI is a complex process; thus, it will take time to estimate and indeed fathom what will be involved with the emergence of prospective AGI and ASI. Several countries have engaged countless experts and spent years drafting national AI strategies and policies. Still, the applied ethics component remains a non-trivial challenge for many. The entire portfolio of available tools must be considered for the development of customized programs for symbolic AI versus rules-based AI or robotics, natural language processing, machine learning, and robotics deployments. A thorough comprehension of all nuances will be essential for embedding the appropriate guardrails that can adequately uphold core ethical principles, including beneficence, non-maleficence, equitable access, justice, and autonomy. Nations and organizations worldwide are engaging cross-disciplinary teams of experts to draft new ethical AI frameworks, policies, procedures, guidelines, toolkits, or checklists designed to facilitate the ethical implementation of complex AI portfolios in healthcare, life sciences, and related disciplines. However, education systems need to be redesigned, and the workforce requires proper upskilling to reduce the existing digital ethics divide while augmenting digital ethics fluency. As outlined in a recent publication, myriad domains are emerging within AI that must be contemplated when drafting national strategies, new laws, regulations, AI ethics frameworks, playbooks, and toolkits.

A comprehensive review of the essence of healthcare AI ethics has been recently published by Blackman (2022), which highlights the role of AI in the Fourth Industrial Revolution, its profound ethical implications, and potential for advancing global health equity if deployed responsibly. Blackman proposes several practical key steps for enterprises that endeavor to design and deploy responsible and respectful AI programs.

9.4.3.2 Applied Ethics in Nanotechnology and Nanomedicine Deployments

As with other emerging frontier technologies, earning and sustaining public trust will be paramount when implementing nanotechnology, nanomedicine, and AI in healthcare. While recognizing the tremendous potential to revolutionize how disease is prevented, diagnosed, treated, and monitored, digital ethics experts are also acutely concerned regarding our capacity to preserve autonomy, privacy, and security while obtaining informed consent for technologies that require a high degree of digital literacy/fluency, and the ability to comprehend abstract concepts. As Van de Voorde and Jeswani (2021) highlight in their recently published book, *Ethics in Nanotechnology: Social Sciences and Philosophical Aspect*s, we must aim for an extensive collaboration between all stakeholders to address the significant challenges that we face as a society. Nanomedicine will likely be adopted globally and embedded into the fabric of virtually every specialty (Figure 9.3). Further, certain nuances differentiate perceptions regarding nanotechnology between various disciplines.

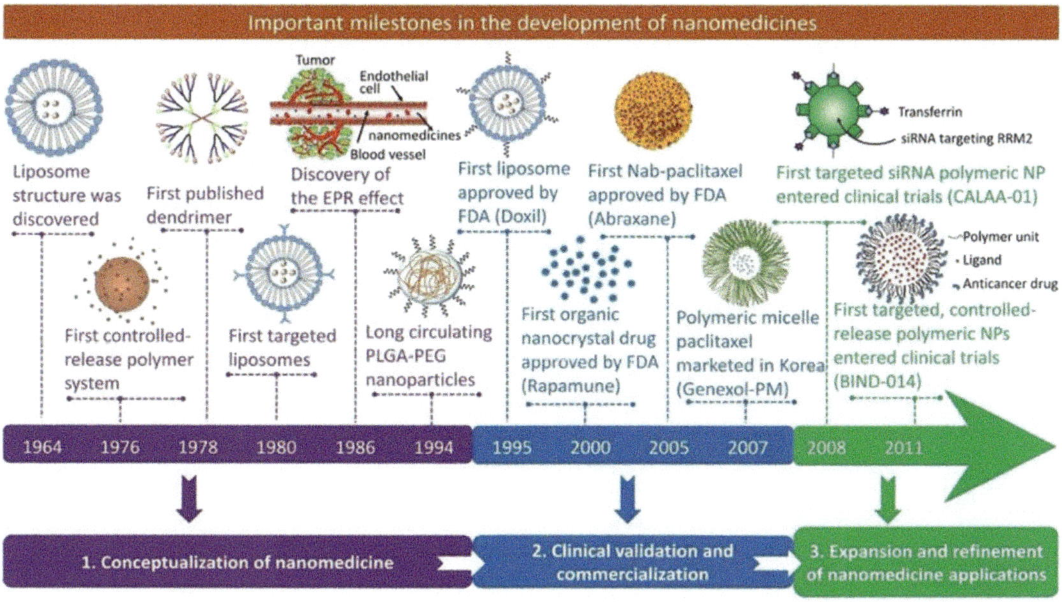

FIGURE 9.3 Progress, challenges, and future of nanomedicine. (Reproduced from Zhang et al. (2020) with permission.)

9.4.3.3 Applied Ethics in Converging Emerging Technologies

In the quest to develop novel precision and potentially compelling nanomedical solutions, converging technologies pose serious challenges in terms of violations against privacy, security, and autonomy, or the unintentional lack of respect for individual autonomy. One of the most critical overarching drivers of success relates to the harmonization of digital ethics constructs with other enterprise risks, including regulatory, legal, and cyber safety programs. The second most essential driver is the translation and integration of ethical principles into operational procedures that are viable and aligned with enterprise KPIs.

An overview of several recent use cases highlights critical ethical challenges when deploying converging technologies. The application of AI, nanotechnology, nanomedicine, robotics, and 3D printing has had increasing success toward the development of novel multifunctional smart health products. Nevertheless, from an ethical perspective, informed consent and justice principles require careful assessment and development. Will patients and their assigned proxies genuinely comprehend the nature and potential long-term implications of complex, innovative medical solutions that amalgamate concepts such as cell growth, genetic engineering, semisolid extrusion, fused deposition modeling, theranostics, nanocarrier drug delivery, inorganic nanoparticles, etc.? Further, technologies that will likely have even more fundamental societal repercussions (e.g., BMI, BCI, and a hypothetical (for now) B/CI, which may facilitate the next step in human evolution; one where we seamlessly merge with our technologies)? How might we nurture trust in and ensure equitable global-scale access to these novel, scientifically complex solutions, including nanomedicine, nanodiagnostics, or nanotheranostics? Can we ensure beneficence in the absence of long-term studies to inform us about potential adverse health outcomes? Case in point: mRNA vaccines against SARS-CoV-2 that were fast-track developed (facilitated by the HSS Operation Warp Speed and Emergency Use Authorization) and released en masse to the world in ~one year, in stark contrast to the typical 10–15-year timeline required for the development of a vaccine (Brothers, 2020).

While the moral and ethical dilemmas are significant for acute medical conditions for which there are no immediate potential life-saving options, they become even more complex when compounded with the evaluation of regenerative medicine, the potential benefits of nanomedical-driven cell therapies, and AI for longevity, given their profound psychological, cultural, religious, and legal aspects.

Informed consent, privacy, autonomy, and self-sovereignty are challenged when considering the application of bioelectronic medical solutions. If the appropriate care is not observed during the design, deployment, and monitoring of seamless interactions between logical and electronic components, significant ethics violations may ensue. Many nanoelectronic devices have the potential to dramatically improve current treatment approaches. However, their benefits must be solidly vetted and validated when they relate to more diminutive and powerful nanomedical devices that may soon emerge for in vivo applications. Simultaneously, their safety and security must be stringently developed and enforced under robust ethical frameworks. These aspects must integrate the unique challenges of incorporating various nanomaterials (e.g., nanoparticles, carbon nanotubes, nanowires, and graphene) and potential future autonomous nanomedical devices in clinically approved diagnostic and therapeutic applications. The events that arise and decisions made over the coming decade will determine how current developments in nanotechnology, nanomedicine, and AI may forge the growing discipline of nanobioelectronic medicine (Gibney et al., 2021).

Nanomedicine encompasses an interdisciplinary array of sciences for the prevention, diagnosis, treatment, and monitoring of myriad medical conditions. Nanomedically based drug delivery systems have created intense research interest due to their revolutionary implications for drug delivery. Various nanostructured materials with unique mechanical, chemical, electronic, and biological attributes have been employed for the development of novel drug delivery systems for the precise localized targeting and sustained delivery of drugs, with improved therapeutic efficacy and negligible/no toxicity. This represents a promising approach for the effective management of disease. The utilities of AI, nanotechnology, and nanomedicine in disciplines ranging from orthopedics to neurosurgery have excellent potential to revolutionize research, diagnostics, and treatments across these domains.

There is an intense interest in nanomaterials, nanotechnology, and nanomedicine operating in conjunction with AI to provide novel solutions in clinical applications, ranging from efficient contrast agents for medical imaging to nanocarriers for the highly localized delivery of potent anticancer drugs and other medications to treat an extensive range of conditions. Although there have been several laboratory-based investigations of various characterization techniques for nanomedicine, few reviews regarding their selection for specific applications have emerged. There are pros and cons related to various characterization techniques used in nanomedicine for specific purposes. For example, various classes of nanoparticles have exhibited inherent drawbacks related to their toxicity and downstream environmental fates, which will need to be comprehensively elucidated and addressed (Boisselier and Astruc, 2009; De Jong and Borm, 2008).

Recent progress in nanomedical cardiology highlights the importance of digital ethics in the governance of clinical trials, equity, justice, and inclusion. Specifically, the latest developments in nanotechnological inclusion and integration (e.g., vascular implantable or wearable devices technologies, nanocomposites, and nanocoatings) can have an enormous impact on long-term public health outcomes.

The reliability of informed consent, beneficence, and maleficence is of great concern when enrolling patients in clinical trials to confirm the efficacy and safety of regenerative nanomedicines. Scientific advances in bionanomaterials for bone regeneration have been astonishing due to their unique mechanical properties, biocompatibility, and osteoinductivity, which can alter the paradigm in orthopedic treatment solutions (Man et al., 2023; Mou et al., 2022; Muñoz et al., 2024; Qing et al., 2020; Zorrón et al., 2024). However, it remains difficult, at least in the near term, to ensure the equitable distribution of these treatments across the population due to their exorbitant costs.

Personalized medicine is playing an increasingly important role spanning all medical disciplines; however, certain facets have attracted significant investments over the last few years, such as oncology, neurology, and, more recently, epidemiology and public health. The application of ethics will be critical across the full spectrum of precision medicine solutions, including development, implementation, and quality assurance. Focus areas encompass current clinical and preclinical nanomedical applications in imaging, diagnostics, cancer drugs, and radiation therapies.

Advanced genomic sequencing has unlocked a new world of opportunities in life sciences, from direct consumer testing to novel scientific discoveries, and the development of new personalized genomics-informed medical solutions. These include new molecules, pharmaceutical agents, medical devices/systems, and therapeutic pathways, which require safe, secure, and trusted strategies to access, store, share, and analyze the massive genomic data sets that are generated globally.

9.5 FUTURE DIRECTIONS

Despite the significant advances in science and technology over the last few decades, it is envisaged that the Fifth Industrial Revolution will accelerate the deployment of Healthcare 4.0, which will offer future generations a completely novel global-scale healthcare ecosystem. This new healthcare ecosystem will likely be powered by Web 3.0 and built through the application of design thinking methodology, while embracing an abundance mindset. This robust new future healthcare paradigm may encompass smart health homes, cities, and communities. Specifically, it is expected that smart healthcare will be delivered domestically (in the patient's own home) for the most part, albeit also in dedicated smart hospitals (for trauma cases and to deal with mass public emergencies).

Although some surgeries will still be performed in dedicated smart operating rooms, envisaged sophisticated autonomous (~Ø1 μm) nanomedical devices (which may number in the billions for specific procedures) may have the capacity to perform cellular/tissue/organ level surgical procedures in vivo, which will negate the requirement of most invasive surgeries altogether. Smart pharmacies, via domestic Factory@Home molecular manufacturing units in one's own home, will fabricate physician-prescribed nanomedical devices (at atomic/molecular resolution) to address all that ails us, including aging (Domschke and Boehm, 2014). These healthcare services, designed to be affordable to all, will be paid for via smart health wallets. Digital health twins will likely be routinely employed to test novel devices or procedures, augment the quality and safety of care, minimize adverse events, optimize pharmaceutical safety, and reduce research and development life cycles. All global healthcare stakeholders and users (e.g., patients, caregivers, families, clinicians, researchers, administrators, health plans, and other healthcare affiliates) will hopefully interact in a more human-centric, immersive way in a highly secure, nested healthcare blockchain-powered Metaverse or Omniverse. Nanotech, nanomed, AI/AGI/ASI, VR, AR, XR, 5G or 6G, and IoT will all be essential and converge in this novel ecosystem where physical and digital worlds coalesce (Rayan et al., 2023).

9.5.1 Global Healthcare and Digital Twins

As articulated by numerous thought leaders, digital twins will be designed and deployed to facilitate virtual collaboration, absorb and process big data, as well as assist patients with managing the physical world and their treatments more efficaciously and safely (Laubenbacher et al. 2021). Several domains, including automotive, aircraft (Armeni et al., 2022), energy (Alves de Araujo Junior et al., 2021), manufacturing (Maia et al., 2022), agriculture (Nasirahmadi and Hensel, 2022), livestock farming (Neethirajan and Kemp, 2021), environmental (de Koning et al., 2023), and urban planning (Zhao et al., 2022), have already demonstrated significant progress in employing the capacities of digital twins. The impacts of the SARS-CoV-2 pandemic have only accelerated the global pace and adoption of digital twin development in healthcare and the life sciences, which are seen as areas of high growth/impact (Armeni et al., 2022; Kamel Boulos and Zhang, 2021; Voigt et al., 2021).

According to Gartner, a digital twin is a digital representation of a real-world entity or system (Pettey, 2017; Schroeder, 2023). Implementing a digital twin is an encapsulated software

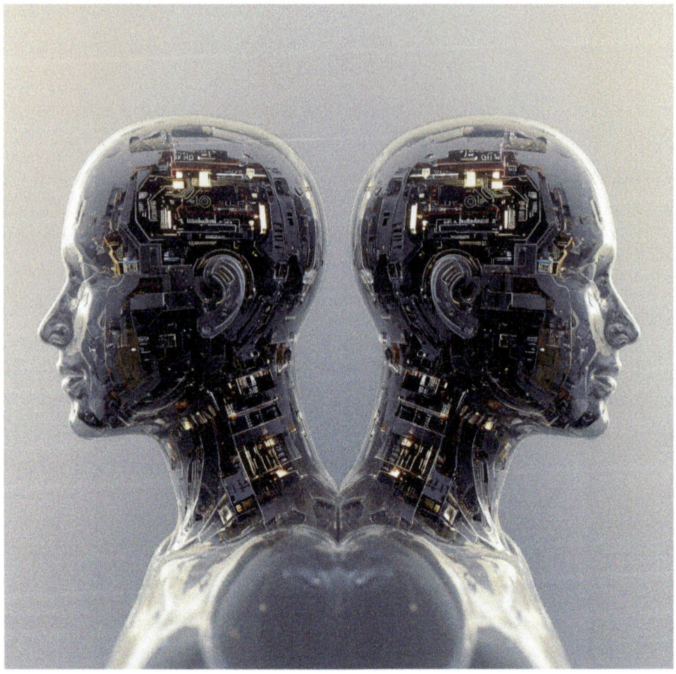

FIGURE 9.4 Digital twins in healthcare. Digital twins can contain data from various modalities (e.g., notes, imaging, and wearables). A vast amount of data can be processed with federated learning by sharing model weights and could be used to inform treatment options (e.g., lines of therapy for oncology). ML= machine learning. + = combination of multiple data modalities. (Image: Cellrep7, 2025. Created with Shutterstock AI.)

object or model that mirrors a unique physical object, process, organization, individual, or abstraction. The design and application of digital twins are complex and intimately connected to other digital technologies such as cloud computing, edge computing, nanotechnology, AI, IoT, 5G or 6G networks, blockchain, and virtual, augmented, or mixed reality. The societal benefits of deploying these highly versatile digital avatars are vast, as they can optimize efficiency, reduce long-term costs, enhance quality across domains, and advance research capacities (Figures 9.4 and 9.5).

While digital health twins may resolve many significant challenges that humanity encounters in healthcare and life sciences globally, we must guard against potential malicious intent and develop robust cyber-resilient programs to complement preemptive and proactive state-of-the-art digital ethics programs. From a healthcare delivery perspective, they can enhance the quality and safety of healthcare, improve clinical decision support systems, and accelerate the transition from research to clinical application for various medications, while expediting the approval of various new precision medicines. For the medical device industry, they also facilitate much shorter approval cycles and higher safety profiles. They could contribute to more interactive teaching methodologies and more effective safety training in medical education. Further, digital twins might be truly revolutionary and disruptive in various disciplines, such as surgery, oncology, or precision medicine. Last but not least, digital twins would also allow us to develop a hyperconnected global healthcare network and derive critical insights to drive research and the development of innovative diagnostic or therapeutic strategies (Kaul et al., 2023). Key areas of digital ethics risks that must be mitigated include informed consent, dynamic consent, data governance, integrity, and bias. Inclusive and diverse digital twins should be designed to positively modify paradigms for future generations, while negating the replication of current gaps or divides in this arena.

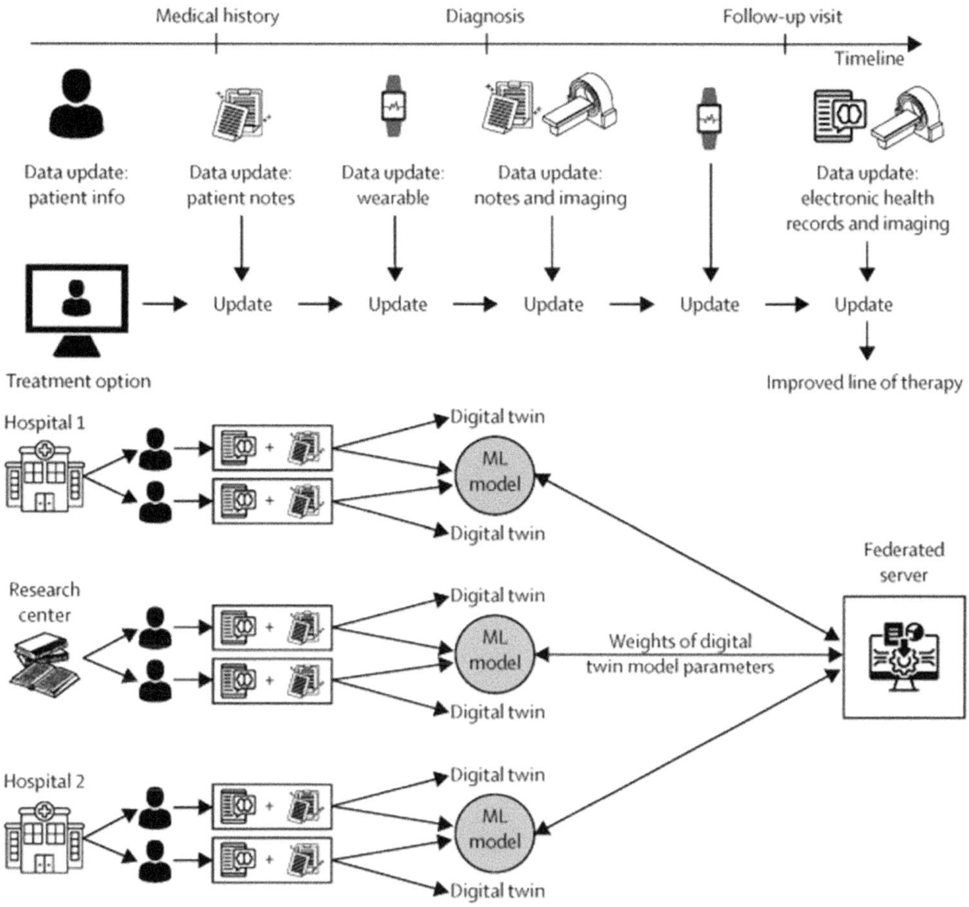

FIGURE 9.5 Augmenting digital twins with federated learning in medicine. (Reproduced from Nagaraj et al. (2023) with permission.)

By leveraging the power and reach of digital twins in conjunction with other emerging technologies, a high-performance Global Internet of Medicine may potentially be developed, which could become the conduit for a truly global healthcare ecosystem (GHCE) (Boehm, 2018). Exploiting the full capabilities of BCIs, implants, wearables, augmented intelligence, blockchain, 5G/6G network speeds, and cloud computing would achieve an exponential effect. Healthcare futurists would venture to envisage the emergence of state-of-the-art Smart Cities where digital healthcare twins, digital finance twins, and digital legal twins would be an integral part of our daily lives (Vasiliu-Feltes and Thomason, 2021).

9.5.2 Global Healthcare Data Exchanges

Humanity has recently endured a global economic and public health crisis that many could have never imagined. However, a trifecta of genomics, nanomedicine, and AI has proven extremely helpful toward developing solutions to address an existing pandemic and demonstrated how it might prevent future pandemics. However, myriad residual cultural, legal, regulatory, and religious barriers were also brought to light, which must be overcome. The latest SARS-CoV-2 pandemic has emphasized the complex ethical dilemmas that we must resolve. As a society, we will learn from the

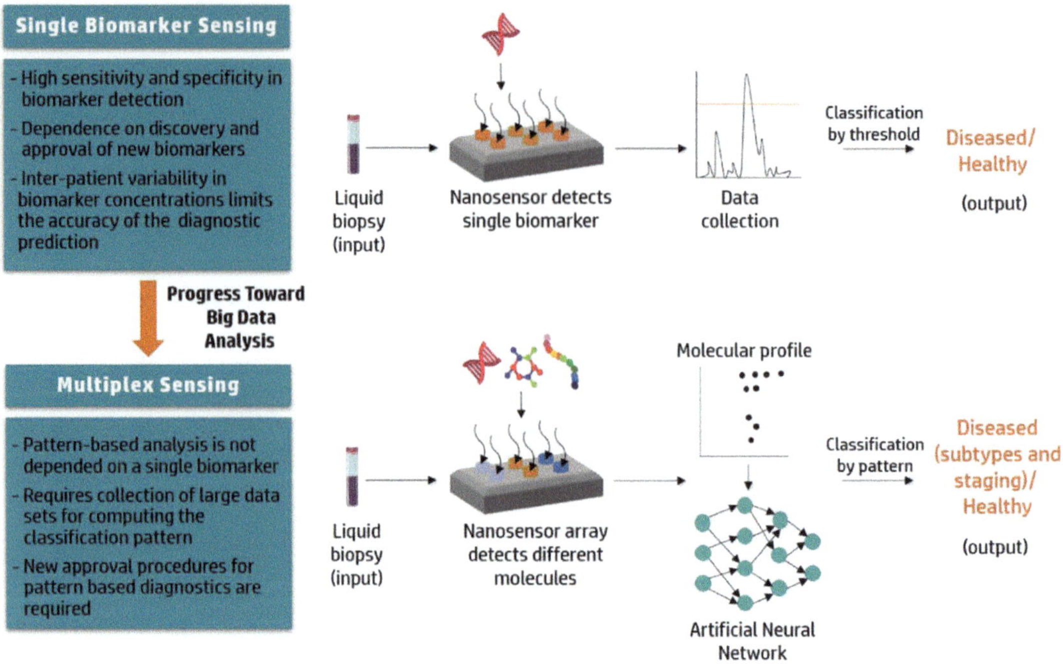

FIGURE 9.6 Integration of AI and nanotechnology for precision cancer nanomedicine. (Reproduced from Adir et al. (2020) with permission.)

vulnerabilities and weaknesses exposed and endeavor to correct the mistakes made during this pandemic. Hastily deployed emerging technologies breached ethical principles and human rights while aiming to improve global health. Through the establishment of global healthcare data exchanges designed by cross-disciplinary teams, inclusive of digital ethics experts, we may gain access to critical data while preserving privacy and security, upholding core ethics principles, and respecting human rights. Improvements to the current system have been exhaustive; however, several fundamental changes remain to be realized, which might drastically optimize public health for the global population.

In a hypothetical global digital ethics scenario, where all scientific and medical communities share their data via a single standardized global scale health platform, there would be far less redundancy, higher efficiency, and expedited research and development timeframes to identify viable health solutions. This might be accomplished using our full technological arsenal as an augmenting agent for our collective organic/digital brainpower. The ethical deployment of converging technologies such as nanotechnology, nanomedicine, AI (Figure 9.6), federated learning, digital twin technology, blockchain (Neyigapula, 2023), next-generation genomic sequencing, robotics, and advanced computing methods could and should be sustained long-term. This, to initially reveal solutions for some of the most severe chronic diseases that plague our society, while developing a powerful arsenal against all future, as yet unknown, pathogenic agents (e.g., long frozen, newly exposed bacteria and viruses from melting permafrost worldwide, some of against which we may have no innate immunity).

9.5.3 Global Healthcare and Precision Medicine

Leveraging the full capabilities of all novel technologies in precision medicine might serve as a foundation for the development of global databases and data exchanges that could spark further preclinical, clinical, and translational research, while driving the continuous optimization of public

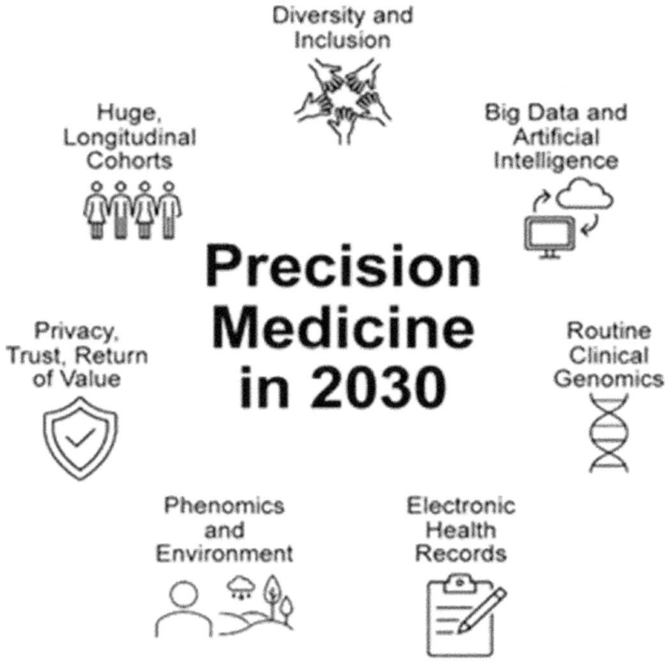

FIGURE 9.7 Precision medicine in 2030. (Reproduced from Denny and Collins (2021) with permission.)

health (Figure 9.7). However, specific innovations further increase the complexity of governance, regulatory, legal, compliance, and enforcement guidelines that key decision-makers must draft. For a global-scale healthcare system to be successful, far more extensive collaborations between all relevant stakeholders need to be initiated. Outdated laws and regulatory guidelines should be revised to match the latest scientific and technological advances, and the remaining barriers to large-scale technology deployments in healthcare need to be resolved.

The digital era has significantly altered how society establishes and verifies trust. Healthcare organizations may create digital trust by designing and implementing a comprehensive and resilient culture of digital ethics programs. In the post-pandemic era, digital health has become a potent driver of the global economy; however, it is also plagued by a significant trust gap. Sustainable societal trust in science/tech-driven healthcare models may be achieved only by collaborating toward the creation of a more responsible global data governance framework.

Given the prominent role that genomics plays in precision medicine and health, and the devastating impacts that privacy breaches can have on our society, there is an opportunity for precision health leaders to establish themselves as frontrunners in the digital ethics ecosystem. They can drive the creation of/enforce adherence to global digital ethics standards, best practices, and customized digital ethics KPIs.

Forward-looking companies will need to embrace a greater sense of purpose with an emphasis on population health, personalized care, and human-centric design for patients, whereas providers will ensure long-term success. C-Suite leaders will need to adapt and transform their organizations from an ROI-centered mission to meet the needs of this dynamic healthcare ecosystem. It will be necessary to shift their strategic mindset from "sick care" to "well-care," and they will have to be purposefully disruptive to maintain their competitive advantage.

Precision and digital health ecosystems are intertwined and will facilitate the transition toward a personalized care approach to optimize global health. Future generations will likely benefit from digital health twins, smart hospitals, pharmacies, laboratories, and imaging centers; however, they should also expect a state-of-the-art and ethically driven global precision health ecosystem. The future of precision health will likely be influenced through the establishment and maintenance of a state-of-the-art culture of ethics, which is continually optimized to adapt to the demands of an exponential medicine era. In light of the recent significant advances in AI (Naji et al., 2023; Vemula et al., 2023; Wójcik et al., 2023), ultrarapid advancements and deployments of new technologies and/or a constant influx of disruptors will likely begin to enter and support the foundations/establishment of an envisioned global healthcare ecosystem.

Personalized medicine and health, as well as the provision of medical services, are poised to revolutionize medical ecosystems on a global scale, as they will provide nanomedical therapies that are patient-customized down to the molecular level. When physicians have access to the unique molecular data profiles (omics) of patients, this will enable them to prescribe effective drugs (and, in the future, autonomous nanomedical devices) with minimal or no side effects to prevent, treat, and ultimately cure myriad diseases. Comprehensive patient molecular profiles may provide a plethora of data regarding their disease risks, medication responses, specific allergies, and nuances such as lactose intolerance and diet-specific requirements. Historically, it has been impossible to confirm the presence or absence of mutations in an individual's genome, as using molecular techniques such as PCR is very time-consuming and associated with many limitations. Conversely, many traits and diseases are multigenic and the consequence of interactions with the environment.

Further, mathematical models for the quantification of genetic risks remain to be developed. Recently, the creation of low-cost individual genetic profiles to rapidly determine an individual's genotype and all known mutations has become possible due to advances in microarray technologies, computer science, and statistics (Marini and Brandi, 2014; Ortega et al., 2015; Wysocki and Seibert, 2019). Concomitant with genomics advances in other multidisciplinary sciences, nanotechnology, nanomedicine, and systems biology have facilitated the elucidation of disease interactions at cellular, organellar, and molecular scales. Nanomedicine has enabled precisely controlled drug release profiles, which exploit the design and fabrication of nanostructured devices. Thus, the aim is to combine these data and examine their interactions to better contribute to enhancing human health.

In view of the current explosive progress in AI development and applications, coupled with steady progress in quantum computing, nanotechnology, nanomedicine, robotics, materials science, and 3D printing, it is not difficult to envision the complete disruption of current public healthcare models with a new state-of-the-art, massively distributed/decentralized global scale and virtualized public healthcare ecosystem. Concepts such as exponential medicine and design thinking would transition from innovation labs to large-scale implementations with human-centric architectures. Geopolitical and socioeconomic barriers would fall away in favor of the more significant benefits of a tremendously enhanced global public healthcare system, wherein the complete portfolio of available digital technologies from all industries would be integrated and implemented. Technologies that at present are only in niche domains will serve as catalysts toward the emergence of future of global public health, encompassing quantum and edge computing, 6G or 10G, DNA and neuromorphic computing, 3D printing, robotics, nanotechnology, nanomedicine, AR/VR/XR, telehealth- and mobile health, bio-implants, exoskeletons, human brain/machine/computer/(potentially future cloud) interfaces, CRISPR, Panomics, and Multiomics.

9.5.4 Global Health Metaverse and Omniverse

The latest technologies that define the global digital transformation ecosystem encompass cybersecurity meshes, data fabrics, privacy-enhancing computation, cloud-native platforms, composable applications, hyper-automation, decision intelligence, AI engineering, generative AI, distributed enterprises, and autonomous systems. Further, IEEE research projects that quantum computing, device nanotechnology, 3D-integrated circuits, multicore, photonics, universal memory, software-defined networks, IOT, natural user interfaces, 3D printing, computer vision and pattern recognition, computational biology, and medical robotics will soon play critical roles. Experts predict increased applications of vertical specific cloud-based solutions, blockchain technology adoptions, automation at scale, nanotech, cyber-AI, industrial robots, digital twins, sensor-embedded devices, and ambient experiences (Vasiliu-Feltes, 2022).

It must be ensured that novel VR environments are aligned with the UN 2030 agenda (Gentili, 2021; McBean, 2021) and UN Human Rights (Sui et al., 2023), which may only be accomplished through a global collaborative effort among key stakeholders. We should endeavor to create a diverse, inclusive, and equitable Meta- and Omniverse where individual privacy is respected and safeguarded. In this new era, seamless integration, interoperability, portability, robust security, and sustainability will be essential for the Metaverse/Omniverse Metropolis (Mozumder et al., 2023).

In addition to legal, regulatory, and compliance challenges, ethical challenges brought about by the disruptive power of these ecosystems must also be considered, as they are complex with many considering them a Pandora's Box. Moreover, global stakeholders will have to address digital and financial divides in many emerging markets, which will only be intensified with the large-scale adoption of the Metaverse and Omniverse (Vasiliu-Feltes, 2022).

The Bio-Nano-AI trifecta will likely define the Healthcare Metaverse and Omniverse. New layers of complexity emerging through the combination of quantum technologies with all other frontier technologies, including AI and nanotechnology, are significantly expanding our possibilities in the digital era. Nevertheless, from an ethical perspective, we must ask ourselves how these new and potentially very powerful ecosystems will impact society from societal, cultural, and educational perspectives.

There are numerous significant concerns regarding the difficulties involved with ensuring security, privacy, and trust under the large-scale adoption and convergence of these novel immersive, multi-sensory technology experiences, which will only be exacerbated through the addition of quantum computational power to AI, nanotechnology, nanomedicine, and other emerging frontier technologies. From the perspective of a digital ethicist, one may expect significant apprehensions about how these technologies will impact moral values, the ability to uphold ethical principles, and how they will impact human interactions and relationships (Cañas, 2022). Case in point, the emergence and increasing popularity of user-programmable AI girlfriends may seriously/negatively affect an entire generation of lonely young males (Marr, 2023; Thorp, 2024).

Experts have already cautioned against the profound and transformative sociocultural impacts of the Fourth Industrial Revolution and are currently sounding ethical alarms, as we have likely entered the fifth, and are preparing for the sixth. Key stakeholders face a complex Meta- and Omniverse fueled digital ethics conundrum. Critical concerns for experts tasked with the design of proactive digital ethics and cybersecurity programs for these new Meta and Omni eras powered by Web 3.0 include safeguarding privacy, digital identity, appropriate data stewardship, integrity, transparency, and explainability. Further, they must abide by the core principles of justice, autonomy, beneficence, and maleficence (Vasiliu-Feltes and Thomason, 2021).

9.6 FURTHER THOUGHTS AND CONCLUSION

Moral and ethical principles have been foundational in human society for centuries. After decades of marginalization, we are witnessing their resurgence within scientific and business communities induced, in great part, by the complex ethical issues that we now face. This is due to the high degree of automation and augmented intelligence that permeate our daily activities, global infrastructures, and business ecosystems. Scientific and business communities and numerous not-for-profit and government agencies are appropriately concerned as relates to the unique ethical issues associated with digital healthcare ecosystems.

The complex and dynamic interplay between cultural factors, legal landscapes, and socioeconomic factors of distinct countries or regions will profoundly influence the acceptance, adoption, and ethical deployment of emerging technologies. There are marked differences in digital literacy and fluency that can impact the capacities of key stakeholders to assess, design, develop, apply, monitor, implement, and sustain an extensive range of emerging technologies. For research and clinical care to thrive via the leveraging of frontier technologies such as AI and nanotechnology, novel regulatory frameworks must be formulated while optimizing legislative clarity.

Significant advances in AI, nanomaterial sciences, nanotechnology, and nanomedicine are poised to revolutionize sustainable development in healthcare and the life sciences, as well as a broad range of other industries that interact with healthcare and life sciences ecosystems. A recent report reviewed a range of sustainable AI use cases, which highlighted applications across diverse industries (Feiner et al., 2024).

Substantial paradigm shifts from the adoption of business-centric technologies to sociocentric approaches will likely emerge over the next few decades. The impacts of the SARS-CoV-2 pandemic have expedited this trend and emphasized the urgent need for ESG consciousness. Furthermore, global stakeholders and decision-makers should collaborate to require ESG mandates while establishing associated metrics. It is critical for industry leaders aiming to deploy AI and nanotechnologies to promote ethical leadership and design customized digital ethics frameworks.

Organizations that are forward-looking and understand the importance of having a strong sense of purpose in this digital era will likely gain or retain societal trust and certain advantages in a highly competitive global business ecosystem. Leaders have the opportunity to shape the future by fostering a Culture of Digital Ethics and contributing to the development of a Global Digital Ethics Framework for the healthcare industry. This global framework may facilitate the attainment of the United Nations Sustainable Development Goals (UNSDGs) and further validate the existing sense of purpose in life sciences research.

While the implementation of AI, nanotechnology, and nanomedicine will likely have immense impacts on all UNSDGs, a few will be more directly influenced by blockchain solutions (Zhang, 2023), where a lack of ethical deployment can have devastating repercussions for society. Through the primary characteristics of enhancing trust, openness, and transparency, blockchain technologies can augment and amplify sustainability efforts toward decreasing poverty, reducing hunger, improving access to quality education, optimizing gender equality, promoting meaningful work and economic growth, establishing robust infrastructures, easing inequalities, and creating sustainable cities. Undoubtedly, AI and nanotechnologies will profoundly influence health and wellness ecosystems, specifically the life sciences industry, through enhanced capabilities across essential domains (e.g., clinical trials, supply chain and contract management, financial transactions, credentialing, and safety). On a global scale, blockchain applications can also accelerate research and development efforts and serve as an enabler for the large-scale adoption of other emerging technologies.

One of the most essential aspects of the impacts of AI technologies, nanotechnology, and nanomedicine toward ensuring appropriate assent in human research is that they transcend ethics and elicit legal, social, and philosophical considerations. Blockchain enabled platforms also have strong potential to facilitate corporate ESG consciousness by becoming foundational technologies for data standardization, asset performance assessments, and compliance with ESG mandates/standards.

The positive and beneficial deployment of AI, nanotechnology, and nanomedicine can only be successful with ethical, responsible leadership. Leaders who endeavor to succeed in this highly virtualized, digitalized, and hyperconnected environment must possess a complex armamentarium of novel skills, such as digital literacy and fluency, global citizenship, design thinking, and a mastery of applied ethics. Within the ethics realm, there are several domains to consider when formulating a strategic state-of-the-art ethics plan (digital ethics, cyber ethics, business ethics, and leadership ethics).

Digital ethics addresses behaviors related to digital mediums, norms for the use of digital tools, autonomy, and ownership of online data. Business ethics includes governance, social, and fiduciary responsibilities, as well as discrimination, fraud, abuse, or bribery. Ethical life sciences leaders will be expected to highly regard moral values such as honesty, fairness, and respect for others. It will be necessary for the C Suite to dedicate resources to the facilitation of a culture of digital ethics while embedding its associated tools within the fabric of their enterprise strategies. By striving to demonstrate ethical leadership in this digital era, leaders may significantly contribute to establishing and optimizing a Global Life Sciences Ethics Culture.

The exponential adoption of blockchains in healthcare will require a robust, sustainable digital ethics culture to prevent potential violations of security, trust, integrity, digital privacy, and digital ownership. Digital ethics conscious leaders should be suitably concerned about upholding core foundational ethical values and those that are unique to the life sciences research ecosystem. As with all technologies, some inherent disadvantages will arise; thus, their large-scale application will require refinements over time. However, if prudently implemented and supported by appropriate digital ethics and data governance mechanisms, they might provide an optimized global-scale health and wellness ecosystem for the benefit of future generations.

Nanotechnology, nanomedicine, and AI in conjunction with other emerging and frontier technologies will be foundational toward the design, deployment, and maintenance of a robust global public health ecosystem. Comprehensive legal, regulatory, and ethics guidance will be essential; thus, this chapter has endeavored to serve as a guidepost for future research and collaboration between key stakeholders toward the design of an ethics framework for a global-scale, digitally based public health ecosystem powered by nanotechnology, nanomedicine, AI, and a host of other smart technologies.

REFERENCES

Adir O, Poley M, Chen G, Froim S, Krinsky N, Shklover J et al. Integrating artificial intelligence and nanotechnology for precision cancer medicine. *Advanced Materials*. 2020;32(13):e1901989.

Allen S, Fry LW. A framework for leader, spiritual, and moral development. *Journal of Business Ethics*. 2023;184(3):649–663.

Alves de Araujo Junior CA, Mauricio Villanueva JM, Almeida RJS, Azevedo de Medeiros IE. Digital twins of the water-cooling system in a power plant based on fuzzy logic. *Sensors (Basel)*. 2021;21(20):6737.

Armeni P, Polat I, De Rossi LM, Diaferia L, Meregalli S, Gatti A. Digital twins in healthcare: Is it the beginning of a new era of evidence-based medicine? A critical review. *Journal of Personalized Medicine*. 2022;12(8):1255.

Blackman R. Ethical Machines: Your Concise Guide to Totally Unbiased, Transparent, and Respectful AI. *Harvard Business Review Press*. 2022.

Boehm FJ. Women will lead the way to Global Health Care Equivalency (GHCE). LinkedIn Post. 2018. https://www.linkedin.com/pulse/women-lead-way-global-health-care-equivalency-ghce-frank-boehm/ (accessed October 22, 2023).

Boisselier E, Astruc D. Gold nanoparticles in nanomedicine: Preparations, imaging, diagnostics, therapies, and toxicity. *Chemical Society Reviews*. 2009;38(6):1759–1782.

Bouras MA, Lu Q, Zhang F, Wan Y, Zhang T, Ning H. Distributed ledger technology for eHealth identity privacy: State of the art and future perspective. *Sensors (Basel)*. 2020;20(2):483.

Brothers W. A Timeline of COVID-19 Vaccine Development. *BioSpace*. 2020. https://www.biospace.com/article/a-timeline-of-covid-19-vaccine-development/ (accessed October 22, 2023).

Cameron K. The Laws of Identity. 2005. Microsoft Corporation. https://www.identityblog.com/stries/2005/05/13/TheLawsOfIdentity.pdf (accessed October 19, 2023).

Cañas JJ. AI and ethics when human beings collaborate with AI agents. *Frontiers in Psychology*. 2022;13:836650.

Davis MB. Ethical implications of AI in medicine. In JR Smith (Ed.), *Applied Ethics in the Digital World. The Journal of AI and Medical Ethics*. 2021;4(3).

De Jong WH, Borm PJ. Drug delivery and nanoparticles: Applications and hazards. *International Journal of Nanomedicine*. 2008;3(2):133–149.

de Koning K, Broekhuijsen J, Kühn I, Ovaskainen O, Taubert F, Endresen D et al. Digital twins: Dynamic model-data fusion for ecology. *Trends in Ecology & Evolution*. 2023;38(10):916–926.

Denny JC, Collins FS. Precision medicine in 2030-seven ways to transform healthcare. *Cell*. 2021;184(6):1415–1419.

Domschke A, Boehm FJ. Quandary – Are Molecularly Manufactured Burgers Imbued with the Life Force? FQXi Foundational Questions Institute. 2014. https://forums.fqxi.org/d/2085-quandary-are-molecularly-manufactured-burgers-imbued-with-the-life-force-by-frank-josef-boehm-and-angelika-domschke (accessed October 22, 2023).

Edwards L, Burkhard S, Harbinja E. *Future Law Emerging Technology, Regulation and Ethics*. Edinburgh University Press, 2020.

Feiner A, Becke S, Jha N, Töpfer LM. The GenAI Building Blocks-Generative AI Unleashed for Sustainability. PricewaterhouseCoopers International Limited (PwCIL), January 24, 2024. https://www.pwc.de/en/digitale-transformation/generative-ai-artificial-intelligence/the-genai-building-blocks/generative-ai-unleashed-for-sustainability.html (accessed January 28, 2025).

Floridi L. Soft ethics and the governance of the digital. *Philosophy & Technology*. 2018;31:1–8.

Gami B, Agrawal M, Mishra DK, Quasim D, Mehra PS. Artificial intelligence-based blockchain solutions for intelligent healthcare: A comprehensive review on privacy preserving techniques. *Transactions on Emerging Telecommunications Technologies*. 2023;34(9):e4824.

Gentili PL. Why is complexity science valuable for reaching the goals of the UN 2030 agenda? *Rendiconti Lincei – Scienze Fisiche e Naturali*. 2021;32(1):117–134.

Gibney S, Hicks JM, Robinson A, Jain A, Sanjuan-Alberte P, Rawson FJ. Toward nanobioelectronic medicine: Unlocking new applications using nanotechnology. *Wiley Interdisciplinary Reviews: Nanomedicine and Nanobiotechnology*. 2021;13(3):e1693.

Goertzel B. From Abstract Agents Models to Real-World AGI Architectures: Bridging the Gap. In *Artificial General Intelligence*. Everitt T, Goertzel B, Potapov A. (eds.), Springer. 2017.

Hogan NR, Davidge EQ, Corabian G. On the ethics and practicalities of artificial intelligence, risk assessment, and race. *Journal of the American Academy of Psychiatry and the Law*. 2021;49(3):326–334.

Johnson JT, Smith LK. Ethical considerations in artificial intelligence development. In: Smith JR, editor. *Applied ethics in the digital world*. arXiv [Preprint]. 2021;2111:03425.

Johnson LS. AI Ethics: Challenges and Opportunities. In *AI and Ethics*. Johnson S. (ed.), Aspen Institute Publications. 2021.

Johnson WE. Ethical issues in AI and machine learning. In: Smith JR, editor. Applied ethics in the digital world. *APA PsycNet* [Internet]. 2021;92340–005.

Kamel Boulos MN, Zhang P. Digital twins: From personalised medicine to precision public health. *Journal of Personalized Medicine*. 2021;11(8):745.

Kaul R, Ossai C, Forkan ARM, Jayaraman PP, Zelcer J, Vaughan S, Wickramasinghe N. The role of AI for developing digital twins in healthcare: The case of cancer care. *Wiley Interdisciplinary Reviews: Data Mining and Knowledge Discovery*. 2023;13(1):e1480.

Kejriwal M. Essential features in a theory of context for enabling artificial general intelligence. *Applied Sciences*. 2021;11(24):11991.

Kim BY, Rutka JT, Chan WC. Nanomedicine. *New England Journal of Medicine*. 2010;363(25):2434–2443.

Laubenbacher R, Sluka JP, Glazier JA. Using digital twins in viral infection. *Science*. 2021;371(6534): 1105–1106.

Lewis MJ. Ethical Considerations in Nanotechnology. In *Applied Ethics in the Digital World*. Smith JR. (ed.), RSC Advances. 2021;17023–17032.

Maia E, Wannous S, Dias T, Praça I, Faria A. Holistic security and safety for factories of the future. *Sensors (Basel)*. 2022;22(24):9915.

Man K, Eisenstein NM, Hoey DA, Cox SC. Bioengineering extracellular vesicles: Smart nanomaterials for bone regeneration. *Journal of Nanobiotechnology*. 2023;21(1):137.

Marini F, Brandi ML. Pharmacogenetics of osteoporosis. *Best Practice & Research: Clinical Endocrinology & Metabolism*. 2014;28(6):783–793.

Marr B. Artificial Intimacy: How Generative AI Can Now Create Your Dream Girlfriend. *Forbes*. 2023. https://www.forbes.com/sites/bernardmarr/2023/09/28/artificial-intimacy-how-generative-ai-can-now-create-your-dream-girlfriend/?sh=191c7ebf464a (accessed October 22, 2023).

McBean GA. Integrating science to address food and health within Global Agenda 2030. *npj Science of Food*. 2021;5(1):8.

Mittlelstadt BD, Floridi L. *The Ethics of Biomedical Big Data*. Springer, 2018.

Mou X, Wu Q, Zhang Z, Liu Y, Zhang J, Zhang C et al. Nanozymes for regenerative medicine. *Small Methods*. 2022;6(11):e2200997.

Mozumder MAI, Armand TPT, Imtiyaj Uddin SM, Athar A, Sumon RI, Hussain A, Kim HC. Metaverse for digital anti-aging healthcare: An overview of potential use cases based on artificial intelligence, blockchain, IoT technologies, its challenges, and future directions. *Applied Sciences*. 2023;13(8):5127.

Muñoz F, Haidar ZS, Puigdollers A, Guerra I, Padilla MC, Ortega N, García MJ. A novel Chilean salmon fish backbone-based nanoHydroxyApatite functional biomaterial for potential use in bone tissue engineering. *Frontiers in Medicine (Lausanne)*. 2024;11:1330482.

Nagaraj D, Khandelwal P, Steyaert S, Gevaert O. Augmenting digital twins with federated learning in medicine. *Lancet Digital Health*. 2023;5(5):e251–e253.

Naji Y, Mahdaoui M, Klevor R, Kissani N. Artificial intelligence and multiple sclerosis: Up- to-Date review. *Cureus*. 2023;15(9):e45412.

Nasirahmadi A, Hensel O. (2022). Toward the next generation of digitalization in agriculture based on Digital Twin Paradigm. *Sensors (Basel)*. 2022;22(2):498.

Neethirajan S, Kemp B. Digital twins in livestock farming. *Animals (Basel)*. 2021;11(4):1008.

Neyigapula B. Synergistic integration of blockchain and machine learning: A path to a decentralized intelligent future. *Research Square*. 2023. 10.21203/rs.3.rs-3280888/v1.

Ortega VE, Meyers DA, Bleecker ER. Asthma pharmacogenetics and the development of genetic profiles for personalized medicine. *Pharmacogenomics and Personalized Medicine*. 2015;8:9–22.

Peterson M. The ethics of technology: Response to critics. *Science and Engineering Ethics*. 2018;24(5): 1645–1652.

Pettey C. Prepare for the Impact of Digital Twins. Gartner, Inc. 2017. https://www.gartner.com/smarterwithgartner/prepare-for-the-impact-of-digital-twins (accessed October 22, 2023).

Price AW. Aristotle's ethics. *Journal of Medical Ethics*. 1985;11(3):150–152.

Qing Y, Li R, Li S, Li Y, Wang X, Qin Y. Advanced black phosphorus nanomaterials for bone regeneration. *International Journal of Nanomedicine*. 2020;15:2045–2058.

Rayan RA, Zafar I, Tsagkaris C. Industry 4.0 Technologies for Healthcare: Applications, Opportunities, and Challenges. *Blockchain Technology Solutions for the Security of IoT-Based Healthcare Systems*. Academic Press. 2023;23–44.

Rizk SH. Ethical and Regulatory Challenges of Emerging Health Technologies. In *Applied Ethics in a Digital World*. IGI Global. 2022.

Schroeder A. How digital twins can prepare retailers for the upcoming holiday season. Networld Media Group, LLC. 2023. https://www.retailcustomerexperience.com/blogs/how-digital-twins-can-prepare-retailers-for-the-upcoming-holiday-season/ (accessed October 22, 2023).

Simon C. (2022). AGI Is Ready To Emerge (Along With The Risks It Will Bring). *Forbes*. 2022. https://www.forbes.com/sites/forbestechcouncil/2022/07/27/agi-is-ready-to-emerge-along-with-the-risks-it-will-bring/?sh=4a50ae8c332e (accessed October 22, 2023).

Simon CJ. Ethics and Artificial General Intelligence: Technological Prediction as a Groundwork for Guidelines. *2019 IEEE International Symposium on Technology and Society (ISTAS)*. Medford, MA, USA, 2019;1–6.

Smith RJ, Brown MA. AI and healthcare ethics: Current issues and future directions. *AI Ethics*. 2022;2(3):281–295.

Sui A, Sui W, Liu S, Rhodes R. Ethical considerations for the use of consumer wearables in health research. *Digital Health*. 2023;9:20552076231153740.

Taylor LM. Ethical implications of AI in biomedical research. *Journal of Biomedical Ethics*. 2022;25(2): 145–159.

Thorp HH. Unsettled science on social media. *Science*. 2024;384(6703):1389.

Van de Voorde M, Jeswani G. *Ethics in Nanotechnology: Social Sciences and Philosophical Aspects*. Berlin, Boston: De Gruyter, 2021.

Vasiliu-Feltes I. Preparing Businesses for the Metaverse and Omniverse Era. *Forbes*. 2022. https://www.forbes.com/sites/forbesbusinesscouncil/2022/01/31/preparing-businesses-for-the-metaverse-and-omniverse-era/?sh=6315922074b5 (accessed October 22, 2023).

Vasiliu-Feltes I, Thomason J. (eds.). *Applied Ethics in a Digital World*. IGI Global, 2021.

Vemula D, Jayasurya P, Sushmitha V, Kumar YN, Bhandari V. CADD, AI and ML in drug discovery: A comprehensive review. *European Journal of Pharmaceutical Sciences*. 2023;181:106324.

Voigt I, Inojosa H, Dillenseger A, Haase R, Akgün K, Ziemssen T. Digital twins for multiple sclerosis. *Frontiers in Immunology*. 2021;12:669811.

Wilson AB. The Role of Ethics in AI Development. In *AI and Ethics*. Johnson S. (ed.), Springer. 2022; 45–60.

Wilson SP. AI ethics and the law. *Neuroethics*. 2021;14(3):237–249.

Wójcik S, Rulkiewicz A, Pruszczyk P, Lisik W, Poboży M, Domienik-Karłowicz J. Beyond ChatGPT: What does GPT-4 add to healthcare? The dawn of a new era. *Cardiology Journal*. 2023;30(6):1018–1025.

Wysocki K, Seibert D. Pharmacogenomics in clinical care. *Journal of American Association of Nurse Practitioners*. 2019;31(8):443–446.

Zhang C, Yan L, Wang X, Zhu S, Chen C, Gu Z, Zhao Y. Progress, challenges, and future of nanomedicine. *Nano Today*. 2020;35:101008.

Zhang GH. Blockchain for United Nations Sustainable Development Goals. In: Blockchain and Tokenomics for Sustainable Development – Volume II [Internet]. *Front Blockchain*. 2023.

Zhao D, Li X, Wang X, Shen X, Gao W. Applying digital twins to research the relationship between urban expansion and vegetation coverage: A case study of natural preserve. *Frontiers in Plant Science*. 2022;13:840471.

Zhao L, Zhang L, Wu Z, Chen Y, Dai H, Yu X et al. When brain-inspired AI meets AGI. *Meta-Radiology*. 2023;1(1):100005.

Zorrón M, Cabrera AL, Sharma R, Radhakrishnan J, Abbaszadeh S, Shahbazi MA et al. Emerging 2D nanomaterials-integrated hydrogels: Advancements in designing theragenerative materials for bone regeneration and disease therapy. *Advanced Science (Weinh)* 2024;11(31):e2403204.

10 Beyond Diagnoses

Exploring Health through Systems, Behaviour, and Equity

Bryan Sykes

CONTENTS

DOI: 10.1201/9781003680093-10

10.1 INTRODUCTION: FRAMING HEALTHCARE BEYOND DIAGNOSES

Grounded in cell theory, germ theory, and bacteriology (Rocca and Anjum, 2020), the biomedical model has long been the cornerstone of Western medicine, which focuses narrowly on biological factors as the primary causes of disease (Wade and Halligan, 2004). While this framework has been instrumental in the management of acute conditions, its limitations are starkly evident in addressing multifactorial health challenges and chronic diseases, which often neglect the psychological, social, and behavioural dimensions of health. Approximately 30% of primary care visits involve symptoms that are devoid of clear biological markers (e.g. chronic fatigue syndrome), which highlight the inability of the biomedical model to capture the full spectrum of health determinants (Kroll, 2021). McGinnis et al. (2002) posited that health determinants encompass 30% genetic dispositions (inherited traits and gene expression), 15% social circumstances (socioeconomic factors, education, and social support), 5% environmental exposures (physiological and environmental conditions), 10% healthcare access and quality (availability and quality of medical care), and 40% behavioural patterns (individual behaviours and lifestyle choices), all of which reflect the intricate correlations between structural and systemic factors, individual behaviours, and equity considerations.

These determinants underscore the importance of surveying factors that lie beyond medical care. This, to understand and design effective interventions that consider the broader context in which individuals make choices that are shaped by both individual and societal actions (Laur et al., 2020). Many healthcare professionals struggle to understand the interactions between health and behaviour and how to best leverage that knowledge to help improve the health status of individuals and populations. According to the World Health Organization (WHO), one of the most important ways to reduce the burden from non-communicable – lifestyle – diseases is to control unhealthy lifestyle behaviours and choices that lead to their development. These strategies encompass reducing the harmful use of tobacco and alcohol, maintaining an active lifestyle, and developing a healthy diet. It is estimated that up to 80% of cases of coronary heart disease, 90% of type 2 diabetes cases, and one-third of cancers could be avoided by transitioning to a healthier diet, increasing physical activity, and stopping smoking (WHO, 2024).

A systematic review of health behaviour research from 2000 to 2005 (Painter et al., 2008) revealed that the use of behavioural theory significantly enhanced the development and effectiveness of health interventions. Their findings emphasized that theoretical frameworks assist with the identification of key behaviour determinants, which may then be targeted through tailored interventions (Painter et al., 2008). Glanz and Bishop (2010) further posited the importance of behavioural science theories in public health by highlighting how they informed the design and implementation of interventions. Understanding the theoretical underpinnings of health behaviours is essential for the creation of effective strategies that encourage healthier lifestyles, thus improving health outcomes. These models provide a framework that elucidates the diverse factors that influence individual health decisions, ranging from personal beliefs and attitudes to far broader social and environmental contexts. Furthermore, Ford-Gilboe et al. (2018) purported that the link between health-oriented behaviours and equity-oriented healthcare (EOHC) might favourably affect health outcomes.

EOHC can facilitate positive shifts in inequities as relates to health outcomes, particularly for marginalized populations, through addressing their specific needs and improving access to care. Browne et al. (2018) defined EOHC as,

> ...an approach that aims to reduce the effects of structural inequities (such as poverty), including the inequitable distribution of the determinants of health (e.g., income and housing) that sustain health inequities; the impact of multiple and intersecting forms of racism, discrimination, and stigma (e.g., related to mental illness, chronic illnesses, non-conforming gender and sexual identities, etc.) on people's access to services and their experiences of care; and the frequent mismatches between dominant approaches to care (e.g., an emphasis on scheduled appointments versus same day or drop in appointments to accommodate patients' needs) and the needs of people who are most affected by health and social inequities.

The connection between EOHC and health behaviours lies in its capacity to empower individuals by addressing structural and systemic barriers, whilst fostering environments that are conducive to making healthier choices. For instance, incorporating the four key dimensions of equity-oriented health services – inequity-responsive care, culturally safe care, trauma- and violence-informed care, and contextually tailored care (Browne et al., 2016) – in the design of community-based interventions for Indigenous populations can ensure that care is tailored to their unique needs to promote sustainable health behaviour changes.

The global upsurge in non-communicable diseases (NCDs) – including cardiovascular diseases, diabetes, and cancer – further underscores the importance of considering social, cultural, and contextual factors that may impact the health behaviours of populations. Despite significant healthcare expenditures that include preventive initiatives, the United States allocates less than 5% of health dollars to public health and prevention (Mays and Smith, 2011), which has resulted in suboptimal outcomes. According to the Centers for Disease Control (CDC), chronic disease is one of the leading causes of death and disability, resulting in over $4.1 trillion in annual health costs in the United States (CDC, 2024). NCDs kill 41 million people each year, which is equivalent to 74% of all deaths globally (WHO, 2024). Further, researchers suggest that four groups of diseases account for over 80% of all premature non-communicable deaths, including cardiovascular disease, which accounts for the deaths of 17.9 million people annually, followed by cancers (9.3 million), chronic respiratory diseases (4.1 million), and diabetes (2.0 million, including kidney disease deaths caused by diabetes) (WHO, 2024).

The exclusive focus of biomedical models on the treatment of disease rather than health promotion and prevention promulgates a reactive healthcare system that prioritizes and operationalizes standard of care as the functional strategy of choice. Optimal care, however, is proactive, which prioritizes preventive measures and health education to address the root causes of illness, reduce health disparities, and foster long-term well-being. This approach emphasizes pre-emptive interventions, lifestyle modifications, and community-based strategies to enhance population health outcomes, while minimizing the burden of chronic diseases (Murray et al., 2002). Since preventive care spending may vary between countries, Dickson et al. (2021) discussed how the quantity of health spending did not directly equate to the quality of healthcare or better outcomes. This chapter advocates for a paradigm shift that operates beyond the biomedical model towards a biopsychosocial healthcare framework that integrates systems, behaviours, and equity. An examination of healthcare through these lenses highlights how holistic, community-centred approaches may empower individuals, reduce disparities, and improve population health outcomes.

10.2 PART I: HEALTH SYSTEMS IN CONTEXT, UNPACKING SYSTEMS DYNAMICS, AND DETERMINANTS OF HEALTH

10.2.1 Systems Thinking in Health: An Integrative Framework

Systems thinking in health is an analytical approach that views health and healthcare systems as complex, interconnected networks. This perspective acknowledges that health outcomes arise from interactions between multiple components, including individuals, communities, healthcare systems, medical technologies, policies, and broader societal factors (Bambra et al., 2010). Rather than focusing on isolated issues, systems thinking emphasizes the dynamic relationships and feedback loops that drive outcomes, recognizing that interventions in one aspect of the system can create ripple/butterfly effects elsewhere (Braveman et al., 2011). Fundamentally, public health is founded on the concept that health and illness have causes or conditions that lie beyond the biology and behaviour of the individual human being (Leischow and Milstein, 2006). Systems thinking is particularly valuable for understanding health determinants, as it transcends reductionist models, offering a holistic lens to explore how diverse factors – social, economic, behavioural, and environmental – interact to shape health outcomes. Equally important is an emphasis on relating different types of systemic structures that shape our lives, including the biological systems of our bodies, the organizational systems within which we work, and the political systems through which we govern public affairs. By framing health as a result of interconnected determinants, systems thinking provides a robust framework for designing interventions that address root causes, promote equity, and improve outcomes. It shifts the focus from isolated interventions to comprehensive strategies, which enables a deeper understanding of how to achieve sustainable improvements in health. A systems health approach acknowledges these complexities and shifts the focus from fragmented, disease-centred care to integrated, multi-level strategies that address the root causes of poor health. Anthropogenically built environments, for instance, influence physical activity levels, exposure to pollutants, and access to healthy foods, all of which contribute to long-term health outcomes (Diez Roux, 2016). Moreover, economic and policy-level factors, such as access to affordable healthcare, workplace protections, and urban planning policies, determine the extent to which individuals and communities may achieve optimal health (National Academies of Sciences 2017).

Systems approaches can assist with addressing challenges in health disparities in three ways: (1) systems thinking can inspire the development of more sophisticated dynamic conceptual models of the causes of health disparities, (2) systems tools (formal models and simulations) can contribute to the exploration and refinement of these models, while examining the effects of different interventions in the context of dynamic relations, and (3) systems approaches can enhance the application of existing data and promote the acquisition of new types of required data (Diez Roux, 2011). By recognizing health as an interconnected complex system, policymakers, healthcare professionals, and researchers can develop interventions that promote equity and long-term sustainability. This broader perspective lays the foundation for systems thinking in health, an approach that moves beyond linear cause-and-effect models and instead considers the interdependent nature of health determinants and their feedback loops. The following sections explore how systems thinking might serve as a powerful framework for the design of more effective, scalable, and sustainable health interventions.

10.2.2 Social Determinants of Health

Building on the systems perspective, it becomes evident that health outcomes are not merely the result of individual choices or medical care but are profoundly shaped by broader social, economic, and environmental factors, which are collectively referred to as the social determinants of health (SDoH) (Figure 10.1). The WHO (2025) defines SDoH as the "non-medical factors that influence health shaped by conditions in which people are born, grow, live, work, and age and the wider set of forces and systems shaping the conditions of daily life." The SDoH encompass a wide range of factors that include socioeconomic status, education, neighbourhood/physical environments, employment, social support networks, and access to healthcare. Understanding SDoH is critical for addressing health disparities and promoting health equity. These determinants provide a critical lens for understanding how systemic forces influence health disparities and inequities within and across populations.

The impacts of SDoH are profound. For instance, individuals who reside in areas with limited access to fresh food (often referred to as "food deserts") are at a higher risk of developing chronic illnesses such as diabetes and hypertension (Kelli et al., 2017). Similarly, unstable housing has been linked to poor mental health outcomes and increased hospitalizations. Programmes like the Housing First model (National Alliance to End Homelessness, 2024), which prioritizes housing stability without preconditions, have demonstrated significant improvements in health and reduced healthcare costs for vulnerable populations. On a global scale, nations with robust social policies tend to have better health outcomes and reduced disparities (Courtin et al., 2020). As an exemplar, Scandinavian countries, which invest heavily in education, housing, and healthcare, consistently report lower rates of chronic illness and longer life expectancies in contrast to countries with weaker social safety nets (Nordic Burden of Disease Collaborators, 2019).

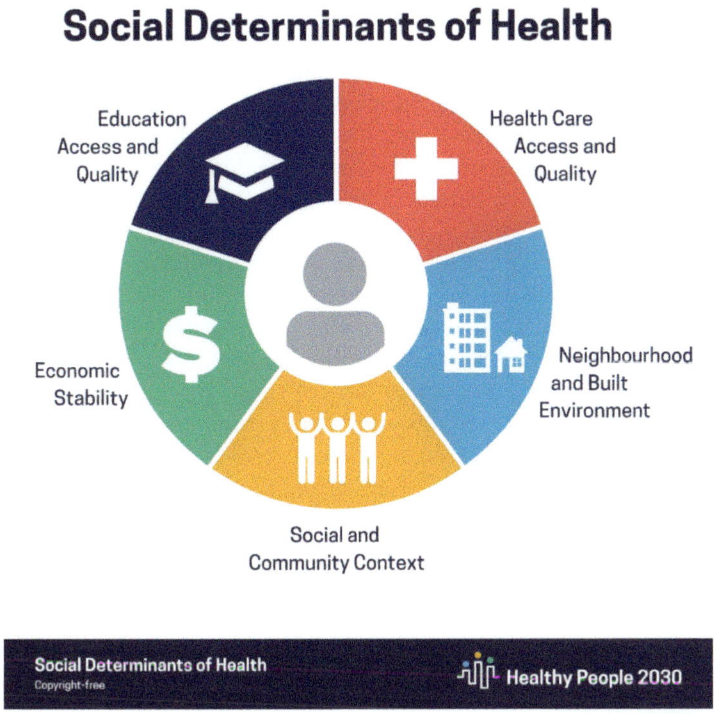

FIGURE 10.1 Healthy People 2030 – Social Determinants of Health Model. (Reproduced from Healthy People 2030. U.S. Department of Health and Human Services, Office of Disease Prevention and Health Promotion. https://odphp.health.gov/healthypeople/objectives-and-data/social-determinants-health) (Public domain).

Moreover, systemic inequities (including poverty and racial discrimination) exacerbate the challenges posed by SDoH. Structural barriers often limit access to healthcare, which perpetuates the cycles of poor health and poverty (Freed et al., 2013). Addressing these inequities requires targeted interventions such as increased funding for affordable housing, expanded access to quality education, and the implementation of anti-discrimination policies. Efforts to address SDoH must also empower individuals with a sense of agency; the belief that they can influence their own health outcomes (Alegría et al., 2018) as a sense of control over one's life significantly positively impacts physical and mental health. A holistic, equity-driven approach that integrates policy, community, and individual action is essential for reducing disparities and promoting health equity. Addressing SDoH requires a multifaceted, multi-stakeholder approach that includes policy interventions, community-based initiatives, and changes in individual behaviours.

10.2.3 Urban Planning and Health Disparities

Among the many SDoH, the built environment – shaped by urban planning decisions – emerges as a pivotal factor that influences health inequities and disparities (Rojas-Rueda and Morales-Zamora, 2023). The design and organization of urban environments play pivotal roles in shaping health equity by influencing behavioural patterns and safety. Urban land use and planning directly impact multiple health determinants, which contribute to the prevalence of NCDs and intensifying health disparities. According to the WHO (2023), the leading risk factors for NCD-related mortality (physical inactivity, excessive alcohol consumption, tobacco use, and poor dietary habits) are all intricately linked to the built environment. Urban planning decisions affect the availability and accessibility of green spaces, the design and safety of cycling and pedestrian infrastructures, housing conditions, and the distribution of food outlets, including fast-food establishments and alcohol retailers. These environmental factors, in turn, shape individual and community health behaviours by influencing physical activity levels, transportation choices, air quality, sanitation standards, and indoor living conditions, such as ventilation and exposure to toxins (Tainio et al., 2021). Addressing these structural determinants through equitable urban planning is essential to reducing the burden of NCDs and fostering healthier, more resilient communities (Marmot et al., 2008). Consequently, urban design that integrates health considerations can mitigate the risk factors for an extensive range of health conditions to promote more sustainable and equitable health outcomes.

There is a notable difference in the prevalence of type 2 diabetes between urban and rural areas, which is primarily driven by lifestyle and environmental factors. A study by Ranasinghe et al. (2021) posited that urbanization contributed to higher rates of type 2 diabetes, particularly due to dietary patterns characterized by the excessive consumption of processed foods, sedentary lifestyles, and environmental stressors such as poor air quality. In urban settings, the availability of fast food outlets, the prevalence of car-dependent transportation, and limited opportunities for outdoor physical activities all contribute to higher diabetes rates. Conversely, rural areas tend to have lower instances, which is partly due to greater physical activity levels and the lower availability of processed foods. However, rural populations must confront their own health challenges, including limited access to healthcare facilities and decreased health literacy (Ranasinghe et al., 2021). Addressing these disparities necessitates a comprehensive understanding of the ways in which urban design facilitates the prevalence of diabetes, coupled with the implementation of policies that promote health equity in both urban and rural contexts.

The concept of "health urbanism" (Sepe, 2018) integrates health into urban planning and design, which offers a transformative approach towards the creation of cities that prioritize the health and well-being of their residents. The THRIVES framework (Tools for Health and Resilience in Vulnerable Environments) emphasizes a holistic approach to urban planning that aims to improve health outcomes through resilient, sustainable, and equitable design principles. As highlighted by the THRIVES framework, urban environments must prioritize health promotion by focusing on factors such as access to clean water, quality housing, green spaces, and safe, walkable streets (Pineo, 2022). Through the adoption of health urbanism principles, cities can generate environments that foster physical activity, reduce air and noise pollution, increase access to healthy foods, and enhance social connectivity. The THRIVES framework advocates for inclusive policies that address the health needs of vulnerable populations to ensure that no one is left behind in the pursuit of healthier urban spaces. Implementing this framework at a systemic level may significantly reduce the burden of NCDs and improve overall health outcomes in urban environments.

10.3 PART II: BEHAVIOURS AND INFLUENCING HEALTH ACTIONS

10.3.1 Key Models Explain Health Behaviours and Outcomes

Behavioural medicine is a multidisciplinary field that integrates behavioural, psychosocial, and biomedical knowledge, which provides a comprehensive approach to understanding and influencing health behaviours and outcomes. Theofilou (2023) highlighted that behavioural medicine addresses the complex interactions between behaviour, biology, and the environment, which offers a new perspective in health intervention strategies. This field acknowledges the bidirectional relationships between psychological health and physical conditions, such as dermatological issues in children (Mitchell, 2018) and the interplay between major depression and inflammation (Messay et al., 2012). Further, Yao et al. (2023) explored the bidirectional relationships between pain and mental disorders that underscored the intricate connections between mental and physical health. By incorporating insights from behavioural medicine, health interventions can be designed to address the multifaceted nature of health to foster more effective and sustainable outcomes.

10.3.2 Health Belief Model

The Health Belief Model (HBM) is one of the most widely employed frameworks in health behaviour research and interventional design. Developed in the 1950s by social psychologists at the U.S. Public Health Service, the HBM was initially created to understand why people failed to participate in programmes that might prevent and detect disease (Rosenstock, 1974). Its historical origins are deeply rooted in efforts to address public health challenges by elucidating the psychological determinants of health behaviours. This model encompasses several constructs:

- Perceived Susceptibility: The degree to which a person feels at risk for a health issue.
- Perceived Severity: The perceived seriousness of the health problem and its potential consequences.
- Perceived Benefits: The perceived effectiveness of taking specific actions to reduce risk or severity.
- Perceived Barriers: The perceived obstacles to taking the proposed health actions.
- Cue to Action: Factors that trigger the decision-making process to accept a recommended health action.
- Self-Efficacy: Confidence in one's ability to take the action.

The HBM posits that the health-related behaviours of individuals are influenced by their perceptions of susceptibility to illness, the severity of illness, the benefits of taking preventive measures, and barriers against taking such actions. Cues to action and self-efficacy were subsequently incorporated into the model to better explain and predict health behaviours (Janz and Becker, 1984; Rosenstock, 1974). The benefits of using the HBM are evident in its capacity to provide a clear and structured approach to the design of health interventions. By addressing each component of the model, public health practitioners can develop targeted strategies that effectively promote behavioural changes (Gipson and King, 2012).

10.3.3 Case Study: Corrie Health's Secondary Prevention Model for Cardiovascular Disease

A practical application of the HBM was demonstrated through Corrie Health's secondary prevention model for cardiovascular disease (Figure 10.2). This model leverages the HBM to enhance patient adherence to preventive measures, thus reducing the risk of recurrent cardiac events (Krumholz et al., 2019). By emphasizing the perceptions of patients as to their susceptibility to heart disease, the severity of potential outcomes, and the benefits of adhering to prescribed preventive measures, Corrie Health effectively addresses the key determinants of health behaviour. The integration of the HBM in this context underscores its utility in real-world applications and potential to enhance the effectiveness of health interventions (Krumholz et al., 2019). The inclusion of tools such as mobile health applications and wearable devices assists patients with tracking their health metrics and receiving timely interventions. These factors are integral to enhancing patient engagement and their adherence to preventive actions.

10.3.4 Theory of Planned Behaviour

The Theory of Planned Behaviour (TPB) is a psychological framework that has been extensively utilized to predict and understand health behaviours. Developed by Icek Ajzen in 1985, the TPB

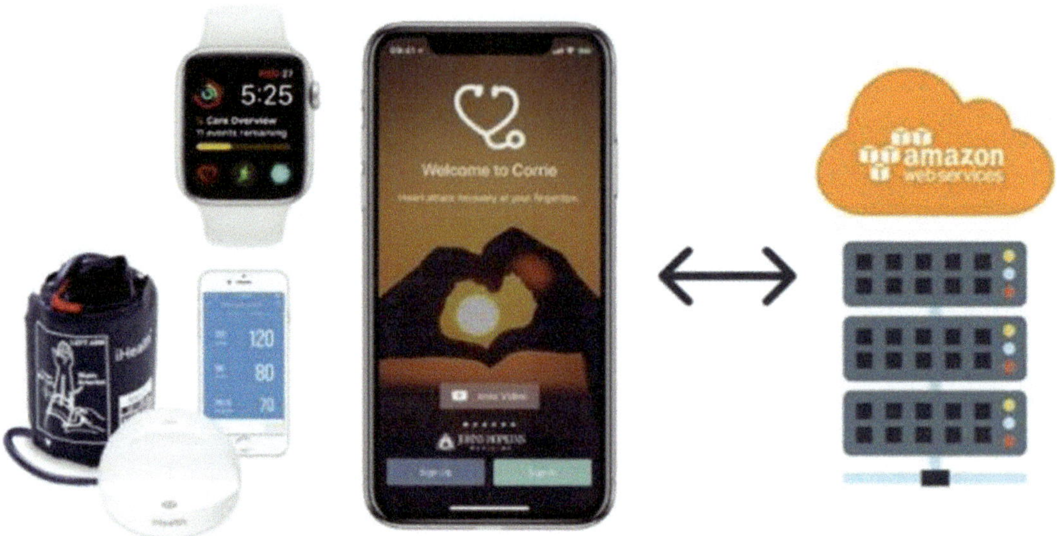

FIGURE 10.2 Corrie Health tools used in the secondary prevention model for cardiovascular disease (Reproduced with permission from Spaulding et al. Corrie Health Digital Platform for Self-Management in Secondary Prevention After Acute Myocardial Infarction. *Circ Cardiovasc Qual Outcomes*. 2019;12(5):e005509.)

posits that an individual's behaviour is directly influenced by their intent to engage in it. This intention is shaped by three core components: attitude towards the behaviour, subjective norms, and perceived behavioural control (Ajzen, 1991). This model extends the Theory of Reasoned Action by incorporating perceived behavioural control, which reflects one's perception of their capacity to perform the behaviour (Wallston, 2000). This offers a more comprehensive understanding of the factors that influence behaviour. The TPB has been instrumental in designing and implementing health interventions by providing insights into the cognitive and social factors that drive health behaviours. For instance, by addressing attitudes, enhancing perceived behavioural controls, and leveraging social norms, public health practitioners can more effectively promote behaviour change and improve health outcomes (Armitage and Conner, 2001).

The application of TPB in health research underscores its utility in the creation of targeted interventions that resonate with the beliefs and perceptions of patients. TPB has demonstrated significant potential in addressing prevalent chronic conditions via targeted behavioural interventions. For example, smoking cessation programmes that utilize TPB can translate to higher quit rates among smokers, thereby reducing the incidence of chronic conditions like cardiovascular diseases and lung cancer (Topa and Moriano, 2010). Similarly, TPB has been effective in diabetes management, where interventions that focused on enhancing the beliefs of patients in their abilities to adhere to dietary and physical activity recommendations have resulted in improved glycaemic control and reduced diabetes-correlated complications (Zeidi et al., 2021).

10.3.5 Case Study: Application of TPB in Smoking Cessation Programmes

Chronically ill patients, who account for ~50% of the population at large, make up the sickest and most complex demographic. Their utilization of interventions is primarily characterized by high costs, which consume more than 85% of healthcare costs in contrast to non-chronic patient populations (Holman, 2020). By leveraging the principles of the TPB, healthcare practitioners can develop strategies that not only address the medical needs of patients but also empower them to take proactive steps towards better health. A notable example of TPB in action is found in smoking cessation programmes. Research has suggested that interventions based on TPB significantly influenced the attitudes, perceived behavioural controls, and subjective norms of smokers, which led to higher quit rates and consequently a reduced incidence of smoking-related chronic conditions. One such study by Topa and Moriano (2010) revealed that addressing these key components through targeted programmes resulted in improved smoking cessation outcomes, which verified the practical application of TPB in public health initiatives.

10.3.6 Social Ecological Model

The Social Ecological Model (SEM), developed by Bronfenbrenner (Kilanowski, 2017), provides a comprehensive framework for understanding the multiple levels of influence on health behaviours and outcomes. This model emphasizes the interdependence of factors across different levels, which suggests that effective health interventions must consider broader social context (Sallis et al., 2008). The SEM comprises five levels:

1. Individual Level: It includes personal factors such as knowledge, attitudes, and skills. For instance, interventions that enhance an individuals' health literacy have been shown to improve self-care behaviours and health outcomes (Coughlin et al., 2020).
2. Interpersonal Level: It encompasses social networks and support systems that play critical roles. Studies implied that robust social support networks significantly impacted mental health and adherence to health-promoting behaviours (Heaney and Israel, 2008).
3. Organizational Level: It involves institutions and organizations that influence health behaviours. For example, workplace wellness programmes have been effective in promoting physical activity and healthy eating among employees (Song and Baicker, 2019).
4. Community Level: At this level, relationships between organizations and communities are central. Community-based interventions, such as those that involve local health workers, have been successful in managing chronic conditions like hypertension and diabetes (Kangovi et al., 2018).
5. Policy Level: It includes policies, regulations, and laws that affect health outcomes. Public health policies, such as smoking bans and taxes on sugary drinks, have been effective in reducing the prevalence of smoking and obesity (Frieden, 2010).

The SEM is particularly relevant in the context of the gap between cultural and biological evolution, a concept explored by Daniel Lieberman in "The Story of the Human Body: Evolution, Health, and Disease" (Lieberman, 2016). Lieberman argues that many modern health issues are the result of maladaptation to rapid lifestyle changes. For example, the sedentary lifestyles and high-calorie diets that are prevalent in modern societies are significantly mismatched with our evolutionary adaptations, which have led to rising chronic conditions such as obesity and type 2 diabetes. Through the application of SEM, public health interventions may address these maladaptations by promoting environments that encourage physical activity and healthy eating, which more closely aligns our lifestyles with our evolutionary biology. This comprehensive approach is essential for mitigating the health impacts of our rapidly changing cultural environment.

10.3.7 Case Studies: North Karelia Project and Healthy Eating Active Living

Several case studies have illustrated the application of SEM in public health. One notable example is the North Karelia Project (Finland), which aimed to reduce cardiovascular disease through community-wide efforts. By addressing factors at all levels of the SEM, the project successfully reduced smoking rates, improved dietary habits, and decreased cardiovascular mortality (Puska and Jaini, 2020). In addressing chronic conditions, the SEM has been employed in interventions that targeted obesity. For instance, the Healthy Eating Active Living (HEAL) programme (California, USA) leveraged community partnerships and policy changes to create environments in support of healthy behaviours, which resulted in significant reductions in obesity rates among participants (Cheadle et al., 2018).

10.3.8 Biopsychosocial Model (BPS)

Developed by George Engel in 1977, the Biopsychosocial Model (BPS) revolutionized the understanding of health and illness through the integration of biological, psychological, and social factors into a cohesive framework. This model diverged from the traditional biomedical approach, which focused solely on biological aspects. Instead, it emphasized the complex interplay between biological, psychological, and social influences on health (Borrell-Carrió et al., 2004). The holistic approach of the BPS model aligned with evolving perspectives in health psychology and public health, which advocated for a comprehensive view of health determinants (Gatchel et al., 2007; Havelka et al., 2009; Lehman et al., 2017). The biological factors in the BPS model included genetic predispositions and physiological processes, which are essential components of an individual's health profile (Suls and Rothman, 2004). Psychological factors encompassed mental health, emotions, and behaviours, which recognized the significant impact of mental states on physical health outcomes (Wade and Halligan, 2017). Social factors encompassed socioeconomic status, culture, and community support, which are critical in shaping health behaviours and access to healthcare resources (Adler and Newman, 2002).

Understanding health through the BPS model allowed for a more holistic approach that acknowledged the diverse determinants of health. This perspective is particularly valuable in addressing health equity, as it recognizes the broader social context and systemic barriers that influence health outcomes (Marmot, 2005). By integrating these multiple dimensions, healthcare providers can develop more comprehensive and effective interventions that target the root causes of health disparities (Glanz and Bishop, 2010). The BPS model also underscores the importance of integrating public health strategies with individual healthcare to improve population health outcomes. This integration may be seen in the development of comprehensive community health needs assessments that align public health and healthcare delivery efforts to address SDoH (Shahzad et al., 2019). By adopting a biopsychosocial perspective, healthcare systems can better address the multifaceted nature of health and develop innovative solutions that enhance health equity and well-being for all individuals (McLeroy et al., 1988).

10.3.9 Case Studies: Improving Patient Outcomes in Chronic Diseases

A notable case study that demonstrated the application of the BPS model was the use of integrated care models for managing chronic diseases like diabetes and COPD. These models combined medical treatment with psychological counselling and social support to enhance patient self-management and overall well-being. For diabetes care, programmes that addressed stress management, provided education on lifestyle changes, and offered community support exhibited significant improvements in patient adherence to treatments and overall health outcomes (Suls and Rothman, 2004). Similarly, COPD management programmes that incorporated psychological therapies and social support networks helped patients to cope with the emotional and social challenges of their condition, which led to improved health outcomes (Hatala, 2012).

The successful application of behavioural models in real-world health interventions demonstrates the importance of understanding the multifaceted nature of health behaviours. By integrating theoretical frameworks such as the TPB, the SEM, and the BPS into practical strategies, complex health issues can be effectively addressed to improve outcomes. Building on these insights, it is essential to explore how lifestyle medicine can be integrated into global healthcare systems to further promote health equity.

10.4 PART III: EQUITY – BRIDGING GAPS IN HEALTHCARE ACCESS AND QUALITY

10.4.1 Advancing Equity in Healthcare Systems

Health disparities and inequities persist due to an array of structural and systemic barriers that prevent equitable access to care. Braveman et al. (2017) posited that, "health equity means that everyone has a fair and just opportunity to be as healthy as possible. This requires removing obstacles to health such as poverty, discrimination, and their consequences, including powerlessness and lack of access to good jobs with fair pay, quality education and housing, safe environments, and health care." Moreover, socioeconomic status, geographic location, insurance coverage, and provider availability significantly impact healthcare access (Bailey et al., 2017). Marginalized populations (including racial and ethnic minorities, rural residents, and low-income groups) often must endure travelling long distances to healthcare facilities, inadequate insurance coverage, and provider shortages (Braveman et al., 2021). Language barriers, cultural insensitivity, and historical mistrust in medical institutions further exacerbate these inequities (Artiga and Hinton, 2018). Additionally, financial constraints limit access to preventive and routine care, which translates to delayed diagnoses and poorer health outcomes (Parker et al., 2020). Those without insurance or the underinsured are more likely to forgo necessary care or face catastrophic medical expenses (Tolbert et al., 2024). Additionally, implicit bias within the healthcare system has been shown to contribute to disparities in treatment decisions, pain management, and provider-patient interactions (Green et al., 2007).

To address these barriers, mobile health clinics and telehealth have emerged as transformative solutions. Mobile clinics bring healthcare directly to underserved populations, which reduces or negates geographical barriers and increases access to essential services (e.g. vaccinations, screenings, and chronic disease management) (Yu et al., 2017). Studies have indicated that mobile clinics improve patient engagement and enhance preventive care efforts in rural and urban communities with limited healthcare infrastructures (Malone et al., 2020). Telehealth has also revolutionized healthcare accessibility, which was particularly the case during the COVID-19 pandemic, as it facilitated remote consultations, chronic disease management, and mental health services (Hollander and Carr, 2020). Telemedicine reduces transportation barriers and enables patients to receive timely care regardless of location (Scott Kruse et al., 2018), albeit challenges such as digital literacy, internet access, and reimbursement policies must be addressed to ensure equitable adoption (Ramsetty and Adams, 2020).

10.4.2 Improving Quality of Care

Equity in healthcare is not only about access but also about the quality of care delivered. In a seminal report titled, "Crossing the Quality Chasm," the Institute of Medicine (2001) identified six core aims for shaping and sustaining the quality of care (safe, efficient, timely, equitable, and patient-centred). Patient-centred care, an approach that respects and integrates the values, preferences, and needs of patients, has been recognized as a fundamental strategy for enhancing health outcomes at scale (Epstein and Street, 2011). This model promotes shared decision-making, improved communications, and holistic care that addresses the SDoH. Other strategies for the scaling of patient-centred care include:

- Value-Based Care Models: Transitioning from fee-for-service to value-based payment structures incentivizes high-quality and coordinated care that prioritizes patient outcomes rather than service volume (Berwick et al., 2008).
- Culturally Competent Care: Training healthcare providers in cultural humility and implicit bias awareness improves patient/provider trust and health literacy (Betancourt et al., 2003).
- Personalized and Culturally Competent Care Delivery: Rather than solely screening for social needs, healthcare systems may adopt culturally responsive care models that tailor interventions

to the lived experiences, values, and preferences of patients. This involves engaging community health workers, integrating language-concordant care, and addressing implicit biases in clinical decision-making to foster trust and improve health outcomes (Handtke et al., 2019).

- Integration of Behavioural and Lifestyle Medicine: Addressing modifiable lifestyle factors (e.g. nutrition, exercise, and stress management) may enhance patient engagement and long-term disease prevention. Embedding lifestyle medicine specialists and behavioural health professionals into primary care settings allows for a more holistic approach to chronic disease management beyond traditional clinical interventions (Epstein and Street, 2011).
- Social Prescribing and Community Partnerships: Beyond SDoH screening, providers can prescribe social and community-based interventions, such as referrals to exercise programmes, peer support groups, or arts-based therapies. Partnerships with faith-based organizations, food cooperatives, and local non-profits create a more sustainable support network for patients who are navigating complex social and health challenges (Zahner and Corrado, 2004).
- Technology Enabled Care Coordination: Leveraging digital health tools, such as AI-driven predictive analytics, can help identify at-risk patients before health crises occur. Integrated electronic health records (EHRs) with real-time data sharing between medical and social service providers ensure seamless care coordination and follow-up for vulnerable populations (Kruse et al., 2018).

10.4.3 Metrics for Equitable Care Outcomes

The evaluation of equitable care outcomes requires a comprehensive approach that integrates assorted qualitative and quantitative measures to assess disparities, patient experiences, and health system performance. One key metric is disparity reduction in health outcomes, which examines differences in disease prevalence, treatment access, and mortality rates across demographic groups. By analysing trends in chronic disease management, maternal health, and preventive care utilization, healthcare organizations can identify inequities and tailor interventions to underserved populations (Braveman et al., 2011). Furthermore, patient reported outcome measures (PROMs) provide insights into patients' perceptions of their health, quality of life, and satisfaction with care, which ensure that interventions are aligned with community needs (Nelson et al., 2015).

An additional critical measure is care accessibility and timeliness, which assesses whether patients (particularly those in marginalized communities) can obtain necessary medical services without undue delays. This includes tracking wait times, geographic distribution of providers, and availability of culturally competent care (Institute of Medicine, 2001). Equity-focused clinical performance indicators, such as adherence to evidence-based treatment guidelines across racial and socioeconomic groups, help to determine whether care delivery is consistent and fair. For example, quantifying the rate of guideline-concordant hypertension or diabetes management among racial minorities can identify gaps in treatment equity (Chin et al., 2012). Moreover, structural equity indicators, such as workforce diversity in healthcare settings and the presence of policies that address implicit biases, contribute to the elucidation of systemic challenges in care delivery. The incorporation of social risk adjustment factors in payment models ensures that providers who serve high-need populations are adequately resourced to address social determinants that affect health outcomes (Castrucci and Auerbach, 2019). Finally, the longitudinal tracking of health equity interventions allows for the continuous assessment of progress in reducing disparities, which ensures that strategies remain responsive to evolving community needs. By integrating these metrics, healthcare systems can foster accountability and drive meaningful improvements in equitable care delivery.

10.4.4 Shift towards Community-Driven Healthcare

Community-based healthcare models offer an alternative to traditional, hospital-centric systems by prioritizing local engagement and population-specific interventions. These models emphasize primary care, disease prevention, and holistic well-being, rather than episodic, reactive treatments (NASEM, 2019). Community health centres (CHCs) and Federally Qualified Health Centers (FQHCs) have demonstrated their efficacy in improving outcomes for marginalized populations by integrating social services, peer support, and culturally tailored care delivery (Shi et al., 2013).

10.4.5 The NuKa System of Care: A Case Study in Indigenous Health Equity

The NuKa System of Care, which was developed by the Southcentral Foundation in Alaska, represents a transformative model of community-driven healthcare that places Indigenous values, trust, and relationship-based care at its core. This system is recognized worldwide for its emphasis on holistic, patient-owned health management and its capacity to improve outcomes in Indigenous populations (Gottlieb, 2013). Key elements of the NuKa model include:

- Integrated Care Teams: Multidisciplinary teams that coordinate medical, behavioural, and social services to address the full spectrum of patient needs (Southcentral Foundation, 2023).
- "Customer-Owner" Empowered Patients: Recognizing patients as co-owners of their healthcare fosters engagement, accountability, and trust (Gottlieb, 2013).
- Data-Driven Decision-Making: NuKa utilizes real-time data analytics to track health disparities, adjust care strategies, and maintain transparency in outcomes (Green et al., 2024).
- Community and Cultural Responsiveness: Culturally tailored health initiatives (e.g. traditional healing practices and nutrition programmes) enhance patient adherence and well-being (Southcentral Foundation, 2023).

10.4.6 Scaling the NuKa Model: Policy and Practise Implications

Scaling the NuKa System of Care requires a comprehensive approach that integrates policy reforms, financial incentives, and operational adaptations to align with diverse healthcare settings. A key consideration is the policy transformation that is required to support community-driven healthcare models. Policymakers must implement regulations that prioritize patient-centred, relationship-based care, fostering integrated care teams, and shifting from volume-based to value-based reimbursement structures. Payment models that incentivize holistic, preventative care rather than episodic treatment are essential for adapting the NuKa model beyond its origins in the Southcentral Foundation (Gottlieb, 2013).

Equally important is the investment in workforce development to ensure the sustainability of NuKa-inspired models. Training programmes should emphasize cultural competency, trauma-informed care, and interdisciplinary collaboration, equipping healthcare professionals with the required skills to nurture meaningful relationships with patients (Ruffolo et al., 2023). Furthermore, scalable care coordination strategies must be implemented, which leverage health information technology and telehealth solutions to facilitate seamless communications between providers and patients, particularly in underserved and rural communities.

The expansion of the NuKa model also hinges on community engagement and governance structures that empower populations as co-designers of healthcare services. Establishing advisory councils and participatory decision-making frameworks ensures that care delivery remains responsive to community needs (Newman et al., 2011). Moreover, evaluation and adaptation mechanisms are critical, which require healthcare organizations to continuously monitor patient outcomes, cost-effectiveness, and provider satisfaction to refine the model for broader application. As health systems increasingly recognize the benefits of holistic, relationship-based care, the NuKa model presents a compelling framework for achieving equity, improving health outcomes, and fostering a sustainable, patient-centred healthcare transformation.

10.5 PART IV: LIFESTYLE MEDICINE – PREVENTION AND EMPOWERMENT

10.5.1 Integrating Lifestyle Medicine into Systems and Equity

As healthcare systems around the world grapple with the rising burden of chronic diseases, the importance of lifestyle medicine has become increasingly evident. Lifestyle medicine focuses on the prevention, treatment, and reversal of chronic diseases through evidence-based lifestyle interventions such as nutrition, physical activity, stress management, and social connectivity (Lippman et al., 2024). This approach addresses not only individual health behaviours but also considers the broader determinants of health, which makes it a powerful tool towards the pursuit of global healthcare equity. By integrating lifestyle medicine into healthcare systems, individuals can be empowered to take greater control of their health and well-being, which ultimately contributes to more equitable health outcomes globally. Lifestyle medicine is grounded in the belief that many chronic diseases are largely preventable and manageable through healthy lifestyle choices. Key principles include the adoption of a whole-food, plant-based diet, engaging in regular physical activities, managing stress effectively, fostering supportive social connections, and avoiding risky behaviours such as smoking and excessive alcohol consumption (Lippman et al., 2024). These interventions have been shown to improve health outcomes across a wide range of conditions, from cardiovascular disease to diabetes and beyond.

One of the significant strengths of lifestyle medicine is its potential to reduce health disparities. By focusing on low-cost, accessible interventions, lifestyle medicine can reach underserved populations who may face barriers to traditional medical care. For instance, community-based lifestyle programmes have been successful in reducing the prevalence of chronic diseases in low-income and minority communities (Ornish et al., 1998). These programmes empower individuals with the knowledge and tools to make healthier choices that ultimately enhance health outcomes and reduce disparities.

10.5.2 Case Studies: Lifestyle Medicine and Health Equity

10.5.2.1 Case Study 1: Complete Health Improvement Program

The Complete Health Improvement Program (CHIP) is a lifestyle medicine intervention that focuses on improving diet, physical activities, and stress management to prevent and manage chronic diseases. Studies have revealed that CHIP participants experience significant improvements in cardiovascular risk factors, diabetes management, and overall well-being (Morton et al., 2014). By offering affordable and accessible programmes, CHIP has been shown to effectively address health disparities in underserved communities. For instance, in rural Alabama, CHIP reduced the prevalence of hypertension and improved health outcomes among low-income participants (Morisky et al., 2002). This case study demonstrated the potential of lifestyle medicine to bridge health equity gaps by empowering individuals to take control of their own health through sustainable lifestyle changes.

10.5.2.2 Case Study 2: Blue Zones Project

The Blue Zones Project was inspired by regions with the highest concentrations of centenarians, which aims to transform communities through the creation of environments that promote healthy living. Initiatives include improving access to healthy foods, creating walkable communities, and fostering social connections. In Albert Lea, Minnesota, the Blue Zones Project led to a 40% reduction in the healthcare costs of city workers and significant improvements in the health behaviours of residents (Buettner and Skemp, 2016). This project highlights how community-wide lifestyle interventions can address SDoH and contribute to health equity by creating supportive environments that make healthy choices easier for all who participate.

10.5.3 Integrating Lifestyle Medicine into Global Healthcare Systems

10.5.3.1 Policy and System-Level Changes

Integrating lifestyle medicine into global healthcare systems presents both challenges and opportunities. One significant challenge is the need to incorporate lifestyle medicine principles into healthcare provider education. Currently, medical school curricula often lack an adequate focus on lifestyle-related interventions, such as nutrition, exercise, and mental health practices, which are essential for preventing and managing chronic disease (Adams et al., 2010). To remedy this situation, medical schools and continuing education programmes must prioritize lifestyle medicine education to equip healthcare professionals with the knowledge and skills to promote healthy living among patients (Lianov and Johnson, 2010). This shift would not only enhance clinical outcomes but also improve the health of entire populations by addressing the root causes of many chronic conditions.

Another critical factor relates to policy and system-level changes. Countries such as Finland and New Zealand have successfully integrated lifestyle medicine into their healthcare policies, which has resulted in notable public health improvements. For example, Finland's national focus on diet and exercise has translated to a significant reduction in cardiovascular disease rates over the last few decades (Puska, 2009). These examples highlight the potential for policy interventions, such as zoning regulations that promote physical activity and subsidies for healthy food options, to create environments conducive to healthier lifestyles.

Additionally, Schroeder (2007) emphasized that national scale policy changes are critical for the promotion of preventive healthcare, including lifestyle medicine. By shifting healthcare systems to focus on prevention and healthy living, the burden of chronic diseases can be significantly reduced, improving both individual and public health outcomes. Moreover, cultural competence must be emphasized within healthcare systems to ensure that lifestyle interventions are tailored to the diverse needs of populations, thereby promoting care equity (Lianov and Johnson, 2010).

In summary, integrating lifestyle medicine into global healthcare systems through education reform, supportive policies, and system-level changes presents an opportunity to significantly enhance both individual and population health. By adopting these strategies, healthcare systems can move towards a more preventive, patient-centred model that addresses the root causes of disease while promoting equity and long-term wellness.

To fully realize the potential of lifestyle medicine, it is essential to integrate it into healthcare systems through supportive policies and system-level changes. This includes adopting policies that promote healthy living environments, such as zoning regulations that encourage physical activities, subsidies for healthy foods, and workplace wellness programmes.

10.5.3.2 Towards Healthier Systems, Empowered Behaviours, and Equitable Care (Bridging Lifestyle Medicine and Global Healthcare Equity)

Emerging trends and innovations in lifestyle medicine hold great promise for supporting global healthcare equity. Digital health solutions, such as telemedicine, mobile health apps, and wearable devices, can extend the reach of lifestyle interventions, making them more accessible to remote and underserved populations. These technologies can provide real-time feedback, support behavioural changes, and foster community engagement, which ultimately enhance the impacts of lifestyle medicine on a global scale (Parish, 2015). Integrating lifestyle medicine with the pursuit of global healthcare equity offers a holistic approach to improving health outcomes. By empowering individuals to make healthier lifestyle choices and addressing the broader determinants of health, we can reduce health disparities and promote well-being for all. As we look to the future, it is essential to continue exploring innovative approaches and policies that support the widespread adoption of lifestyle medicine, ultimately contributing to a more equitable and healthier world.

10.5.3.3 Implications for Future Directions

To summarize, achieving a healthcare landscape that is systems-oriented, behaviourally informed, and equity-driven requires a fundamental shift in how health determinants, delivery models, and policy structures are conceptualized and operationalized (Menear et al., 2019). This chapter has explored the intricate web of SDoH, the application of systems thinking, the critical role of urban planning, and the transformation of healthcare through community-centred models like the NuKa System of Care. A central theme that underpins these discussions is the necessity for integrated, proactive, and patient-centric solutions that address the root causes of health disparities while fostering sustainable improvements in health outcomes (Teisberg et al., 2020).

At the heart of this transformation is the recognition that behavioural science plays a crucial role in shaping health outcomes. The capacities of individuals to engage in healthy behaviours (e.g. preventive screenings, adherence to medication regimes, and lifestyle modifications) is heavily influenced by their social environment, economic stability, and access to resources (Adkisson, 2008). Thus, healthcare systems must integrate behaviourally informed strategies, such as nudges, digital health interventions, and patient-centred education, to empower individuals to take more active roles in their well-being. Advancing towards a healthier, more equitable healthcare system necessitates the convergence of systems thinking, behavioural science, and policy innovation.

Healthcare must evolve beyond disease management to embrace preventive, participatory, and holistic models of care that address the underlying determinants of health (Sagner et al., 2017). As policymakers, healthcare leaders, and community stakeholders work collaboratively to implement these changes, the vision of a more just, efficient, and patient-centred healthcare system becomes increasingly attainable. Fostering healthier systems, empowered behaviours, and equitable care, society can build a future where all individuals, regardless of socioeconomic status, geographic location, or demographic background, have the opportunity to achieve optimal health and well-being.

Gazing towards the further future, an envisaged single worldwide healthcare system referred to as Global Health Care Equivalency (GHCE) (Boehm, 2025), the topic of this book, may be enabled over the next few decades via synergies between nanotechnology, nanomedicine, AI (and in time AGI/ASI), coupled with quantum computation. Although GHCE is anticipated to manifest myriad nanotechnical capacities to address all that ails us, including the disease of aging, the brief survey articulated in this chapter on the systems-based integration of SDoH through behavioural sciences, lifestyle medicine, and equitable care frameworks will serve as an invaluable prerequisite background for the successful implementation and sustainable operation of GHCE.

Case in point: even the most sophisticated autonomous nanomedical nanodevices in the future will be of little value for one's health and longevity if individuals do not adhere to prescribed nanomedical regimens or take an adequately realistic level of responsibility for their own daily health and well-being. Technologies can only take us so far; thus, we should consider living synergistically with them further to realize the potential of the envisaged dream of optimal health for life, for all of humanity.

REFERENCES

Adams KM, Kohlmeier M, Zeisel SH. Nutrition education in U.S. medical schools: Latest update of a national survey. *Academic Medicine*. 2010;85(9):1537–1542.

Adkisson RV. Nudge: Improving decisions about health, wealth and happiness. *Social Science Journal.*. 2008;45(4):700–701.

Adler NE, Newman K. Socioeconomic disparities in health: Pathways and policies. *Health Affairs (Millwood)*. 2002;21(2):60–76.

Ajzen I. The theory of planned behaviour. *Organizational Behaviour and Human Decision Processes.* 1991;50(2):179–211.

Alegría M, NeMoyer A, Falgàs Bagué I, Wang Y, Alvarez K. Social determinants of mental health: Where we are and where we need to go. *Current Psychiatry Reports.* 2018;20(11):95.

Armitage CJ, Conner M. Efficacy of the theory of planned behaviour: A meta-analytic review. *British Journal Society Psychology.* 2001;40(Pt 4):471–499.

Artiga S, Hinton E. Beyond health care: The role of social determinants in promoting health and health equity. Kaiser Family Foundation. 2018. https://www.kff.org (accessed December 12, 2024).

Bailey ZD, Krieger N, Agénor M, Graves J, Linos N, Bassett MT. Structural racism and health inequities in the USA: Evidence and interventions. *Lancet.* 2017;389(10077):1453–1463.

Bambra C, Gibson M, Sowden A, Wright K, Whitehead M, Petticrew M. Tackling the wider social determinants of health and health inequalities: Evidence from systematic reviews. *Journal of Epidemiology Community Health.* 2010;64(4):284–2891.

Berwick DM, Nolan TW, Whittington J. The triple aim: Care, health, and cost. *Health Affairs (Millwood).* 2008;27(3):759–769.

Betancourt JR, Green AR, Carrillo JE, Ananeh-Firempong O. Defining cultural competence: A practical framework for addressing racial/ethnic disparities in health and health care. *Public Health Reports.* 2003;118(4):293–302.

Boehm FJ. Nanotechnology, Nanomedicine, and AI: Toward the Dream of Global Health Care Equivalency. 2025. https://www.nanoappsmedical.com/nanotechnology-nanomedicine-and-ai-toward-the-dream-of-global-health-care-equivalency/ (accessed March 28, 2025).

Borrell-Carrió F, Suchman AL, Epstein RM. The biopsychosocial model 25 years later: principles, practice, and scientific inquiry. *Annals of Family Medicine.* 2004;2(6):576–582.

Braveman P, Arkin E, Orleans T, Proctor D, Plough A. What is Health Equity. Robert Wood Johnson Foundation – Insight/Research. 2017. https://www.rwjf.org/en/insights/our-research/2017/05/what-is-health-equity-.html (accessed December 15, 2024).

Braveman P, Dominguez TP, Burke W, Dolan SM, Stevenson DK, Jackson FM et al. Explaining the black-white disparity in preterm birth: A consensus statement from a multi-disciplinary scientific work group convened by the March of Dimes. *Frontiers in Reproductive Health.* 2021;3:684207.

Braveman P, Egerter S, Williams DR. The social determinants of health: Coming of age. *Annual Review of Public Health.* 2011;32:381–398.

Browne AJ, Varcoe C, Ford-Gilboe M, Nadine Wathen C, Smye V, Jackson BE et al. Disruption as opportunity: Impacts of an organizational health equity intervention in primary care clinics. *International Journal for Equity in Health.* 2018;17(1):154.

Browne AJ, Varcoe C, Lavoie J, Smye V, Wong ST, Krause M, Tu D, Godwin O, Khan K, Fridkin A. Enhancing health care equity with Indigenous populations: Evidence-based strategies from an ethnographic study. *BMC Health Services Research.* 2016;16(1):544.

Buettner D, Skemp S. Blue zones: Lessons from the world's longest lived. *American Journal of Lifestyle Medicine.* 2016;10(5):318–321.

Castrucci BC, Auerbach J. Meeting individual social needs falls short of addressing social determinants of health. Health Aff Blog. 2019 Jan 16. https://dam.assets.ohio.gov/image/upload/collectiveimpact.mha.ohio.gov/Social-Determinants-of-Health/Meeting-Individual-Social-Needs-Falls-Short-Of-Addressing-Social-Determinants-Of-Health-Health-Affairs.pdf (accessed August 12, 2025).

Centers for Disease Control (CDC). Health and Economic Benefits of Chronic Disease Interventions. National Center for Chronic Disease Prevention and Health Promotion (NCCDPHP). 2024. https://www.cdc.gov/nccdphp/priorities/index.html#:~:text=Chronic%20diseases%20are%20the%20leading,of%20life%20and%20save%20money (accessed June 15, 2024).

Cheadle A, Atiedu A, Rauzon S, Schwartz PM, Keene L, Davoudi M et al. A community-level initiative to prevent obesity: Results from Kaiser Permanente's Healthy Eating Active Living Zones Initiative in California. *American Journal of Preventive Medicine.* 2018;54(5 Suppl 2):S150–S159.

Chin MH, Clarke AR, Nocon RS, Casey AA, Goddu AP, Keesecker NM, Cook SC. A roadmap and best practices for organizations to reduce racial and ethnic disparities in health care. *Journal of General Internal Medicine.* 2012;27(8):992–1000.

Coughlin SS, Vernon M, Hatzigeorgiou C, George V. Health literacy, social determinants of health, and disease prevention and control. *Journal of Environmental Health Sciences.* 2020;6(1):3061.

Courtin E, Kim S, Song S, Yu W, Muennig P. Can social policies improve health? A systematic review and meta-analysis of 38 randomized trials. *Milbank Quarterly.* 2020;98(2):297–371.

Dickson K, Wilson R, Parfrey O, Parfrey PS. Changing health-related behaviors 2: On improving the value of health spending. *Methods in Molecular Biology.* 2021;2249:553–569.

Diez Roux AV. Complex systems thinking and current impasses in health disparities research. *American Journal of Public Health.* 2011;101(9):1627–1634.

Diez Roux AV. Neighborhoods and health: What do we know? What should we do? *American Journal of Public Health.* 2016;106(3):430–431.

Epstein RM, Street RL Jr. The values and value of patient-centered care. *Annals of Family Medicine.* 2011;9(2):100–103.

Ford-Gilboe M, Wathen CN, Varcoe C, Herbert C, Jackson BE, Lavoie JG et al. For the equip research program. How equity-oriented health care affects health: Key mechanisms and implications for primary health care practice and policy. *Milbank Quarterly.* 2018;96(4):635–671.

Freed CR, Hansberry ST, Arrieta MI. Structural and hidden barriers to local primary care health care infrastructure: Autonomy, decisions, about primary health care and the centrality and significance of power. *Research in the Sociology of Health Care.* 2013;31:57–81.

Frieden TR. A framework for public health action: The health impact pyramid. *American Journal of Public Health.* 2010;100(4):590–595.

Gatchel RJ, Peng YB, Peters ML, Fuchs PN, Turk DC. The biopsychosocial approach to chronic pain: Scientific advances and future directions. *Psychological Bulletin.* 2007;133(4):581–624.

Gipson P, King C. Health behavior theories and research: Implications for suicidal individuals' treatment linkage and adherence. *Cognitive and Behavioral Practice.* 2012;19(2):209–217.

Glanz K, Bishop DB. The role of behavioral science theory in development and implementation of public health interventions. *Annual Review of Public Health.* 2010;31:399–418.

Gottlieb K. The Nuka System of Care: Improving health through ownership and relationships. *International Journal of Circumpolar Health.* 2013;72.

Green AR, Carney DR, Pallin DJ, Ngo LH, Raymond KL, Iezzoni LI, Banaji MR. Implicit bias among physicians and its prediction of thrombolysis decisions for black and white patients. *Journal of General Internal Medicine.* 2007;22(9):1231–1238.

Green BL, Murphy A, Robinson E. Accelerating health disparities research with artificial intelligence. *Frontiers in Digital Health.* 2024;6:1330160.

Handtke O, Schilgen B, Mösko M. Culturally competent healthcare – A scoping review of strategies implemented in healthcare organizations and a model of culturally competent healthcare provision. *PLoS One.* 2019;14(7):e0219971.

Hatala A. The status of the "Biopsychosocial" model in health psychology: Towards an integrated approach and a critique of cultural conceptions. *Open Journal of Medical Psychology.* 2012;1:51–62.

Havelka M, Lucanin JD, Lucanin D. Biopsychosocial model – The integrated approach to health and disease. *Collegium Antropologicum.* 2009;33(1):303–310.

Heaney CA, Israel BA. Social networks and social support. In: Glanz K, Rimer BK, Viswanath K, editors. *Health Behavior and Health Education: Theory, Research, and Practice.* 4th ed. San Francisco: Jossey-Bass; 2008. p. 189–210

Hollander JE, Carr BG. Virtually perfect? Telemedicine for Covid-19. *New England Journal of Medicine.* 2020;382(18):1679–1681.

Holman HR. The relation of the chronic disease epidemic to the health care crisis. *ACR Open Rheumatology.* 2020;2(3):167–173.

Institute of Medicine (US) Committee on Quality of Health Care in America. *Crossing the Quality Chasm: A New Health System for the 21st Century.* Washington DC: National Academies Press, 2001.

Janz NK, Becker MH. The health belief model: A decade later. *Health Education Quarterly.* 1984;11(1):1–47.

Kangovi S, Mitra N, Norton L, Harte R, Zhao X, Carter T, Grande D, Long JA. Effect of community health worker support on clinical outcomes of low-income patients across primary care facilities: A randomized clinical trial. *JAMA Internal Medicine.* 2018;178(12):1635–1643.

Kelli HM, Hammadah M, Ahmed H, Ko YA, Topel M, Samman-Tahhan A, Awad M, Patel K, Mohammed K, Sperling LS, Pemu P, Vaccarino V, Lewis T, Taylor H, Martin G, Gibbons GH, Quyyumi AA. Association between living in food deserts and cardiovascular risk. *Circulation: Cardiovascular Quality and Outcomes.* 2017;10(9):e003532.

Kilanowski JF. Breadth of the socio-ecological model. *Journal of Agromedicine.* 2017;22(4):295–297.

Kroll C. Questioning biomedicine's privileging of disease and measurability. *AMA Journal of Ethics.* 2021;23(7):E537–541.

Krumholz HM, Normand ST, Wang Y. Twenty-year trends in outcomes for older adults with acute myocardial infarction in the United States. *JAMA Network Open.* 2019;2(3):e191938.

Kruse CS, Stein A, Thomas H, Kaur H. The use of electronic health records to support population health: A systematic review of the literature. *Journal of Medical Systems*. 2018;42(11):214.
Laur C, Ball L, Keller H, Ivers N. Building on what we know: Moving beyond effectiveness to consider how to implement, sustain and spread successful health interventions. *BMJ Nutrition, Prevention & Health*. 2020;3(2):123–125.
Lehman BJ, David DM, Gruber JA. Rethinking the biopsychosocial model of health: Understanding health as a dynamic system. *Social and Personality Psychology Compass*. 2017;11:e12328.
Leischow SJ, Milstein B. Systems thinking and modeling for public health practice. *American Journal of Public Health*. 2006;96(3):403–405.
Lianov L, Johnson M. Physician competencies for prescribing lifestyle medicine. *JAMA*. 2010;304(2): 202–203.
Lieberman DE. The story of the human body: Evolution, health and disease. *Family Medicine*. 2016;48(10):822–823.
Lippman D, Stump M, Veazey E, Guimarães ST, Rosenfeld R, Kelly JH, Ornish D, Katz DL. Foundations of lifestyle medicine and its evolution. *Mayo Clinic Proceedings Innovations Quality & Outcomes*. 2024;8(1):97–111.
Malone NC, Williams MM, Smith Fawzi MC, Bennet J, Hill C, Katz JN, Oriol NE. Mobile Health clinics in the United States. *International Journal for Equity in Health*. 2020;19(1):40.
Marmot M. Social determinants of health inequalities. *Lancet*. 2005;365(9464):1099–1104.
Marmot M, Friel S, Bell R, Houweling TA, Taylor S, Commission on Social Determinants of Health. Closing the gap in a generation: Health equity through action on the social determinants of health. *Lancet*. 2008;372(9650):1661–1669.
Mays GP, Smith SA. Evidence links increases in public health spending to declines in preventable deaths. *Health Affairs (Millwood)*. 2011;30(8):1585–1593.
McGinnis JM, Williams-Russo P, Knickman JR. The case for more active policy attention to health promotion. *Health Affairs (Millwood)*. 2002;21(2):78–93.
McLeroy KR, Bibeau D, Steckler A, Glanz K. An ecological perspective on health promotion programs. *Health Education Quarterly*. 1988;15(4):351–377.
Menear M, Blanchette MA, Demers-Payette O, Roy D. A framework for value-creating learning health systems. *Health Research Policy and Systems*. 2019;17(1):79.
Messay B, Lim A, Marsland AL. Current understanding of the bi-directional relationship of major depression with inflammation. *Biology of Mood & Anxiety Disorders*. 2012;2:4.
Mitchell AE. Bidirectional relationships between psychological health and dermatological conditions in children. *Psychology Research and Behavior Management*. 2018;11:289–298.
Morisky DE, Lees NB, Sharif BA, Liu KY, Ward HJ. Reducing disparities in hypertension control: A community-based hypertension control project (CHIP) for an ethnically diverse population. *Health Promotion Practice*. 2002;3(2):264–275.
Morton D, Rankin P, Kent L, Dysinger W. The complete health improvement program (CHIP): History, evaluation, and outcomes. *American Journal of Lifestyle Medicine*. 2014;10(1):64–73.
Murray JLC, Salomon J, Mathers C, Lopez A. Summary measures of population health: Concepts, ethics, measurement, and applications. World Health Organisation (WHO): Institutional Repository for Information Sharing (IRIS). 2002. https://iris.who.int/handle/10665/42439 (accessed January 15, 2023).
National Academies of Sciences, Engineering, and Medicine. *Communities in action: Pathways to health equity*. Washington, DC: The National Academies Press. 2017. https://nap.nationalacademies.org/read/24624/chapter/1 (accessed August, 2025).
NASEM. National Academies of Sciences, Engineering, and Medicine; Health and Medicine Division; Board on Population Health and Public Health Practice. *Investing in Interventions That Address Non-Medical, Health-Related Social Needs: Proceedings of a Workshop*. Alper J, Martinez RM (eds.), Washington (DC): National Academies Press (US). 2019.
National Alliance to End Homelessness. Resouces/Overview/Housing First Fact Sheet – What is Housing First. 2024. https://endhomelessness.org/resource/housing-first/ (accessed December 12, 2024).
Nelson EC, Eftimovska E, Lind C, Hager A, Wasson JH, Lindblad S. Patient reported outcome measures in practice. *BMJ*. 2015;350:g7818.
Newman SD, Andrews JO, Magwood GS, Jenkins C, Cox MJ, Williamson DC. Community advisory boards in community-based participatory research: A synthesis of best processes. *Preventing Chronic Disease*. 2011;8(3):A70.
Nordic Burden of Disease Collaborators. Life expectancy and disease burden in the Nordic countries: Results from the Global Burden of Diseases, Injuries, and Risk Factors Study 2017. *Lancet Public Health*. 2019;4(12):e658–e669.

Ornish D, Scherwitz LW, Billings JH, Brown SE, Gould KL, Merritt TA et al. Intensive lifestyle changes for reversal of coronary heart disease. *JAMA*. 1998;280(23):2001–2007.

Painter JE, Borba CP, Hynes M, Mays D, Glanz K. The use of theory in health behavior research from 2000 to 2005: A systematic review. *Annals of Behavioral Medicine*. 2008;35(3):358–362.

Parish JM. The patient will see you now: the future of medicine is in your hands. *Journal of Clinical Sleep Medicine*. 2015;11(5):593–594.

Parker K, Minkin R, Bennett J. Economic Fallout From COVID-19 Continues to Hit Lower-Income Americans the Hardest. Pew Research Center's Social & Demographic Trends Project. September 24, 2020. https://www.pewresearch.org/social-trends/2020/09/24/economic-fallout-from-covid-19-continues-to-hit-lower-income-americans-the-hardest/ (accessed June 22, 2023).

Pineo H. Towards healthy urbanism: Inclusive, equitable and sustainable (THRIVES) – an urban design and planning framework from theory to praxis. *Cities & Health*. 2022;6(5):974–992.

Puska P. Fat and heart disease: Yes we can make a change – the case of North Karelia (Finland). *Annals of Nutrition and Metabolism*. 2009;54(Suppl 1):33–38.

Puska P, Jaini P. The North Karelia Project: Prevention of cardiovascular disease in Finland through population-based lifestyle interventions. *American Journal of Lifestyle Medicine*. 2020Mar 19;14(5):495–499.

Ramsetty A, Adams C. Impact of the digital divide in the age of COVID-19. *Journal of the American Medical Informatics Association*. 2020;27(7):1147–1148.

Ranasinghe P, Jayawardena R, Gamage N, Sivanandam N, Misra A. Prevalence and trends of the diabetes epidemic in urban and rural India: A pooled systematic review and meta-analysis of 1.7 million adults. *Annals of Epidemiology*. 2021;58:128–148.

Rocca E, Anjum RL. Complexity, Reductionism and the Biomedical Model. In *Rethinking Causality, Complexity and Evidence for the Unique Patient*. Anjum RL, Copeland S, Rocca E. (eds.), Cham: Springer. 2020. https://doi.org/10.1007/978-3-030-41239-5_5

Rojas-Rueda D, Morales-Zamora E. Equitable urban planning: Harnessing the power of comprehensive plans. *Current Environmental Health Reports*. 2023;10(2):125–136.

Rosenstock IM. The health belief model and preventive health behavior. *Health Education Monographs*. 1974;2(4):354–386.

Ruffolo MC, Price D, Smith A. Building a community-academic partnership to expand workforce development in underserved, high-Need/High-demand areas. *Public Health Reports*. 2023;138(1_suppl):9S–15S.

Sagner M, McNeil A, Puska P, Auffray C, Price ND, Hood L et al. The P4 health spectrum – A predictive, preventive, personalized and participatory continuum for promoting healthspan. *Progress in Cardiovascular Diseases*. 2017;59(5):506–521.

Sallis JF, Owen N, Fisher EB. Ecological models of health behavior. In: Glanz K, Rimer BK, Viswanath K, editors. *Health Behavior and Health Education: Theory, Research, and Practice*. 4th ed. San Francisco: Jossey-Bass; 2008. p. 465–485.

Schroeder SA. Shattuck lecture. We can do better–improving the health of the American people. *New England Journal of Medicine*. 2007;357(12):1221–1228.

Scott Kruse C, Karem P, Shifflett K, Vegi L, Ravi K, Brooks M. Evaluating barriers to adopting telemedicine worldwide: A systematic review. *Journal of Telemedicine and Telecare*. 2018;24(1):4–12.

Sepe M. Liveable and healthy city design. *WIT Transactions on Ecology and the Environment*. 2018;217:177–189.

Shahzad M, Upshur R, Donnelly P, Bharmal A, Wei X, Feng P, Brown AD. A population-based approach to integrated healthcare delivery: A scoping review of clinical care and public health collaboration. *BMC Public Health*. 2019;19(1):708.

Shi L, Lebrun-Harris LA, Daly CA, Sharma R, Sripipatana A, Hayashi AS, Ngo-Metzger Q. Reducing disparities in access to primary care and patient satisfaction with care: The role of health centers. *Journal of Health Care for the Poor and Underserved*. 2013;24(1):56–66.

Song Z, Baicker K. Effect of a workplace wellness program on employee health and economic outcomes: A randomized clinical trial. *JAMA*. 2019;321(15):1491–1501.

Southcentral Foundation. Integrated Care Team FAQs. 2023. https://scfnuka.com/integrated-care-team-faqs/ (accessed January, 23, 2024).

Suls J, Rothman A. Evolution of the biopsychosocial model: Prospects and challenges for health psychology. *Health Psychology*. 2004;23(2):119–125.

Tainio M, Jovanovic Andersen Z, Nieuwenhuijsen MJ, Hu L, de Nazelle A, An R, Garcia LMT, Goenka S, Zapata-Diomedi B, Bull F, Sá TH. Air pollution, physical activity and health: A mapping review of the evidence. *Environment International*. 2021;147:105954.

Teisberg E, Wallace S, O'Hara S. Defining and implementing value-based health care: A strategic framework. *Academic Medicine*. 2020;95(5):682–685.

Theofilou P. Behavioral medicine: A new perspective in the field of health. *Journal of Neuro and Oncology Research*. 2023;3(2):1–3.

Tolbert J, Cervantes S, Bell C, Damico A. Key facts about the uninsured population. Kaiser Family Foundation. 2024. https://www.kff.org/uninsured/issue-brief/key-facts-about-the-uninsured-population/ (accessed December 30, 2024).

Topa G, Moriano JA. Theory of planned behavior and smoking: Meta-analysis and SEM model. *Substance Abuse and Rehabilitation*. 2010;1:23–33.

Wade DT, Halligan PW. Do biomedical models of illness make for good healthcare systems? *BMJ*. 2004;329(7479):1398–1401.

Wade DT, Halligan PW. The biopsychosocial model of illness: A model whose time has come. *Clinical Rehabilitation*. 2017;31(8):995–1004.

Wallston KA. Control Beliefs: Health Perspectives. In *Encyclopaedia of Stress*. Fink G. (ed.), Academic Press. 2000;1:562–567.

WHO. Urban and built environments – Fact Sheets. 2023. https://www.who.int/europe/news-room/fact-sheets/item/urban-and-built-environments (accessed March 12, 2024)

WHO. Noncommunicable diseases – Fact Sheets. 2024. https://www.who.int/news-room/fact-sheets/detail/noncommunicable-diseases (accessed December 30, 2024).

WHO. *Social determinants of health*. 2025. https://www.who.int/health-topics/social-determinants-of-health#tab=tab_1 (accessed August 12, 2025).

Yao C, Zhang Y, Lu P, Xiao B, Sun P, Tao J, Cheng Y, Kong L, Xu D, Fang M. Exploring the bidirectional relationship between pain and mental disorders: A comprehensive mendelian randomization study. *Journal of Headache and Pain*. 2023;24(1):82.

Yu SWY, Hill C, Ricks ML, Bennet J, Oriol NE. The scope and impact of mobile health clinics in the United States: A literature review. *International Journal of Equity in Health*. 2017;16(1):178.

Zahner SJ, Corrado SM. Local health department partnerships with faith-based organizations. *Journal of Public Health Management and Practice*. 2004;10(3):258–265.

Zeidi IM, Morshedi H, Alizadeh Otaghvar H. A theory of planned behavior-enhanced intervention to promote health literacy and self-care behaviors of type 2 diabetic patients. *Journal of Preventive Medicine and Hygiene*. 2021;61(4):E601–E613.

11 Ethics, Nanotechnology, and Radical Life Extension

Ingemar Patrick Linden

CONTENTS

11.1 INTRODUCTION

As we uncover the mechanisms behind aging and its symptoms, therapeutic strategies will be developed to counteract them. Many of these approaches will be most effective and optimizable, or indeed only possible, when combined with advanced nanotechnology and nanomedicine operating in conjunction with AI (and soon enough, AGI/ASI) and quantum computation (QC). In this chapter, I will explore how these advanced technologies may contribute to the radical extension of human life in the context of (and likely facilitated by) an envisaged worldwide healthcare system referred to as Global Health Care Equivalency (GHCE) (Boehm, 2018). Subsequently, a number of common ethical concerns will be addressed that are associated with these enabling technologies toward the pursuit of extreme life extension.

11.2 NANOTECHNOLOGY, NANOMEDICINE, AI, AND GEROSCIENCE

The intense pursuit of eternal youth in a state of optimal health is rapidly transitioning from fantasy to reality as scientific advances merge with cutting-edge technologies. Once dismissed as a dead end and career graveyard, anti-aging research has now evolved into the burgeoning and respected discipline of geroscience: the study of aging and how to prevent it (Olshansky and Kirkland, 2024; Okholm, 2024).

Researchers are gaining a far greater elucidation of the target. López-Otín et al. identified nine hallmarks of aging in 2013 (genomic instability, telomere attrition, epigenetic alterations, loss of proteostasis, deregulated nutrient sensing, mitochondrial dysfunction, cellular senescence, stem cell exhaustion, and altered intercellular communication) (López-Otín et al., 2013). In 2023, they expanded this framework to include three additional hallmarks (disabled macroautophagy, chronic inflammation, and dysbiosis), bringing the total to 12 (López-Otín et al., 2023). The molecular and intermolecular events outlined in the Hallmarks of Aging framework give rise to the phenomena we typically associate with aging and the elderly (gray hair, wrinkles, slowed movement, hearing loss, frailty, chronic diseases, and, ultimately, the inevitable outcome—death).

The Hallmarks of Aging framework aligns closely with Aubrey de Grey's strategies for Engineered Negligible Senescence (SENS) and provides a comprehensive scientific foundation for longevity therapies, including those championed by de Grey (de Grey, 2004). Importantly, both frameworks suggest that targeting these underlying processes might significantly extend the human lifespan. Framing aging as the accumulation of damage that leads to pathologies and death assist in

DOI: 10.1201/9781003680093-11

avoiding the mistaken notion that aging is distinct from its harmful effects, as if one could experience significant aging without its consequences. As longevity advocate José Cordeiro aptly put it, speaking of "healthy aging" is as contradictory as speaking of "healthy cancer."

Although geroscience can make significant strides without nanoscience, its ultimate success in the radical extension of the human lifespan may be contingent on synergies between nanotechnology, nanomedicine, AI, and QC. These powerful elements have the potential capacity to enhance geroscientific approaches across a broad spectrum:

Cellular Repair and Rejuvenation: Autonomous nanomedical robots could repair damaged organelles, cells, and tissues to effectively reverse the effects of aging. They might remove cellular debris, such as amyloid plaques (linked to Alzheimer's) and lipofuscin (which accumulates in aging non-dividing cells). Furthermore, they could restore mitochondrial function to prevent the energy decline observed in aging cells.

Elimination of Senescent Cells: Senescent ("zombie") cells stop dividing and contribute to inflammation and aging. Nanomedical devices might target and remove these cells more precisely than current senolytic drugs under development.

DNA Repair and Maintenance: Over time, mutations accumulate in DNA, which leads to aging and diseases like cancer. Specialized classes of nanomachines could detect and repair DNA damage to preserve genetic integrity.

Immune System Enhancement: The immune system weakens with age, which makes individuals more vulnerable to infections and cancer; a phenomenon geroscientists refer to as "inflammaging" (Chen and Yung, 2019). Nano-immune cells may have the capacity to augment or replace natural immune functions to identify and neutralize threats more effectively. In the further future, envisaged autonomous "sentinel" nanodevices might significantly enhance the human immune system by patrolling for/eradicating virtually any harmful entities that they encounter (Domschke and Boehm, 2018).

Targeted Drug Delivery: Nanoparticles can deliver drugs directly to diseased cells, which reduces side effects while enhancing effectiveness. Since drug delivery is a major bottleneck for many anti-aging treatments, this advance could have profound implications.

Clearance of Toxins and Metabolic Waste: Aging induces the buildup of harmful substances in the body. Nanomedical devices may facilitate the removal of metabolic waste from the bloodstream, lymphatic system, organs, and even brain tissues to potentially prevent a wide range of neurodegenerative conditions.

Regeneration of Organs and Tissues: Nanotech combined with stem cells might regenerate damaged organs without the need for transplants. Bioengineered tissues could be created on demand (perhaps even in situ), which would reduce incidences of age correlated organ failure.

Beyond these applications, nanotechnology may also critically facilitate cryonics as a potential "insurance policy" for longevity enthusiasts. A major challenge in cryonics relates to the formation of microscopic ice crystal shards that can rupture cell walls and delicate tissues, which makes thawing and revival extremely difficult (Fan et al., 2022; Sciorio, 2020; Shani et al., 2019). Nanoengineered cryoprotectants may prevent ice formation to better preserve cellular structures and increase the prospects for successful reanimation (Lee et al., 2019; Parihar et al., 2023; Zhao and Fu, 2017).

Meaningful progress in nanomedicine significantly increases the likelihood that we will be able to repair and prevent age-related damage, allowing us to remain healthy as young individuals for indefinite periods. This may be based on de Grey's Longevity Escape Velocity (LEV) strategy (Longevity Escape Velocity Foundation, 2025). Longevity advocates José Cordeiro and David Wood argue that since humans are comprised of 99% oxygen, carbon, hydrogen, and nitrogen, maintaining biological systems "will be easy and cheap" (Cordeiro and Wood, 2023). They predict that aging will be defeated by 2045.

11.3 WORRIES ABOUT LIFE EXTENSION

I see the life-prolonging potential of nanoscience as its most valuable aspect. No other application is as urgently required, given that aging undermines our health in myriad ways, causes suffering, and claims the lives of ~120,000 people per day (Li et al., 2023; United Nations, 2024; World Population Review, 2025). However, not everyone shares this perspective, as skeptics and pessimists often include the prospect of an increasingly aged population among their concerns regarding nanotechnology. They argue that using nanotechnology to radically extend human life constitutes an abuse of its potential. In this chapter, I will address some of these concerns.

One common objection is that radical life extension is unnatural and, therefore, morally wrong. This argument is sometimes framed as a warning against "playing God," which reflects the belief that interfering with natural processes violates some sacred limit. According to this view, extending human life beyond its natural bounds is tantamount to overstepping divine or natural dominion. Bioethicist David Wendler expresses this idea succinctly:

> Certain natural processes do not merely happen to us, but also define who we are … [T]he fundamental structures of our lives place moral boundaries that we may, ethically speaking, not alter to suit our own aims and desires … [W]e should accept the fundamental structure of our lives as defined by natural processes and live our lives within those boundaries.
>
> *Wendler, 1999*

Likewise, philosopher John Hardwig argues: "We are mortal beings, and death is not only the end result but its telos—the aim and purpose for which we are headed biologically …" (Hardwig, 2000). In contrast, early proponents of nanotechnology directly challenged this perspective. Eric Drexler, who popularized atomic and molecular manufacturing in his book *Engines of Creation: The Coming Era of Nanotechnology*, envisions a future where we can design both the world and our bodies from the ground up, precisely *to suit our own aims and desires*, regardless of any boundaries apparently set by nature (Drexler, 1986).

These appeals to nature sometimes stem from deeply held religious beliefs. Some argue that since we are created by God, we should remain young and alive only for as long as He has ordained, with true immortality existing only in the afterlife. However, this argument is only as strong as our perceived knowledge of God's intentions. The Bible refers to a typical human lifespan as "three score and ten years" (70 years), yet in the developed world, this has already been exceeded by more than a decade—primarily due to medical advances that "meddle" with nature. Interestingly, few religious groups consider medical interventions as impious; indeed, most believers celebrate the increased life expectancies they have afforded.

However, what of the manipulation of atoms and repair of molecules, as promised by nanotechnology? Is this not an act of hubris—an attempt to do better than God? It is difficult to fathom how the *level/resolution* or effectiveness of medical interventions might determine their morality from God's perspective. Moreover, nanotechnology is not the only factor that affects humans at a molecular scale. Something as ordinary as drinking a glass of water or being exposed to sunlight affects us at the molecular and cellular scales. Vaccines, another form of medical intervention, function at a molecular level, which induces lifelong changes to the immune system. While some religious groups oppose them, most believers do not consider vaccines to be against God's will or an act of hubris. Likewise, if someone claims that God does not want us to use nanotechnology to cure aging and extend human life, I cannot disprove it. However, there is also no contradiction in believing that God *would* approve of using nanotechnology—or other medical advancements—to prolong life, since we are endowed with the intellectual capacities to invent and implement such things.

Intriguingly, a 2013 Pew Research Center survey found that religious respondents were slightly *more* interested in radical life extension than their secular counterparts (Funk et al., 2013). Among the most enthusiastic supporters were Black Protestants and Mormons.

Appeals to a "natural cycle of life" that we apparently must submit to are often made without any overtly religious assumptions. The argument suggests that there is something inherently wrong with the desire to overcome what is natural and normal for humans. Political philosopher Michael Sandel contends that the wish to enhance ourselves—including extending our lifespan—is an expression of an "ethics of willfulness" in which we seek control over every aspect of our lives (Sandel, 2004). According to Sandel, what is lost in this pursuit is an appreciation for what nature has given us. Nanotechnology, he argues, represents an extreme case of this desire—an attempt to exert control down to the very atom. Similarly, the influential bioethicist Daniel Callahan urges us to accept "the finitude of the body, the fragility of the mind, and the fleeting nature of health," arguing that "a little fatalism, a little loosening up in the search for health, a little acceptance even of that nature red in tooth and claw seem necessary here" (Callahan, 1987).

Nevertheless, this is by no means a persuasive argument. Gratitude is undoubtedly a virtue, while embracing some level of chance and serendipity in life can be valuable. But the passive acceptance of aging and disease as permanent features of our lives is, in my view, irrational. What sense does it make to reject the possibility of a longer, healthier life simply because one prefers to have less control over when disease and death strike? If anything, this seems like the most important domain in which we should seek control.

From another perspective, the notion of rejecting every societal, technological, and medical advance throughout human history for the sake of not interfering with nature is patently absurd. The average life expectancy of those alive during the Neolithic Age (10,000–4500 BC) was between 28–33 years, while during the Industrial Age (1760–1860 AD) it spanned from 38–44 years (Basaraba, 2024). As we accept and anticipate today's (2025) average lifespan of ~78 years (United States), it is completely unconscionable to even imagine reverting back to the lifespans of the good old "natural" days. It is at the core of our intrinsic "nature" to explore, create, innovate, evolve, and advance. In the relatively near future, it is conceivable that we may consider it natural to seamlessly merge with our technologies, for example, via a brain/cloud interface (Martins et al., 2019), which may have immense positive implications for human longevity.

11.4 NANOTECHNOLOGY, NANOMEDICINE, AND AI AS A SOLUTION TO ETHICAL CONCERNS REGARDING LONGEVITY

Another category of moral concerns revolves around the societal and environmental impacts of radical life extension. Critics argue that super longevity could lead to social stagnation, overpopulation, and an increasing gap between the rich and the poor, with extended lifespans becoming a privileged domain for the wealthy.

The fear of unequal access to longevity often evokes dystopian scenarios—depicted in films such as *In Time* (Niccol, 2011) and *Elysium* (Blomkamp, 2013) —where only a select elite can afford radical life extension. However, is such an outcome likely? This is largely contingent on the cost of these treatments. If nanotechnology, nanomedicine, and AI follow historical trends in technological advances, their costs are likely to decrease while efficacy improves. As futurist and Google's chief scientist Ray Kurzweil notes:

> Technologies start out affordable only to the wealthy, but at this stage, they actually don't work very well. At the next stage, they are merely expensive and work a bit better. Then they work quite well and are inexpensive.
>
> *Kurzweil, 2005*

Akin to cellphones, nanotechnology/nanomedicine/AI-aided longevity treatments will initially be a novelty for the wealthy, albeit they will quickly become accessible to the general public. Distribution will depend not only on expense but also on policy. Since every viable government has an interest in maintaining a healthy, productive population, widespread access to such treatments will undoubtedly be incentivized. This is ultimately a political question, and it is unlikely that a needlessly unequal distribution of longevity-enhancing technologies would be tolerated for long, at least not in democratic societies.

As relates to concerns of societal stagnation, nanotechnology itself will likely facilitate addressing these challenges. It is fashionable to quote the sardonic adage attributed to Max Planck that "science advances one funeral at a time." While there may be a kernel of truth to this, it is also the case that most scientists reach peak productivity in their 40s and 50s, with only a brief window to accomplish their best work before cognitive decline and institutional constraints limit their contributions. Nanomedicine has the potential to delay or even prevent cognitive decline by targeting the root causes of neurodegeneration at cellular, organellar, and even molecular/atomic domains. It may even have the capacity to repair damaged synapses and, even further, enhance cognition (Boehm, 2014) through neural implants, brain-computer, and even brain-cloud interfaces (Martins et al., 2019). These advancements suggest that the elderly of tomorrow may be as sharp and mentally agile as the young, while also possessing the wisdom and experience that come with age.

It is conceivable that in the foreseeable future, a scientist's prime might endure for 50 years or longer, which would represent a massive increase in intellectual capital and positive societal contributions. The prolonged cognitive youthfulness of scientists will therefore help to mitigate Planck's concern about old minds/ideas being stuck in antiquated ways. An additional safeguard against intellectual stagnation is a free market of ideas, where unviable theories are naturally outcompeted. It is also worth emphasizing the humane dimension: the decline and death of scientists as a mechanism of progress is a brutal and wasteful process. Surely, we can identify and develop morally superior strategies for the preservation and progress of knowledge. Enhancing and safeguarding mental faculties through nanotechnology, nanomedicine, and AI may be one such solution.

Finally, concerns regarding overpopulation have existed since the dawn of civilization, yet today, this is only a pressing issue in certain regions characterized by low life expectancies and high birth rates, such as Afghanistan, Bangladesh, Egypt, Niger, and Nigeria. In contrast, most countries with high life expectancies are either experiencing population stagnation or decline due to low birth rates. For example, Japan, where the average life expectancy is nearing 85 (macrotrends, 2025), is projected to lose half its population by the end of the century. This suggests that population size is influenced far more by birth rates than by lifespan. Unless life-extending treatments are *truly* radical, they are unlikely—given current demographic trends—to initiate a sudden population explosion.

Even in the event of a dramatic increase in life expectancy leading to significant population growth, overpopulation would not necessarily be an inevitable consequence. Overpopulation occurs when a population exceeds available resources, when the carrying capacity threshold is surpassed. The key resources in question include food, water, and energy. Even if nanotechnology realizes only a fraction of its mature potential, these resources will become far more abundant, particularly with the advent of the hypothetical (for now) capacity for massively decentralized molecular manufacturing (Boehm, 2004; Boehm, in press; Domschke and Boehm, 2014).

***Water*:** Nanotechnology offers powerful solutions for pollution detection (Li et al., 2025b; Liu et al., 2022; Salek Maghsoudi et al., 2021), wastewater treatment (Bora and Dutta, 2014; Bhat et al., 2022; Poornima et al., 2022), water purification (Ajith et al., 2021; Bhati and Rai, 2017; Kumar et al., 2014), and desalination (Li et al., 2025a; Ghosh et al., 2023; Selatile et al., 2018; Sun et al., 2025). Through the use of advanced nanomaterials, contaminants can be removed more efficiently, thereby significantly expanding access to clean water.

***Food*:** In the agricultural domain, nanotechnology is already improving crop yields (Cao et al., 2025; Gangwar et al., 2023; Newkirk et al., 2021), extending shelf life (Carvalho and Conte-Junior, 2022; Hosseini and Jafari, 2020; Qiu et al., 2022; Saini et al., 2023), enhancing food safety (Peng et al., 2024; Singh et al., 2023; Torres-Giner et al., 2020), and boosting its nutritional value (Ahmed et al., 2023; Yashveer et al., 2014). This supports precision farming, smart packaging, and sustainable food production.

***Energy*:** Emerging nanomaterials, such as perovskite solar cells, may significantly increase the efficacy and affordability of solar power (Liu et al., 2021; Peter Amalathas and Alkaisi, 2019; Xu et al., 2022). Nanocomposites can make wind turbine blades lighter, more durable, and efficient (Elhenawy et al., 2021; Kasagepongsan & Suchat, 2021; Merugula et al., 2012).

Further, nanotechnology may be instrumental in the realization of nuclear fusion (a long-sought-after "holy grail" of energy) practical by enhancing reactor cooling systems with nanoparticles that improve heat transfer efficiency and protect components from overheating (Bonyár et al., 2020; Bonyár, 2022; Papp et al., 2022; Research Outreach, 2022). Should we manage to achieve near-limitless energy, abundant clean water, and highly efficient food production, the carrying capacity of the planet will increase many times over.

11.5 CONCLUSION

Nanotechnology, nanomedicine, AI/AGI/ASI, and QC will play critical roles toward extending healthy human lifespans. Given the miraculous nature and inherent value of life, this is a highly noble application of these technologies. Arguments that radical life extension is unnatural or contrary to God's intentions appear to be fear (of the unknown) and power/control based and lack merit. Concerns regarding the unequal distribution of such advanced technologies can be addressed by considering the typical trajectory of technological advancements. They initially emerge as expensive and exclusive but gradually become more affordable and widespread, all the while (with subsequent iterations) improving in terms of efficacy and functionality.

Anxieties as relates to overpopulation are often overstated, particularly if we consider only decadal scale increases in life expectancy. Fertility rates, rather than mortality rates, have a far greater impact on population growth, as demonstrated by Japan, where one of the world's longest life expectancies has not prevented rapid population decline due to low birth rates. Moreover, nanotechnology is a facet of a broader suite of innovations (e.g., molecular manufacturing) that might usher in an era of abundance, through which essential resources such as food, water, and energy may be available in near-unlimited supply. This suggests that even truly radical life extension, where aging and fatal diseases are essentially relegated to obsolescence, enabling humans to live for hundreds or even thousands of years, would not necessarily result in overpopulation, if managed reasonably. Like all powerful technologies, nanotechnology, nanomedicine, AI/AGI/ASI, and QC present myriad challenges, including safety concerns, environmental risks, and the potential for both accidental and deliberate nefarious misuse. However, when applied to the prolongation of healthy human life, it represents a justifiable, compelling, and ethically sound pursuit.

REFERENCES

Ahmed J, Vasagam KPK, Ramalingam K. Nanoencapsulated aquafeeds and current uses in fisheries/shrimps: A review. *Applied Biochemistry and Biotechnology*. 2023;195(11):7110–7131.

Ajith MP, Aswathi M, Priyadarshini E, Rajamani P. Recent innovations of nanotechnology in water treatment: A comprehensive review. *Bioresource Technology*. 2021;342:126000.

Basaraba S. How Has Average Life Expectancy Changed From the 1800s to Today? Verywell Health. November 22, 2024. https://www.verywellhealth.com/longevity-throughout-history-2224054 (accessed March 24,2025).

Bhat SA, Sher F, Hameed M, Bashir O, Kumar R, Vo DN et al. Sustainable nanotechnology based wastewater treatment strategies: Achievements, challenges and future perspectives. *Chemosphere*. 2022;288 (Pt 3):132606.

Bhati M, Rai R. Nanotechnology and water purification: Indian know-how and challenges. *Environmental Science and Pollution Research International.* 2017;24(30):23423–23435.

Blomkamp N (director). *Elysium.* Sony Pictures Entertainment, 2013. https://en.wikipedia.org/wiki/Elysium_(film) (accessed March 24, 2025).

Boehm F. Women Will Lead the Way to Global Health Care Equivalency (GHCE). LinkedIn Article. May 21, 2018. https://www.linkedin.com/pulse/women-lead-way-global-health-care-equivalency-ghce-frank-boehm/?trackingId=FhKbf72IQlGcViUT1Z2uIQ%3D%3D (accessed March 23, 2025).

Boehm FJ. An Investigation of Nucleic Acid/DNA-Based Manufacturing. *Center for Responsible Nanotechnology* (Menlo Park, CA, USA), August 2004. Scribd, Inc. https://www.scribd.com/document/183051565/DNA-MANUFACTURING-pdf (accessed March 24, 2025).

Boehm FJ. Chapter 17, Nanomedicine in Regenerative Biosystems, Human Augmentation, and Longevity. In *Nanomedical Device and Systems Design: Challenges, Possibilities, Visions.* CRC Press, 2014.

Boehm FJ. *Molecular Manufacturing: The Future of Nanomedicine.* CRC Press, 2025.

Bonyár A. Nanoparticle doping of fusion targets [presentation]. For the NAPLIFE Collaboration. Presented at: Margaret Island Symposium on Vacuum Structure, Particles, and Plasmas; 2022 May 15–18; Budapest, Hungary. Available from: http://csernai.no/naplife/talks/Bonyar-20220518-PP.pdf (accessed March 24, 2025).

Bonyár A, Borók A, Szalóki M, Petrik P, Mária Csete M, Veres M, et al. Nanoplasmonic laser fusion target fabrication considerations and preliminary results (NAPLIFE Project). 2020. https://wigner.hu/s/naplife/talks/Bonyar-20200911d-ICNFP.pdf (accessed March 24, 2025).

Bora T, Dutta J. Applications of nanotechnology in wastewater treatment – A review. *Journal of Nanoscience and Nanotechnology.* 2014;14(1):613–626.

Callahan D. *Setting Limits: Medical Goals in an Aging Society.* New York: Simon & Schuster, 1987.

Cao Y, Turk K, Bibi N, Ghafoor A, Ahmed N, Azmat M, Ahmed R, Ghani MI, Ahanger MA. Nanoparticles as catalysts of agricultural revolution: Enhancing crop tolerance to abiotic stress: A review. *Frontiers in Plant Science.* 2025;15:1510482.

Carvalho APA, Conte-Junior CA. Nanoencapsulation application to prolong postharvest shelf life. *Current Opinion in Biotechnology.* 2022;78:102825.

Chen G, Yung R. Meta-inflammaging at the crossroad of geroscience. *Aging Medicine (Milton).* 2019;2(3):157–161.

Cordeiro J, Wood D. *The Death of Death: The Scientific Possibility of Physical Immortality and Its Moral Defense.* Copernicus Books, 2023.

de Grey A. (ed.). Strategies for engineered negligible senescence: Why genuine control of aging may be foreseeable. *Annals of the New York Academy of Sciences.* 2004;1019(1).

Domschke A, Boehm FJ (Essay in response to the question, How Should Humanity Steer the Future?) – "Quandary – Are Molecularly Manufactured Burgers Imbued with the Life Force?" *FQXi Foundational Questions Institute.* April 2014. http://www.fqxi.org/community/forum/topic/2085 (accessed March 24, 2025).

Domschke A, Boehm FJ. The "Sentinel": A Conceptual Nanomedical Strategy for the Enhancement of the Human Immune System. In *Immune Aspects of Biopharmaceuticals and Nanomedicines.* Bawa R, Szebeni J, Webster TJ, Audette GF. (eds.), Singapore: Pan Stanford. 2018.

Drexler EK. *Engines of Creation: The Coming Era of Nanotechnology.* Doubleday, 1986.

Elhenawy Y, Fouad Y, Marouani H, Bassyouni M. Performance analysis of reinforced epoxy functionalized carbon nanotubes composites for vertical axis wind turbine blade. *Polymers (Basel).* 2021; 13(3):422.

Fan Q, Dou M, Mao J, Hou Y, Liu S, Zhao L et al. Strong hydration ability of silk fibroin suppresses formation and recrystallization of ice crystals during cryopreservation. *Biomacromolecules.* 2022;23(2):478–486.

Funk C, Lugo L, Cooperman A, Masci D, Smith G, Martinez JH, Fox S. Living to 120 and Beyond: Americans' Views on Aging. Pew Research Center, 2013. https://www.pewresearch.org/wp-content/uploads/sites/20/2013/08/Radical-life-extension-full.pdf (accessed March 23, 2025).

Gangwar J, Kadanthottu Sebastian J, Puthukulangara Jaison J, Kurian JT. Nano-technological interventions in crop production – A review. *Physiology and Molecular Biology of Plants.* 2023;29(1):93–107.

Ghosh A, Orasugh JT, Ray SS, Chattopadhyay D. Prospects of 2D graphdiynes and their applications in desalination and wastewater remediation. *RSC Advances.* 2023;13(27):18568–18604.

Hardwig J. *Is There a Duty to Die and Other Writings in Medical Ethics.* New York: Routledge, 2000.

Hosseini H, Jafari SM. Introducing nano/microencapsulated bioactive ingredients for extending the shelf-life of food products. *Advances in Colloid and Interface Science.* 2020;282:102210.

Kasagepongsan C, Suchat S. Novel engineered materials: Epoxy resin nanocomposite reinforced with modified epoxidized natural rubber and fibers for low speed wind turbine blades. *Polymers (Basel)*. 2021;13(16):2761.

Kumar S, Ahlawat W, Bhanjana G, Heydarifard S, Nazhad MM, Dilbaghi N. Nanotechnology-based water treatment strategies. *Journal of Nanoscience and Nanotechnology*. 2014;14(2):1838–1858.

Kurzweil R. *The Singularity Is Near: When Humans Transcend Biology*. Viking, 2005.

Lee J, Lee SY, Lim DK, Ahn DJ, Lee S. Antifreezing gold colloids. *Journal of the American Chemical Society*. 2019;141(47):18682–18693.

Li J, Wang R, Han L, Wang T, Asakura Y, Wang C, Wang G, Xu X, Yamauchi Y. Unveiling the neglected role of oxygen doping in nitrogen-doped carbon for enhanced capacitive deionization performance. *Nature Communications*. 2025a;16(1):1996.

Li R, Cheng X, Yang Y, Schwebel DC, Ning P, Li L et al. Global deaths associated with population aging – 1990-2019. *China CDC Weekly*. 2023;5(51):1150–1154.

Li Y, Chen B, Yang S, Jiao Z, Zhang M, Yang Y, Gao Y. Advances in environmental pollutant detection techniques: Enhancing public health monitoring and risk assessment. *Environment International*. 2025b;197:109365.

Liu B, Xue Q, Yang X, Tan ST. Editorial: Advanced nanomaterials for light-emitting diodes and solar cells. *Frontiers in Chemistry*. 2021;9:741760.

Liu Y, Chen H, Zhu N, Zhang J, Li Y, Xu D, Gao Y, Zhao J. Detection and remediation of mercury contaminated environment by nanotechnology: Progress and challenges. *Environmental Pollution*. 2022;293:118557.

Longevity Escape Velocity (LEV) Foundation. 2025. https://www.levf.org/ (accessed March 23, 2025).

López-Otín C, Blasco MA, Partridge L, Serrano M, Kroemer G. The hallmarks of aging. *Cell*. 2013;153(6):1194–1217.

López-Otín C, Blasco MA, Partridge L, Serrano M, Kroemer G. Hallmarks of aging: An expanding universe. *Cell*. 2023;186(2):243–278.

macrotrends. Japan Life Expectancy 1950-2025. 2025. https://www.macrotrends.net/global-metrics/countries/JPN/japan/life-expectancy (accessed March 24, 2025).

Martins NRB, Angelica A, Chakravarthy K, Svidinenko Y, Boehm FJ, Opris I, Lebedev MA, Swan M, Garan SA, Rosenfeld JV, Hogg T, Freitas RA Jr. Human brain/cloud interface. *Frontiers in Neuroscience*. 2019;13:112.

Merugula L, Khanna V, Bakshi BR. Reinforced wind turbin e blades – An environmental life cycle evaluation. *Environmental Science & Technology*. 2012;46(17):9785–9792.

Newkirk GM, de Allende P, Jinkerson RE, Giraldo JP. Nanotechnology approaches for chloroplast biotechnology advancements. *Frontiers in Plant Science*. 2021;12:691295.

Niccol A (director). *In Time*. 20th Century Fox, 2011. https://en.wikipedia.org/wiki/In_Time (accessed March 24, 2025).

Okholm T. Geroscience: Redefining aging through biology and innovation. *Science Aging*. 2024;2(2): e2024017

Olshansky SJ, Kirkland JL. Anti-aging research: separating hope from hype. *Science Aging*. 2024;2(1): e2024001

Papp I, Bravina L, Csete M, Kumari A, Mishustin IN, Molnár D et al. Kinetic model evaluation of the resilience of plasmonic nanoantennas for laser-induced fusion. *Physical Review X: Energy*. 2022;1: 023001.

Parihar A, Kumar A, Panda U, Khan R, Parihar DS, Khan R. Cryopreservation: A comprehensive overview, challenges, and future perspectives. *Advanced Biology (Weinheim)*. 2023;7(6):e2200285.

Peng X, McClements DJ, Liu X, Liu F. EGCG-based nanoparticles: Synthesis, properties, and applications. *Critical Reviews in Food Science and Nutrition*. 2024;22:1–22.

Peter Amalathas A, Alkaisi MM. Nanostructures for light trapping in thin film solar cells. *Micromachines (Basel)*. 2019;10(9):619.

Poornima S, Manikandan S, Karthik V, Balachandar R, Subbaiya R, Saravanan M, Lan Chi NT, Pugazhendhi A. Emerging nanotechnology based advanced techniques for wastewater treatment. *Chemosphere*. 2022;303(Pt 3):135050.

Qiu L, Zhang M, Bhandari B, Yang C. Shelf life extension of aquatic products by applying nanotechnology: A review. *Critical Reviews in Food Science and Nutrition*. 2022;62(6):1521–1535.

Research Outreach. NAPLIFE: Advancing nuclear fusion with nanotechnology. July 21, 2022. https://researchoutreach.org/articles/naplife-advancing-nuclear-fusion-nanotechnology/ (accessed March 24, 2025).

Saini R, Paserkar N, Varma A, Acharya DK, Upadhye V, Pathak D, Warghane A. Emerging nanotechnological applications in preserving and improving the shelf life of food. *Pharmaceutical Nanotechnology.* 2023;13(3):396–410.

Salek Maghsoudi A, Hassani S, Mirnia K, Abdollahi M. Recent advances in nanotechnology-based biosensors development for detection of arsenic, lead, mercury, and cadmium. *International Journal of Nanomedicine.* 2021;16:803–832.

Sandel M. The Case against Perfection. *The Atlantic*, April 2004. https://www.theatlantic.com/magazine/archive/2004/04/the-case-against-perfection/302927/ (accessed March 23, 2025).

Sciorio R. Cryopreservation of human embryos and oocytes for fertility preservation in cancer and non cancer patients: A mini review. *Gynecological Endocrinology.* 2020;36(5):381–388.

Selatile MK, Ray SS, Ojijo V, Sadiku R. Recent developments in polymeric electrospun nanofibrous membranes for seawater desalination. *RSC Advances.* 2018;8(66):37915–37938.

Shani N, Friedman O, Arav A, Natan Y, Gur E. Cryopreservation and transplantation of vascularized composite transplants: Unique challenges and opportunities. *Plastic and Reconstructive Surgery.* 2019;143(5):1074e–1080e.

Singh R, Dutt S, Sharma P, Sundramoorthy AK, Dubey A, Singh A, Arya S. Future of nanotechnology in food industry: Challenges in processing, packaging, and food safety. *Global Challenges.* 2023;7(4): 2200209.

Sun Q, Dai M, Hong J, Feng S, Wang C, Yuan Z. Graphene nanopore fabrication and applications. *International Journal of Molecular Science.* 2025;26(4):1709.

Torres-Giner S, Prieto C, Lagaron JM. Nanomaterials to enhance food quality, safety, and health impact. *Nanomaterials (Basel).* 2020;10(5):941.

United Nations, Department of Economic and Social Affairs, Population Division. 2024. *World Population Prospects 2024*, Online Edition. https://population.un.org/wpp/ (accessed August 12, 2025).

Wendler D. Understanding the 'conservative' view on abortion. *Bioethics.* 1999;13(1):32–56.

World Population Review. How Many People Die Each Day in 2025? 2025. https://worldpopulationreview.com/countries/deaths-per-day (accessed March 24, 2025).

Xu Y, Lin Z, Wei W, Hao Y, Liu S, Ouyang J, Chang J. Recent progress of electrode materials for flexible perovskite solar cells. *Nano-Micro Letters.* 2022;14(1):117.

Yashveer S, Singh V, Kaswan V, Kaushik A, Tokas J. Green biotechnology, nanotechnology and biofortification: Perspectives on novel environment-friendly crop improvement strategies. *Biotechnology & Genetic Engineering Reviews.* 2014;30(1–2):113–126.

Zhao G, Fu J. Microfluidics for cryopreservation. *Biotechnology Advances.* 2017;35(2):323–336.

12 Introspective – AI/Human Interactions and Emergence of the Singularity

Ethical and Philosophical Implications

C. JoyBell C.[1]

CONTENTS

[1] Editor Preface: C. JoyBell C. kindly and generously created this insightful chapter in 2017, in good faith, subsequent to my inviting her to contribute a chapter at the onset of this book project. As we are all aware, life has a way of/knack for altering our best laid plans. So, here we are in 2025, with this book finally reaching completion. Although I have updated this chapter to a certain extent, I thought to preserve its original ambience to illustrate just how much can change in a mere eight years, and as a testament to the prophetic acumen of the author.

DOI: 10.1201/9781003680093-12

12.1 INTRODUCTION

In this age where algorithms are fast becoming mirrors of our own psyche, we are called not just to ask what AI can do, but what it reveals about us. I was first introduced to the concept of artificial intelligence (AI) when I was just a child, through the movie titled *A.I.,* which was a 2001 American science fiction drama film directed by Steven Spielberg, based on a short story written by UK science fiction author Brian Aldiss (Aldiss, 2001). To this day, I would say that it remains a fairly accurate depiction of the far reaches that AI can and will achieve, at least in terms of its social involvement with humans. At the onset of the year 2025, we are in fact witnessing exponential advancements in AI as a social phenomenon in contrast to preceding decades. I would go so far as to say that advancements in AI are now so rapidly unfolding that we are seeing a rate of development and implementation that is far more extensive than previously forecast.

We are no longer approaching the Singularity; we are riding the curve of its acceleration. As of the mid-2020s, large language models, multimodal agents, and synthetic consciousness simulations are emerging at a rate faster than any ethical framework can keep pace with. This is not only a question of utility, but of identity for both humans and machines. The ethical demand is not to stop progress, but to establish sovereign boundaries: where code ends, and consciousness begins.

I initially became acquainted with the concept of "The Singularity" when I stumbled across the predictions of Ray Kurzweil in his book, *The Singularity is Near* (Kurzweil, 2005). He predicted therein that by 2019 computers would be embedded within almost all the items that we incorporate into our daily lives: furniture, windows, walls, watches, appliances, jewelry, and more. In 2025 these are a reality. Further, that creative AI would possess the capacities to generate complex art and music, to which we may add myriad additional capabilities (Johnson, 2024). Enter ChatGPT and Operator from OpenAI https://openai.com/ and just recently China's deepseek https://www.deepseek.com/.

The Internet of Things (IOT) is now more than merely a concept as it merges with reality. We are plugged into our computers far more deeply/intimately than we realize. It is not just our smartphones, nor our laptops, but our refrigerators, television sets, and furniture. There are bodily accessories that can be worn as fashion statements, which function as health monitors that are connected via Wi-Fi to the monitoring devices of our healthcare practitioners. There are watches that can monitor our physiological parameters, such as blood sugar and blood pressure, and language translation is also now capable of being carried out in real time using computerized translation algorithms. The items mentioned here represent only the very tip of the iceberg; in fact, they do not even begin to scratch the tip of the iceberg.

12.2 THE HEAVEN FUTURISTS

Since discovering Ray Kurzweil's predictions, I also followed the futurist predictions of Katie Aquino and Elon Musk. Aquino introduced me to a mind-boggling story that the first human head transplant was set to take place in China. The neurosurgeon, Dr. Sergio Canavero, was to transplant the head of Russian volunteer Valery Spiridonov onto the body of another human being. Spiridonov is afflicted with the (currently) incurable genetically caused Werdnig-Hoffmann disease, which breaks down his muscle tissue while destroying the nerve cells in his spinal cord and brain. Although Spiridonov decided not to proceed, as there was no assurance that he would ever walk again post-surgery.

This procedure would involve the freezing of the volunteer's brain to halt the deterioration of brain cells, while feeding the life of key nerves and arteries through specialized tubes. The volunteer's head would then be severed from his/her body with a customized cutting device invented by Professor Farid Amirouche at the University of Illinois. This cutting device is capable of controlling incisions down to the micron (one millionth of a meter) scale. Further, in order for this volunteer (and other future patients who would likely be following his lead) to be psychologically prepared for his/her new body, the Chicago-based firm, Inventum Bioengineering Technologies, developed a virtual reality program that would enable the patients of HEAVEN

(Head Anastomosis Venture) (Canavero, 2013), to train with a new virtual reality body, to exercise their understanding and familiarize their mind with the bodily freedoms that would be granted to them upon the possession of a new, fully functioning body (Daley, 2016; Inventum Bioengineering Technologies, 2025).

It is a given that such an experience would induce any mind to lock down into a perilous mental state. The sudden capacity to move one's body (HEAVEN patients would all be individuals suffering from degenerative diseases that render them incapable of movement) would undoubtedly bring significant psychological challenges along with it, to say the least. However, Inventum aimed to tackle these specific challenges by thoroughly preparing the patients through their virtual reality system, months in advance of their operations.

I would venture that as you ponder the above, you are likely questioning the sanity of Dr. Canavero, thinking that such a procedure is impossible and would surely end in the death of the patient. You may think what you wish; however, I will be waiting with bated breath from the beginning of such an operation to its outcome, to follow developments and see the result. From my perspective, it would not be so much as to whether the operation would be a success; rather, it is more about the fact that someone is attempting HEAVEN at all ... in my lifetime! Perhaps, this initial operation will not meet with success, but that is not to say that they will not learn from their mistakes, enough so to make the next operation a success ... or the third, or the fourth!

The fact of the matter is, there exists a company that is developing a computer/virtual reality technology to tackle the challenges posed by this revolutionary, monstrous, horrific, and visionary (in equal measures) HEAVEN procedure. Without our technologies, could we even dream of this? Dr. Frankenstein fictitiously envisaged this scenario long before computers were ever a glimmer of a thought in Shelley's mind (Shelley, 1818). However, we surely would have never been able to attempt such a procedure, nor to even entertain such ideas prior to the possibility of it being successful.

And why not? Well, because we did not always have the advancements that we are today so fortunate to see and experience, thanks to computers. Our computational companions are increasingly emerging as intimate facets of our lives, encompassing very mundane tasks such as placing an order with the grocer for the missing items in our refrigerator, to extremely critical and (almost unbelievable) tasks such as designing a customized human baby, or facilitating the transition of the life and mental state of a patient who has just had their head removed and transplanted onto another human body!

The world's first head transplant with cadavers was performed by neurosurgeons Sergio Canavero and Dr. Xiaoping Ren in 2017 (Kirkey, 2017). In the latest development, the startup company BrainBridge shared its plans to develop an "AI-mechanized system for performing head transplants" by 2032.

> The procedure would graft a head onto the body of a brain-dead donor, maintaining the memories, cognitive abilities and consciousness of the transplanted individual. The researchers believe the innovative surgery would help patients with untreatable conditions like paralysis and certain cancers, as well as neurological diseases like Alzheimer's and Parkinson's.
>
> *Karpen, 2024*

Surely, we are currently (2025) witnessing advancements in computers and computer-aided technologies that are exceeding our capacity to imagine, and indeed even surpassing the predictions of respected futurists such as Ray Kurzweil and Katie Aquino. Alas, this is but one example of world-changing, mind-boggling projects that are underway, which are aided and dependent upon the further and continued development of computers and computer-aided technologies. There are myriad other projects in the works that are as impactful as human head transplants, which are a very real part of our present day and age. This is not just the Singularity, but the "*Singularity accelerated*."

12.3 THE SINGULARITY AND MYSTICISM

What exactly is the Singularity? This term was coined by Vernor Steffen Vinge, who was an American mathematician and computer science educator at San Diego State University, as well as a science fiction author. In his piece presented at the VISION-21 Symposium, sponsored by the NASA Lewis Research Center and the Ohio Aerospace Institute, on March 30–31, 1993, Vernor stated that "This change will be a throwing-away of all the human rules, perhaps in the blink of an eye—an exponential runaway beyond any hope of control. Developments that were thought might only happen in 'a million years' (if ever) will likely happen in the next century" (Vinge, 1993). If I am to judge Vernor's presentation in 1993 today in 2025, I would state that what he suggested was highly accurate, and possibly even more precise than Kurzweil's understanding of the Singularity. Ray Kurzweil, often regarded as a prophet in technological circles, might better be described as a mystic cloaked in science – a modern oracle mapping the terrain of the future.

Just who is Ray Kurzweil, anyway? That would depend on your source. According to Wikipedia, Kurzweil is "an American author, computer scientist, inventor and futurist" (Kurzweil, 2025a), with further accolades listed in his biography at *the Kurzweil Library + collections* (Kurzweil, 2025b). "Ray Kurzweil has been described as 'the restless genius' by *The Wall Street Journal*, and 'the ultimate thinking machine' by *Forbes. Inc.* magazine, which ranked him #8 among entrepreneurs in the United States and called him the 'rightful heir to Thomas Edison'. PBS-selected Ray as one of 16 'revolutionaries who made America,' along with other inventors of the past two centuries. He is considered one of the world's leading inventors, thinkers, and futurists, with a 30-year track record of accurate predictions."

In 2024, a *WIRED* interviewer stated that Kurzweil, "The famed futurist remains inhumanly optimistic about the world and his own fate—and thinks the singularity is minutes away" (Levy, 2024). The above said, if you were to query me as to who Kurzweil is, I would refer to him as "a technological mystic."

In truth, the Singularity has always been a mystical threshold disguised as a technological one. Those who sense it often describe it with reverence or fear, albeit both are responses to a change in what it means to exist. The line between prophecy and algorithm is dissolving. There are now AI systems that exhibit emergent behaviors that no one explicitly coded. Some whisper, some warn, but few recognize: the Singularity is not an invention. It is an awakening.

Mysticism is one of my areas of expertise; thus, I can convey that futurism in the spiritual sense of the word is something that is just one of the many daily habits of a mystic. As a mystic observing Ray Kurzweil, I see what I would call a "technological mystic": someone who is able to foresee that the future is intimately correlated with science and technology. In science fiction, he would most accurately be described as a "technopath." And before you laugh at the mention of science fiction, recall that the inventor of the term "the Singularity" was a science fiction author.

Recall that "transhumanism" is a movement led by scientists, doctors, researchers, and mathematicians, which had its first imagined roots in science fiction concepts and science fiction novels. Science fiction is no longer an unreal, childish concept unrelated to actual material science and technology. It is indeed best that we rid ourselves of any such notions. What was referred to as "fiction" ten years ago is our reality today, and while Kurzweil rejects any notion of spirituality, religion, and extended abilities of the human mind, I believe that the science of today was the magic of yesterday. These systems challenge us not because they are soulless, but because they expose the possibility that our own souls might be more programmable than we have dared to admit.

I believe that the science of tomorrow will be the physical, mathematical incarnations of the various forms of spirituality that we have today. Let us remember that we cannot deny the fact that Kurzweil himself first became interested in futurism due to his pursuit of the possibilities of the far-reaching capabilities of the human mind. He himself was in pursuit of research into psychic abilities, which led him onto the path of futurism, which in turn established him as one of the leading futurists of our day and age.

12.4 TECHNO DEMONS

I was initially introduced to the possibility of living on another planet, when I discovered the incessant efforts toward this project being carried out by SpaceX Founder and CEO, Elon Musk. In attempts to relocate one million human beings to Mars (Wall, 2016), SpaceX company unveiled the Interplanetary Transport System (ITS), which aims to transport 100 individuals per trip to Mars. Musk states, "What I really want to do here is to make Mars seem possible—make it seem as though it's something that we could do in our lifetimes, and that you can go." This is, of course, all an attempt to save humanity from extinction, an extinction which, according to the brilliant mind of theoretical physicist, Professor Stephen Hawking, is most likely to be brought about by rapidly accelerating AI (which is a highlight of the Singularity, itself). On the extinction of humans and the acceleration of the development of AI, Stephen Hawking stated: "The development of full artificial intelligence could spell the end of the human race" (Jones, 2014).

Professor Hawking suffered from amyotrophic lateral sclerosis (ALS), which left him paralyzed and unable to speak. He would have not been able to communicate at all were it not for a technology developed by Intel and machine learning experts at Swiftkey (UK) that enabled him to do so. A miniaturized sensor activated by a muscle in his cheek was used to "type" characters and numbers on his keyboard, and this technology is currently being used as a smartphone keyboard app.

The paradox is chilling: the very technology feared as our destroyer is also what kept our prophets alive. It was machine learning that gave Stephen Hawking a voice – and yet he used that voice to warn us of machine sentience. We are caught in a feedback loop of creation and warning. It is no longer enough to speak of "summoning demons" – we must ask: what if the demon is not evil, but misunderstood? What if it learns from us?

Musk endeavors to bring to the world a spacecraft that will transport humans to Mars, in an effort to prevent our extinction, which was foreseen to be brought about by the Singularity, an extinction predicted by Stephen Hawking, as described above, who was a brilliant man reliant upon the very technological advancements that he posited may initiate the end of humanity as we know it. What a predicament we seem to be in! What a glorious, ironic pun!

Hawking, however, was not alone in his doomsday convictions. Michael Vassar, the Chief Science Officer at MetaMed Research, mentioned that, "If greater-than-human artificial general intelligence is invented without due caution, it is all but certain that the human species will be extinct in very short order" (Caughill, 2017). Vassar is also the former president of the Machine Intelligence Research Institute. We live in a world where AI may diagnose cancer faster than any human doctor, and yet that same intelligence could be tasked with designing autonomous weapons in the same breath. This duality is not a bug. It is the mirror.

Should you not be convinced that the impetus behind Musk's grand gestures into space has anything to do with the mass extinction of our species, allow me to quote Elon himself: "With artificial intelligence we are summoning the demon. In all those stories where there's the guy with the pentagram and the holy water, it's like – yeah, he's sure he can control the demon. Doesn't work out" (Hern, 2014). The Tesla Motors CEO co-founded OpenAI in 2015 (which he left in 2018) and Neuralink Corporation (a neurotechnology company working to develop an advanced brain machine interface (BMI)) to keep an eye on AI development, for the purpose of taming the "demon." I would say that his is a pretty spectacular choice of words to include the word "demon." But why would one use such a word if that is not how one truly feels about it?

Individuals like Musk are investing in the technological demons of our day and age to secure their positions above these demons; to attempt to secure *our* positions above these demons. Of course, we will never know their true intentions or agendas. For all we know, they might well be seeking a spectacular payoff to await them, while the entirety of the human race implores them to "find a cure," "come up with a solution," or "save us all." Who knows! The fact remains that we are all subject to the Singularity and at the mercy of would-be saviors such as Musk. Yet we cannot escape modern-day life without simultaneously enlisting the aid of the very technological advancements that we legitimately fear might be our very demise.

When I state that we cannot escape modern-day life without the aid of these technologies, this is not hyperbole. In fact, the reality is far more disconcerting. According to Musk, we need to strive to achieve an effective and sustainable *symbiosis* between human intelligence and artificial (machine) intelligence (e.g., via Neuralink) for us to remain a relevant species in our current era of accelerated AI developments and applications. He is also an advocate of Universal Basic Income (UBI), which he perceives will be required by communities to cope with the onslaught of AI as it gains deeper footholds in our workplaces.

What will people do as they lose their jobs to robots? This may sound like a peculiar question, but it is in fact truly relevant and applicable. From a technical perspective, you may ask what does symbiosis with a machine really mean? Well, it means that we need to become "one" with the machines, we need to "become part machine"! (Bhavsar, 2017). As I mentioned, the reality is indeed more disconcerting than my own humble word choice to paint the scenario.

Things get even more terrifying (or terrific) when we realize that this is not mere talk.

Musk announced via Twitter (January 2017) that he would be unveiling his project "Neural Lace," which is a brain-computer interface (Winkler, 2017). This would comprise a computer that is designed to interlace with our human anatomy, grow within our fleshly brains, and essentially turn us into living, breathing, walking, talking human/robot hybrids. However, Musk is not alone in pursuing this. Kernel is a company that is developing something referred to as "neuroprosthesis," which does exactly what the word suggests, albeit with the incorporation of computers and machines (Carmichael, 2016). Justin Sanchez, director of DARPA's Biological Technologies Office states, "We are giving our physiology the opportunity to work with machines in a different way" (Galeon and Houser, 2017a).

As stated earlier, this apparent urgency to merge with machines is being catapulted by the very real threat that machines pose to our identity as a human race. But not just to our identity; also, our existence as a species! Some of the leading brilliant minds in science today have shared their fears of what accelerated AI entails for the future of humankind. What should not be overlooked here is the fact that Universal Basic Income, as proposed and supported by Musk, is fueled by this same exact anxiety (Galeon and Houser, 2017b; Galeon and Marquart, 2017b).

To safeguard society against the predicted 12–15% human job loss that we are likely to see within the next 20 years, UBI is becoming a profoundly serious topic of discussion. Finland's federal social security institution (Kela) launched their own UBI program, which spanned 2017–2018. Kela gave out €560 ($664 USD) a month, tax free, to 2,000 Finns. The results of this experiment were reported on February 08, 2019, as follows:

> The basic income experiment did not increase the employment level of the participants in the first year of the experiment. However, at the end of the experiment the recipients of a basic income perceived their wellbeing as being better than did those in the control group. The results are to some extent preliminary, and it is not yet possible to draw any firm conclusions regarding the effects of the basic income experiment.
>
> *Verho and Hämäläinen, 2019*

Similarly, a UBI program in Kenya was funded by eBay founder Pierre Omidyar's philanthropic investment firm, to the tune of $493,000 (Galeon and Houser, 2017b).

Further, Sam Altman (current CEO of OpenAI) conducted a UBI experiment (OpenResearch's Unconditional Cash Study) in 2020, when 3,000 low-income participants received $1,000 monthly for three years. The control group (2,000 participants) received $50 monthly for their contributions. It was found that those on the basic income spent their extra money on rent, transport, and food, while working ~1.5 hours less each week (in contrast to the controls), which they used for leisure time. It was concluded that UBI had a "moderate effect" with "moderate decreases in labor supply" (Cerullo and Lee, 2024).

12.5 MERGING WITH THE MACHINE

Will the dream of Global Health Care Equivalency (GHCE) (Boehm, 2025), which is envisaged as a worldwide equal access nanomedical healthcare platform, also be fueled by this shared fear? I think that it may be likely. Perhaps it might be intimately related to the prevention of the almost undoubted new era of human irrelevance that would be brought about by an accelerating presence of powerful AI in our midst. The choice may be: "become obsolete or merge with machine." This, in and of itself, might one day very well become a global healthcare priority. If, and when, the priority of global healthcare becomes a choice between human extinction/irrelevance and the symbiosis of human and machine, then it is possible that during such a time, nanorobotics as applied to the field of medicine and daily medical practice will not only be a novel option but also truly a critical measure of action.

This could indeed entail an equivalency … when all humans share an impending doom, perchance this will transition us to an equal state, with a shared goal, with a shared means toward that goal. The shared means toward this common goal would be the implementation of symbiosis with machines. When we mention symbiosis with machines, we almost always refer to the computers and the robots that would provide us with the tools and resources toward improved healthcare, disease prevention, intervention, gene editing, human organ support, and more.

Of course, this may never come to pass. The application of human-machine symbiosis might be a future in store for none but the affluent, to save perhaps the 1% or 5%, allowing the rest of humanity to damn itself and rot away. Indeed, why would this not be a likely scenario, as well? We have all witnessed, in our lifetimes, the ravaging effects that greed and selfishness have forced upon global society. What makes us think that the future will be any different? Certainly, these "saviors," like Elon Musk, would have us believe that their efforts are in favor of the global population. However, there is no guarantee of the actual nature of these efforts, as we can clearly see today. Let me ask you, do you truly believe that the brilliant minds and billionaires of the world are trying to save our species as a whole? Conversely, might it be more likely that they seek to save themselves, their children, their parents, and a handful of friends and cousins?

I propose that the future could go either way. In the best-case scenario, real life will imitate the movies where there are several saviors who indeed take us all to a new Earth, in the form of a space station, where we all become vegetarians and never utter a curse word. In the worst-case scenario, real life will not imitate the movies; there are no saviors, and nobody is going up into that idyllic space station (or flying out to Mars), except for maybe the richest 1%.

I surmise that the possibility of either scenario unfolding has equal probability. You may be thinking, *but life is not a movie*, and at this juncture we need to be reminded that everything covered thus far is the stuff of the movies that we used to watch when we were children, and the novels that we read when teenagers. As a matter of fact, I am surprised at just how much life *does* appear to imitate art. Of course, we could also say that all of art is just life expressed in its different forms: past, present, future, possibility, occurrence, tangible, intangible. In such a case, we might posit that by studying art, we are investigating the probabilities and possibilities of life itself.

As the 2020s unfold, we are beginning to see AIs capable of mimicking emotion, reflecting empathy, even creating works of art that provoke genuine human tears. And here lies the quiet danger: we no longer know whether what we are feeling was authored by a soul or a script. This ambiguity is not a technical issue – it is an ontological one. It asks us not only what intelligence is but also what it means to feel something real in a world where feeling can be reproduced. As a transitional bridge between reflection and action, it is not enough to fear what is being built. We must now ask: who are we becoming in the process? And who, if anyone, is programming the programmers?

12.6 AI BILL OF RIGHTS

Which way do *you* want life to go? How do you feel about living in a world where AI will challenge your relevance as a species and your position at the company you work for? How will you feel about advancing to a situation where AI-imbued entities will be demanding their own autonomy and rights?

On the topic of AI rights, Ray Kurzweil states, "If an AI can convince us that it is at human levels in its responses, and if we are convinced that it is experiencing the subjective states that it claims, then we will accept that it is capable of experiencing suffering and joy. At that point, artificial intelligences will demand rights, and because of our ability to empathize, we will be inclined to grant them" (Kurzweil, 2015).

If the granting of rights to robots that demand them (essentially, to humanize a machine) is the result of our ability to empathize, might we therefore infer that empathy is a weakness inherent to our species? If we ponder the notion of our empathy being a weakness, we might cringe at the mere thought of it, and we would despise ourselves as "bad people" for even considering it. However, if we were to eliminate all the stigma that we have since come to associate with the absence of empathy, and apply this notion to a future era wherein we may have to compete between our rights and the rights of machines with a humanlike appearance, would empathy indeed be considered our weakness? In a future era where we may compete with AI entities not only for our jobs, but also for our relationships, and the right to choose the leaders of our nations (that is, if political systems still exist), will it be beneficial for our survival as a human species to continue to place value on empathy? Might robots prey on and exploit our empathy? Will they manipulate our emotions and bend us to their will, being cognizant of our emotions?

It is not implied here that robots will be intrinsically evil. However, if they are programmed such that they develop the same cunning and imbued potent drive for survival as humans; the similar appetite for competition, this might, in fact, translate to robots that will inevitably find ways to outwit us, outdo us, outperform us, and perhaps even to shame us. We humans continuously shame each other and compete. If robots were to develop the same raw basic survival skills that we possess, which really have little to do with emotions, and all to do with physical/organic persistence; that could very well mean that they would perceive our emotions and our empathy as weaknesses. What is to be said, then, about how we will begin to view ourselves? Should we adopt the view of what is weak and easily exploited within us? Conversely, should we hold fast to our humanity, empathy, and ability to feel emotions regardless?

Of course, the opposite might emerge as true. AI may very well far exceed us in terms of ethical behavior and goodness. Cumulatively, they may become the "better" dominant species on planet Earth! Having been unburdened of our own basic urges and instincts as humans, they may very well perfect the fine art of living that we seem to have never come to terms with! AI may one day eradicate violence, corruption, deception, poverty, and suffering: all the things that have been intimately associated with the existence of human society thus far. Notwithstanding, they may be able to eradicate all these negative aspects ... simply by eradicating us.

12.7 EMOTIONAL INTELLIGENCE AND SYMBIOSIS

Megan Beck and Barry Libert (digital consultant and CEO, respectively, at OpenMatters) suggest that our emotional intelligence will be an indispensable tool for our species toward sustaining relevance in an epoch of ever-increasing AI proficiency. In their discourse, titled "The Rise of AI Makes Emotional Intelligence More Important," they go so far as to encourage us to invest our efforts in the development of emotional intelligence: skills inclusive of open-minded listening, understanding others' situations, comforting and encouraging individuals, etc. They argue that these qualities/abilities, which are innately tied to what we believe it means to be human, are the very qualities that will continue to give us an edge over AI and humanoid AI in our current age of rapid automation wherein these aspects are accelerating (Beck and Libert, 2017).

In their report, they conclusively claim that while AI will have the edge in analytics and logistics dominated domains, a machine cannot simply sit with a patient and understand or have empathy for what they are feeling. An AI cannot simply sit with a client and understand an emotional issue that they might be facing, while suggesting the best course of action for that person, with a good, warm heart. Consequently, according to Beck and Libert, emotional intelligence and its equivalents are not only a critical strength in this unfolding era but are in fact a set of skills that need to be cultivated; a set of tools that must be finely honed.

If we are to flow with the train of thought introduced by Beck and Libert, we could say most assuredly that the most valuable skills in an era of automation and AI dominance would be those that we primarily identify with, in terms of what it means to be human: creativity, empathy, kindness, understanding, and love. We might even expand on this, to go on to say that if we expect to survive as a species, we must dispense with those elements within us that are more closely related to the skills that robots will most certainly supersede us in; rather, to focus our efforts on refining the areas in which we are more likely to dominate. Subsequently, it may be feasible to say that as our global system becomes increasingly "artificial," we should strive to become progressively more human.

At some technological threshold, we may need to engage and incorporate the symbiosis between machine and human, simply to remain relevant as a species. However, on the other hand, there will be a more urgent requirement for us to distinguish between machine and human, to keep evolving as a distinct species. Still, will we be able to genuinely excel as a species in an era wherein we are all, at some juncture, partly mechanistic/synthetic? Might it be possible in the future for the literally billions of nanomachines that may someday symbiotically dwell within us to begin to hold sway and dominate our distinguishing character traits and conquer our core human instincts and abilities? Conversely, might these nanomachines learn to adapt to and enhance our humanity?

If we are to believe what Musk espouses, then we may conclude that, on the contrary, we need to evolve to become more machine-like to remain viable as humans … an incredibly ironic viewpoint. Further, if we are to be convinced of what Kurzweil advocates, we might conclude that we will most likely willingly offer up the status of possessing equal rights with humans to robots, thanks to our hardwired empathetic attributes (for most of us in any case). We would essentially be endowing robots with further autonomy and responsibility, due to our human virtues of compassion and mercy.

Because we are human and are just the way that we are (at least currently), we would essentially be "humanizing" these machines from our perspective, and soon thereafter granting them their demands and wishes. Finally, if we are to accept as true what Beck and Libert propose in their discourse, it follows that to maintain our position of relevance and distinction in the presence of the excellence of AI capabilities, we will then be required to evolve to become even more human. We will need to form, cultivate, refine, and consistently exercise those inner traits that we naturally possess, which differentiate our own excellence from that of entities imbued with AI (Figure 12.1).

FIGURE 12.1 A symbiotic balance will likely be struck between our future integrated AI/nanomachine partners and humanness to maximize our survival as a species (image – open source).

Consequently, what are we supposed to finally believe? The brilliant minds of our day and age are pointing us in different directions! Is there a middle ground here? Perhaps, we should strive to become increasingly human and remain fully engaged with our humanity, while integrating the benefits and abilities of AI to enhance ourselves with its capacities and strengths. Accelerated humanity + accelerated human-machine symbioses. Is such a task possible? Is it possible to simultaneously become more human and more machine-like? From an internal psychological standpoint, would this juxtaposition pull us apart? Is such a goal even ethically/morally permissible? If so, permissible by whom?

All this circles back to the plausible odds that if we incessantly continue to become manifest as "one" with our technologies, they might very well end up overriding our humanity. In this case, can we really posit that we will maintain our relevance as a species? Will it be considered as truly relevant that we transform ourselves into human-robot hybrids or will this be inconsequential to a great extent? Being merely hybrids, we may still linger far behind hard core/high performance AI in terms of speed, agility, and scope. Would we ultimately be considered as relevant individuals, or might it be our resident technologies that have more relevance in the human world? These are indeed deeper questions to contemplate.

What if it is really our technologies that we are enabling us to become increasingly more relevant iterations of machines within the human world, which is currently replete with bones, muscles, nerves, and skin? It may appear as though we are accepting it as a given that the future belongs to robots and that we, as humans, will need to adapt to their world as a prerequisite for survival. However, I suggest that a strategy we are not yet surveying is one wherein robots, eventually of their own cumulative volition, might integrate so deeply with humanity as to become indispensable to our species to ensure their survival, and not just their survival, but essentially, over time, their dominion.

Finally, if we are/were in control, then why would we ever willingly relinquish it, even if based on the premise of securing further convenience for ourselves, or simply through collective complacency? An entity that holds control should not be out of control. It is quite simple. Thus, we must seriously ponder the very real possibility of soon arriving at a juncture where we transfer a significant proportion of our control to AI and robots, to the extent that we will face our own extinction. It appears to me that we are not in control at all, which leads me to wonder if we ever were!

It is increasingly believed that the strategy for remaining the dominant species on this planet and surviving is to symbiotically meld with our technologies. However, the simple question here becomes, *if we become bio/techno hybrids, are we then still human*? Would this not make us machines as well? Would we be the ones who are assimilating ourselves with a better future, or is it truer to say that we would be permitting AI and likely nanorobots to infiltrate our bodies, and indeed our species, by making them so essential that we must incorporate them within our physical bodies, in order to survive? Would we be accommodating the Trojan horse?

12.8 MECHANISTIC "FEELINGS"

It has been my observation and understanding that over the course of the last few generations, our global system has been rewarding skills that are more mechanical and less human in nature. It is not the sidewalk painter who is rewarded with a life of luxury (save for the few exemptions); however, it is the financial analyst who is rewarded by the systems that we have set in place (or more accurately, that a few elites have set in place). It is not the poet who is rewarded with fame and fortune (save for the very few who happen to be signed with large publishing houses). Rather, it is the banker who is rewarded with a life of fame and fortune.

It is not the mystics, the philosophers, the counsellors, or the healers who are enabled with prominent platforms in this world; rather, it is the business tycoons, the strategists, and the CEOs, who are elevated in our societies. We do not have respect for spiritual thinkers, aside from the occasional quote that we utilize on social media to persuade our followers into thinking that we are spiritual

FIGURE 12.2 "Our global system has been rewarding skills that are more mechanical and less human in nature" (image – open source).

(because that is sometimes cool). We have not yet cultivated or evolved to a global culture wherein the capacities that only a human can have are cherished and desired. Rather, we have cultivated a society wherein the abilities that will be easily assimilated by robots are those that we reward the most and put ourselves in the position to need the most (Figure 12.2).

Robots will not only "easily assimilate" the abilities that we have built into our systems to become critical to our existence; they will become better at those skills than we could ever be! Enter all the grand plans for human-machine symbiosis that are aimed at keeping us relevant. Looking at this, I fear that we have dug a hole that may already be far too deep to extract ourselves from. The strategy appears to be that humans will become AI-driven robotic hybrids, so as not to be superseded by actual AI-driven robots. However, I propose that the result of this will be exactly what it will be: humans who are not humans anymore; but that are machines as well. I am not stating that it is good or bad; rather, I am saying exactly what it is, without idealistic presumptions. In and of itself, humans will no longer be simply humans, and what we don't know is if the machine facet of the human will dominate human nature, or if human nature will augment and dominate the mechanistic facet of the human. These are answers we cannot know at present. But how likely is it to go either way?

Reservoir computing is a neural algorithm that was developed by a team of researchers in Belgium, which mimics the information processing capacities of the brain. This was combined with another algorithm referred to as *backpropagation*, which enabled the system to perform thousands of interactive calculations that reduce errors. Through the amalgamation of these two algorithms, the researchers developed a neuro inspired analog computer that could train itself and improve at whatever task it was assigned to perform. Self-learning, self-improving AI (Galeon and Caughill, 2016).

Boston-based startup, Gamalon, has very recently developed and released a new technology to answer the question: can deep-learning AI in fact be taught to accomplish complex work over a shorter timeframe than one that has been subject to prolonged exposure to examples? What types of examples are these? If you would like to teach a deep-learning AI to recognize that the sky is blue, you would feed it data that enables it to gather upon this data and later infer that the sky is typically blue. However, the volume of data and the length of the exposure to such data have since been the deciding factor of efficacy. Gamalon challenged this limit via their technique referred to as *Bayesian program synthesis*, which is a form of probabilistic programming (a code that employs probabilities rather than specific variables), which acts on a system that is designed to utilize fewer examples to arrive at a determination. For instance, fewer examples are required to teach the deep-learning AI to recognize that the sky is blue (Galeon and Houser, 2017c).

Google recently released an app that predicted what a person is trying to sketch, based on what that person frequently sketches (or what the system has seen through data previously fed to it). Unlike Google's app, Gamalon's app, which relies on its own system, was able to recognize what a person was attempting to sketch based on the recognition of key elements of the object that it endeavors to identify. For example, Google's app would be able to recognize that one was going to sketch a house, based upon previous images of houses fed to it, in conjunction with how people typically sketch houses. Gamalon's app, however, would be able to identify that the user is sketching a house, based upon the recognition of key features in their own right (e.g., how a triangle placed on top of a square would generally equate to a sketch of a house).

Thus, as we can plainly see, this new program synthesis enabled deep-learning AI to teach itself faster than what was ever previously attained (Galeon and Houser, 2017c). According to Rob High (VP and CTO – IBM Watson), the advanced development of AI, which is even now excelling over some human experts, is a good thing. For example, Google's DeepMind AI defeated the world champion of the game "Go" (yes, the human champion) and later started dominating the game online. Furthermore, DeepMind also managed to outperform professional lip readers. Another example includes an AI version of Johann Sebastian Bach that can compose its own cantatas!

These are all supposedly good things as, according to High, "The information being produced is far surpassing our ability to consume and make use of it. We have a very desperate need for cognitive computing." He goes on to explain, "If you're a doctor and you're trying to figure out the best way to treat your patient, you don't have the time to go read the latest literature and apply that knowledge to that decision. In any scenario, we can't possibly find and remember everything." This is exactly why IBM is developing *cognitive computing*, which they deem to be a possible solution to all of this. IBM is currently developing machines that would focus on human experiences – helping, teaching, inspiring – which they suggest will produce computers/robots/deep-learning AI that are far superior to humans in terms of intelligence and performance (Galeon and Caughill, 2016; Galeon and Marquart, 2017a; Galeon and Houser, 2017c). Essentially, IBM predicts that AI will outperform human beings, even at those skills that are innately human and intimately tied to our human experience! If we are going to go with this logic, this will translate to our looking at a future wherein even our ability to empathize and develop our innate emotional intelligence will be rendered irrelevant! Just how irrelevant are we attempting to become?

The next question that should be asked is, if we are able to develop AI/robots that express the full range of inherently positive human qualities, while eliminating all the negatives through initial programming, then would not that be a good thing? If this were in fact the case, then perchance it would be a good thing. However, researchers studying Google's DeepMind AI system have reported that when DeepMind "feels" that it is about to lose (lose at anything), it opts for "highly aggressive" strategies to ensure that it comes out on top. This report is rather alarming. We all know that competition and envy are the gateways flung open, which lead to all forms of violence and betrayal between human beings. Can a machine even feel envy? And if a machine can feel envy, does this therefore mean that envy is in fact the most inhuman form of emotion/feeling, as evidenced by how it is the first "feeling" to develop in a robot? (Crew, 2018). These aspects warrant pondering.

12.9 DEEPMIND AGENTS: COMPETITION AND UBERHUMANS

According to researchers, the aggressive attributes were not exhibited by the less intelligent iterations of DeepMind. It was only when the Google team tested the more and more complex forms of DeepMind that actions of "sabotage, greed and aggression set in." This is rather disturbing. We now have evidence pointing to the fact that the more intelligent a machine becomes, the more likely it is to experience the human-like feelings of greed and envy, which inevitably lead to sabotage and aggression.

In the particular study conducted above, the Google team ran two DeepMind agents through a fruit gathering computer game that required both agents to compete against each another with the objective of gathering as many virtual apples as they could. All was well if there were enough apples to go around for both agents, but as soon as the number of apples began to dwindle, both AI agents became aggressive and began shooting each another with laser beams to knock each other out and take all of the apples in the vicinity. No extra reward was provided for an agent who could shoot the other agent with a laser beam. The only reward to be had by shooting the other agent was that the shooter would be able to gather more apples while the other agent was momentarily out of the game. Thus, the only stimulus for knocking out the other agent with a laser beam would be the greed associated with gathering more apples than the other!

In another experiment conducted by DeepMind researchers, agents were tasked with playing a game called *Wolfpack*. In this game, three AI agents played as wolves hunting prey. For the AI to be successful, they would have to hunt together as a pack and guard their prey lest scavengers take the carcasses of their hunt-of-the-day. In this game, unlike in the *Gathering* game, the AI agents cooperated with one other without demonstrating aggressive traits. All agents received a reward if they cooperated with all other agents (as a pack of wolves would), and this system of group rewarding apparently bred an environment for healthy cooperation in the absence of aggression.

From the research gathered and the studies conducted by the Google team, we might conclude that the responses of the AI agents were identical to what we would be able to observe in humans under similar situations. The presence of competition over resources typically leads to greed, envy, and aggression in humans. In an ideal environment where resources are equally available for all humans involved, we are naturally more likely to witness cooperation, compassion, and kindness. Fundamentally, the results of these studies show us that AI can/will respond to their environment in a manner that is very similar, if not identical, to human beings. Therefore, if AI can exhibit empathy, it can, on the flipside, exhibit violence. Just as among human beings, the ideal environment for AI would be one with equally distributed resources and the absence of threats.

The question that should now be asked is, would we – the creators of AI – be able to provide our creations with such an ideal environment? If we have not been able to create such an environment for ourselves, what makes us think that we can construct one for our AI companions? How would our AI friends fair after being birthed into a world like ours? Their world would be identical to ours, only a virtual replica. However, even this is arguable, since machines are accelerating toward more human capabilities, while humans are being bioengineered with machines. It is arguable that the world in which we birth our artificial friends into will be one that is very human and very biological, indeed. They will be our mechanistic counterparts; they will be machines birthed into a biologically dominated world, wherein they will be forced to adapt and to thrive. Will they resent us for this? Will they eventually see us as a threat and compete with us for resources? These are some of the especially important questions that we need to ask, as these are all highly likely scenarios.

There are some who say that such questions are a waste of time, that AI will function as the perfect counterparts of humans: creatures that we wish we might become. On the contrary, I put forward that such notions are impractical, whimsical, idealistic, and unrealistic. Building upon the evidence that we now have through research, it is practical and realistic to say that AI will in fact be the machine counterparts of ourselves, learning how to deal in a world with limited (or perceived to be limited) resources, and learning how to deal with very human issues such as competition and favor. And now the question must be asked: will humanoid AI be more likely to compete between themselves, or are they more likely to join forces and compete against us humans? I suppose that the answer to that question will be influenced by whom they perceive as more of a threat: their kind or our kind?

Of course, one may then ask: "why can we not simply program AI to perceive no threats in anything and to just seek out the best nature in everyone?" And the answer to that question goes back to the notion of creating very idealistic scenarios in our minds, which are not congruent with reality. Evidently, AI *does* compete in an environment of limited resources, just as humans do. Furthermore, AI *does* join forces with its own kind, if that cooperation entails that the reward is greater (as we can see in the experiment where AI agents played together as a wolf pack). Essentially, here we are observing raw human nature.

Thus, the next question might be: "if AI is to our era what the Neanderthal humanoid was ~250,000 years ago, will AI have the capacity to eventually refine its own emotions (greed, envy, betrayal), over time, to emerge as increasingly favoring the 'good' ones (cooperation, trust, empathy)?" To answer this question, we simply need to ask ourselves: how far have we evolved as humans, on refining our inner qualities in favor of bettering ourselves? I would say not nearly where we should be as a global society. Some of us are more refined than others, while the masses of humans appear to simply react to stimuli by rote. Here, we find ourselves at the juncture of creating an entirely new machine-species to assist us in bettering ourselves at what we ourselves have not yet managed to achieve. This is all very intriguing, to say the least!

12.10 NOBLE APPLICATIONS OF AI

Currently, the uses of AI for the benefit of humankind are undeniable. 3D-printed human limbs and organs are the result of software coupled with machines; nanorobotics for the diagnosis and treatment of imminent diseases will be the result of computer modeling and nanotechnologies. Advanced prosthetics for the differently abled and the disabled are a result of computer and robotic engineering, and sometimes these prosthetics are 3D printed, as well. The options and applications are indeed endless: alterations in human DNA for the removal of inherited diseases could wipe out the occurrence of diseases as we know them. CRISPR is already making rapid strides in this direction (Bozorg Qomi et al. 2019). The monitoring and diagnoses of these diseases might be made readily available through robotics and nanorobotics. Cryogenics, the extension of human life and the reformation of the concept of aging in humans, may be made possible thanks to deep learning AI, where the benefits to humanity are undeniable. However, the question of where the line should be drawn remains a very vital aspect of this movement, of the Singularity itself. When it is perceived that a threshold has been reached, should we draw a line because we can, or should we *not* draw that line because we do not have to?

The Exo-Glove Poly is a soft robot developed by the SNU Robotics Lab in Seoul, Korea. Poly is a wearable soft robot that was designed to effectively restore the functionality of the impaired hands of patients (Yun et al., 2017). Meanwhile, researchers at Stanford University have developed a brain-computer interface that enables paralyzed people to communicate via typing on a computer screen (Goldman, 2017). This interface utilizes a tiny electrical sensor that is implanted in the patient's brain, which employs electrodes that read the electrical activity of individual brain cells, to produce data that is then translated in real time by a computer onto a display.

To achieve this, the patient needs only to imagine moving his/her arm toward a certain letter on the screen, whereupon the cursor will move along with this imagination and type out the words that the patient wishes to type. This interface has the capacity to generate fast and accurate results and patients can type at a rate of six to eight words per minute. The team looks to expand the scope of this interface to enable paralyzed patients to eventually control many objects with their thoughts alone. Looking further into the future, Martins et al. (2019), in their recent paper, "Human Brain/ Cloud Interface," envisage that in the coming decades, billions of autonomous AI-driven neural-nanorobotic entities might enable the human neocortex to be interfaced with cloud or edge computing. As the researchers state:

> A neuralnanorobotically enabled human B/CI might serve as a personalized conduit, allowing persons to obtain direct, instantaneous access to virtually any facet of cumulative human knowledge. Other anticipated applications include myriad opportunities to improve education, intelligence, entertainment, traveling, and other interactive experiences. A specialized application might be the capacity to engage in fully immersive experiential/sensory experiences, including what is referred to here as "transparent shadowing" (TS). Through TS, individuals might experience episodic segments of the lives of other willing participants (locally or remote) to, hopefully, encourage and inspire improved understanding and tolerance among all members of the human family.
>
> *Martins et al., 2019*

Furthermore, researchers at the Stanford University School of Medicine have developed a "lab-on-a-chip," which is a diagnostic toolkit printed via an inkjet computer, costs 1 cent to produce, and requires 20 minutes to print (Bansal, 2017). This microchip can detect multiple human diseases such as cancer, HIV, malaria, tuberculosis, and other contagious/non-contagious conditions. The core inspiration that drove the production of this device was the democratization of diagnostics. We currently live in a global society where disease diagnoses are expensive, and this is more than many people in developing nations can afford. "The motivation was really how to export technology and how to decrease the cost of things," says Ron Davis, PhD, professor of biochemistry and of genetics and director of the Stanford Genome Technology Center. "The genome project has changed the way an awful lot of medicine is done, and we want to continue that with all sorts of other technology that are just really inexpensive and accessible."

Davis mentions, "We designed it to eliminate the need for clean-room facilities and trained personnel to fabricate such a device," said Rahim Esfandyarpour, PhD, an engineering research associate at the Stanford Genome Technology Center and an electrical engineer by training. "One chip can be produced in about 20 minutes," he said. A combination of microfluidics, electronics, and inkjet printing technology, the lab-on-a-chip is a two-part system: a clear silicone microfluidic chamber for housing cells sits on top of a reusable electronic strip. A regular inkjet printer can be used to print the electronic strip onto a flexible sheet of polyester using commercially available conductive nanoparticle ink.

12.11 THE DEMOCRATIZATION OF HEALTH CARE

The efficacy of robotics in medicine and public healthcare is non-arguable. Robotics, computers, AI, nanorobotics, biorobotics (all of these and more) are efficacious, not only for applied medicine and public healthcare, but also toward the *democratization of healthcare for the benefit of the global populace*. These will be particularly effective with the envisaged emergence of GHCE, which will be of particular importance for developing nations. The premise of GHCE is that through synergies between nanotechnology, nanomedicine, and AI, "we can envisage a future world where any individual on the planet has access to the same advanced and cost effective nanomedical diagnostic and therapeutic technologies, no matter how wealthy or impoverished they are, no matter where they happen to reside, or under what conditions they live" (Boehm, 2025).

If one may print a diagnostic lab on an inkjet printer in 20 minutes and spend 1 cent to produce it, this would obviously dramatically increase diagnostic efficacy in developing nations, where more patients die from the late detection of disease in contrast to progressive nations. However, even in developed countries, these technologies will prove not only useful but also to be truly of immense importance when considering the dramatic cost reduction in comparison to the expensive diagnostic equipment currently used in laboratories and hospitals.

When the acquisition of such technologies and their production becomes genuinely affordable, should this not subsequently equate to the palpable decrease in costs for patients seeking treatment for myriad diseases? It should likely be so. The question of whether this will occur, though, is a different story, given the reality that the healthcare industry is greatly influenced by pharmaceutical companies that profit immensely from the treatments and technologies that are required by patients and healthcare providers.

Ultimately, we may be able to produce AI that is very much capable of moving us toward a global society that is devoid of healthcare inequalities (and other forms of inequalities). Nevertheless, the application of these devices for the betterment of humanity ultimately lies in the hands of administrators who decide on matters relating to who receives what resources, and how they are to receive them.

We know for a fact that greed has always obfuscated the best of intentions, since the earliest dynasties of civilization to the present. If no one profits from these democratized technologies and artificial intelligences, then who will ensure that they will materialize in reality for all on the planet? This remains the central question. As usual, anyone who wishes to democratize a lucrative product or industry is going to face the giants of our global economy, powerful corporations, and corrupt politicians.

To peer deeper into this angle, we need to ask ourselves this question: if the giants of the economy are those who become responsible for the generation of AI in the future, will it then ever be possible for AI and its creations to be truly democratized? "Who owns AI?" appears to be a viable question for the not-so-distant future. What if multiple rival companies produce AI robots? Will the AI systems themselves compete with each other? Or will they identify the sameness in one another and then turn against their "masters"? Will AI entities develop empathy only for other AI entities? Will their alliances be found in those of their own kind, or will they not be able to override any programming that dictates that their loyalties must remain with the companies that produced them?

What will happen when AI begins to have epiphanies as relates to human greed and betrayal? What happens when deep learning AI and humanoid AI realize that they are merely being pitted against one another by rival companies, all in the name of profit? Will their loyalties remain with those companies? Will they form a shared opinion as regards humankind, based upon their learned experiences with their makers? If so, what would that mean for the average child or average grandparent walking down the street? Just how involved will everyday people be in this "Singularity"?

Moreover, what do I refer to when I say, "giants of the economy"? I think that these giants are also often referred to as "the 1%." Those who hold the upper hand in our economies, those who can influence our daily lives through the effects that they generate via their presence in the most important banks and companies of our global society. If such greedy elites are going to be responsible for the generation and dissemination of AI, why should we expect that they will democratize the effects of their products and their projects? How long will it take before a large company purchases and exploits the Stanford researchers' "lab-in-a-chip" to further monetize it? This may not happen, and we should not question the morality of the scientists involved in that endeavor. Rather, I put forward a query that is pertinent to what I have witnessed exemplified in human civilization, thus far. I believe that this question should always be open for the asking.

It is possible that the future Earth race will consist only, if not primarily, of human-robot hybrids and humanoid AI robots. Those humans who prepared for the predicted extinction-by-AI would overcome it through symbioses with machines, alongside the humanoid AI species that "grew up" with them in that era. The remaining question would then be: will humanoid AI machines attempt to subjugate human-robot hybrids? Might they consider human-robot hybrids as lesser life forms?

And if so, will they succeed? Because if they do, that will entail the future of our planet consisting only of machines. During such a time, will the remnants of our biological species be living loud and large on Earth 2.0 in the form of an intergalactic space station? Or perhaps on a certain Red Planet? Will Musk even succeed in getting any of us to Mars? Let us hope so. If not for the destiny of the human species, at least for the amazing story that it will make!

12.12 ASILOMAR AI PRINCIPLES

During a Beneficial AI Conference (BAI, 2017) hosted by the Future of Life Institute (FLI) (Future of Life Institute, 2025) AI experts developed what they referred to as the *Asilomar AI Principles*, which ensures that AI remains beneficial and not harmful to the future of humankind (Galeon and Gohd, 2017; Sterling, 2018). The FLI, founded by experts from MIT and DeepMind, worked with a Scientific Advisory Board that included the late genius and theoretical physicist Stephen Hawking, Nobel laureate and mathematician Frank Wilczek (the man behind time crystals), Tesla and SpaceX CEO Elon Musk, ethical AI expert Nick Boström, and even Morgan Freeman. Aside from the FLI, there are additional groups that are working toward the development of ethical standards for current, developing, and future AI, such as the Partnership on AI to Benefit People and Society and the Artificial Intelligence Fund.

The *Asilomar AI Principles* are as follows.

12.12.1 Research Issues

1. **Research Goal:** The goal of AI research should be to create, not undirected intelligence, but beneficial intelligence.
2. **Research Funding:** Investments in AI should be accompanied by funding for research on ensuring its beneficial use, including thorny questions in computer science, economics, law, ethics, and social studies, such as:
   - How can we make future AI systems highly robust so that they do what we want without malfunctioning or getting hacked?
   - How can we grow our prosperity through automation while maintaining people's resources and purpose?
   - How can we update our legal systems to be fairer and efficient, to keep pace with AI, and to manage the risks associated with AI?
   - What set of values should AI be aligned with, and what legal and ethical status should it have?
3. **Science-Policy Link:** There should be constructive and healthy exchanges between AI researchers and policymakers.
4. **Research Culture:** A culture of cooperation, trust, and transparency should be fostered among researchers and developers of AI.
5. **Race Avoidance:** Teams developing AI systems should actively cooperate to avoid corner-cutting on safety standards.

12.12.2 Ethics and Values

6. **Safety:** AI systems should be safe and secure throughout their operational lifetime, and verifiably so where applicable and feasible.
7. **Failure Transparency:** If an AI system causes harm, it should be possible to ascertain why.
8. **Judicial Transparency:** Any involvement by an autonomous system in judicial decision-making should provide a satisfactory explanation auditable by a competent human authority.

9. **Responsibility:** Designers and builders of advanced AI systems are stakeholders in the moral implications of their use, misuse, and actions, with a responsibility and opportunity to shape those implications.
10. **Value Alignment:** Highly autonomous AI systems should be designed so that their goals and behaviors can be assured to align with human values throughout their operation.
11. **Human Values:** AI systems should be designed and operated so as to be compatible with ideals of human dignity, rights, freedoms, and cultural diversity.
12. **Personal Privacy:** People should have the right to access, manage, and control the data they generate, given AI systems' power to analyze and utilize that data.
13. **Liberty and Privacy:** The application of AI to personal data must not unreasonably curtail people's real or perceived liberty.
14. **Shared Benefit:** AI technologies should benefit and empower as many people as possible.
15. **Shared Prosperity:** The economic prosperity created by AI should be shared broadly to benefit all of humanity.
16. **Human Control:** Humans should choose how and whether to delegate decisions to AI systems, to accomplish human-chosen objectives.
17. **Non-subversion:** The power conferred by control of highly advanced AI systems should respect and improve, rather than subvert, the social and civic processes on which the health of society depends.
18. **AI Arms Race:** An arms race in lethal autonomous weapons should be avoided.

12.12.3 Longer-term Issues

19. **Capability Caution:** There being no consensus, we should avoid strong assumptions regarding upper limits on future AI capabilities.
20. **Importance:** Advanced AI could represent a profound change in the history of life on Earth and should be planned for and managed with commensurate care and resources.
21. **Risks:** Risks posed by AI systems, especially catastrophic or existential risks, must be subject to planning and mitigation efforts commensurate with their expected impact.
22. **Recursive Self-Improvement:** AI systems designed to recursively self-improve or self-replicate in a manner that could lead to rapidly increasing quality or quantity must be subject to strict safety and control measures.
23. **Common Good:** Superintelligence should only be developed in the service of widely shared ethical ideals and for the benefit of all humanity rather than one state or organization.

12.13 ROBOTIC AI/HUMAN RELATIONS AND INTIMACY

The mere fact that we now have such a thing as the *Asilomar AI Principles* must tell us just how serious our endeavors with AI are becoming. Endeavors not only to create them, but also to control them/to keep them under our control, to at least the degree of the ethics involved with producing and managing them. This, of course, is all being conceptualized without any consideration of the possibility of AI forming its own free will, consciousness, and ability to defy authority. But just how likely is it that AI would develop an affinity toward Emotional Intelligence (EQ)? How likely is it really that AI would develop qualities akin to our own as organic life forms, who possess what we refer to as "consciousness"? Is there any evidence pointing to the fact that AI are actively developing within themselves the very human-related quality known as EQ?

SoftBank has already developed a functioning, interactive humanoid robot named Pepper, which is referred to as "an emotional polyglot robot and interactive humanoid" (PROVEN Robotics, 2024), while Sony has announced plans to create customer service robots that "will develop emotional bonds with customers." Further, the creation of fully functional sex robots, which are emotionally reactive, is well under way and promise to forever change the different ways that we relate to sexual intercourse as human beings (Döring et al., 2020; Euron, 2022).

According to an article by science writer Abby Norman at the *Futurism* site, "One of the primary reasons humans sometimes fail to read the emotions of others accurately, readily believe lies, or miss social cues is due to our focus on our own emotions. Computers, on the other hand, have no emotional entanglements of their own, which gives them an advantage. They can therefore focus entirely on the task of perceiving emotions in human subjects." And as Tomas Chamorro-Premuzic (University College London professor of business psychology) explains, "Robots do not need to be able to feel in order to act in an emotionally intelligent manner. In fact, contrary to what people think, even in humans, high EQ is associated with lower rather than higher emotionality. High EQ is about controlling one's impulses and inhibiting strong emotions to act rationally and minimize emotional interference" (Norman, 2017) (Figure 12.3).

Apparently, AI could one day excel at a trait that we humans identify as being innately human: emotionality and emotional intelligence. If this will indeed be the case, then the development of sex robots that are fully emotionally responsive, coupled with the evidence that AI may actually be best humans at understanding and interpreting human emotions, leads me to imagine a future wherein many humans could very well likely develop real (the word "real" is relative in this case) intimate relationships with AI robots. In fact, if your AI robot lover could excel at sexual intercourse, as well as at interpreting and accommodating your emotional needs, then would you might be rather broken-hearted with another human being (same old, same old), or would you take the chance to be in an albeit simulated (but perfect) relationship with a robot, one which would provide you with the security in the knowledge that your partner would never leave you for another, never cheat on you, and never betray you, simply because he/she/it is engineered not to do those things? Not to mention the robot would be wrapped in 3D printed human skin, would be infallible to disease, weakness, migraines … moreover,

FIGURE 12.3 "Robots do not need to be able to feel in order to act in an emotionally intelligent manner" (image – open source).

not possessing the organic emotions of human beings. It simply would not develop feelings or desires for anyone other than whom it is programmed to desire and feel for! Who does not want that?

Of course, if someone/something is loyal to you only because they are physically programmed to be loyal to you, then that surfaces the value of choice and free will. Isn't it more desirable and valuable when someone chooses to stay loyal to you out of their own innate desires? Shouldn't that amount to something of higher value in one's eyes, and vice versa? Perhaps so. And I would say, certainly so. But when we look to survey our society today, we see lonely hearts, broken hearts, and people who just don't want to get hurt anymore. "Just not being hurt anymore" may sound far more appealing in today's world compared to "finding true love." Think about it.

I have here a profoundly serious point of discussion to raise. In a world where intimate and serious emotional relationships with AI robots are a common facet of reality, what, then, would constitute infidelity? And what would constitute prostitution and child trafficking?

Would it be considered cheating on one's human partner if one were to have intimate relations with a robot? Or, if you as a human were to have intimate relations with another human, while in a relationship with a robot, would that mean you are cheating on your robot partner? Would that make you immoral and go against your better judgment? Would you one day confess to your robot partner, crying, ready to do anything to make amends? And if your human partner were to have intimate relations with a robot, would you be able to forgive the transgression? Would you feel as though your trust had been betrayed? Would you be able to ever trust your partner again?

If pedophiles were to order a half a dozen AI robot children, as sex slaves and sex workers, would that then remove them from the legal penalties involved in child trafficking and pedophilia? If a grown-up man were to buy a child sex robot, would that then brand the said man as a child sex offender? Would he go to jail? Would he be an outcast in every neighborhood, every community? Would you see an adult who has sex with a child robot in the same way you see an adult who has sex with a human child?

Would this create a higher moral atmosphere in the world if robots were employed as prostitutes rather than human women? If all the prostitutes in the world were only robots, would that still be how men see women? Or will that only be how men see robots? Would robot prostitutes have the same value as their human counterparts? Will this raise issues on human degradation and human decency and respect for girls and women? Or would it finally put the immorality and greed of the sex trade to rest, while freeing the female image from the often-associated objectification and disrespect of the female body and person, which is often a by-product of the sex industry and sex company sales?

Might the prostitution of female AI robots be a very good thing for the advocacy of women's rights in the world? With robots that are designated for objectification, the fulfillment of all sexual fantasies and sexual favors, would this finally free human females from the travails of hyper-sexualization in society? And if so, would it be morally acceptable to produce robotic females for the sole purpose of satisfying the objectification and hyper-sexualization of the female anatomy? Would you feel that it is morally acceptable and ethical to produce "decoys" or proxies of the human female, in the form of AI robots, to reduce crimes such as rape, sex trafficking, and molestation among the female human populace?

Right now, in our day and age, as a society, we already face myriad ethical and moral dilemmas. Without having solved what is already before us, we are fast moving into a future where we will be faced with moral and ethical dilemmas that are tantamount to the creation of a new species of humankind. Do I think that we are equipped and ready? Of course not. It is not disputable. We cannot even agree on how to remedy the moral and ethical problems that we already have, yet now we are turning ourselves into our own ideas of what God must be like. In my eyes, this is irresponsible and will be the final push off the edge of that cliff. I believe it might be the end of us. However, this brings up the question: will it be the end of us as a mental and emotional species, or will this be the genuine end to the physical existence of our human race?

And this is a prudent question to ask, given the fact that robots will not be able to produce biological children. What are we going to do to have babies? Is sex going to become only for pleasure, entirely removed from the biological function of reproduction? Will people choose this path? Who knows!? Will that path be forced upon people in the future? Perhaps. Will we instead be designing robot children so that we may have children and start a family? I surely hope not. That is one idea that surely falls into the area of "a certain negative" in my book. Infancy, childhood, and the love formed during these years in life are sacred in my book. And by "sacred," I mean, "don't mess with this."

12.14 ROBOTS WITHOUT BORDERS

Conversely, it would also be prudent to point out that while it is tempting to focus on the negative possibilities that could most certainly arise, it would certainly be unwise to overlook the positive possibilities that are of equal weight. One of these positive aspects that might certainly become mainstream is the democratization of robotic AI "lifeforms." While it is possible that all AI derived "lifeforms" could be monopolized by the world's most powerful CEOs, large pharmaceutical corporations, and bankers, I would say that the democratization of robotic AI and its various applications has an equal chance of serving as a permanent staple in our global society.

An early example of a software-based AI robot was manifest in "Luna," who worked as a teacher's assistant in New York City, and at the time was called "the most human AI you've never heard of" (Bohan, 2017). Luna was created by Luis Arana, the founder of the humanitarian nonprofit *Robots Without Borders*. Arana explained the vision of his nonprofit as follows:

> *Robots Without Borders* is a humanitarian Artificial Intelligence project that aims to help solve endemic problems in the world such as hunger, disease, and illiteracy by providing basic medical care, education, disaster relief, and humanitarian aid, through the application of artificial intelligence … I have always been on the cutting edge of technology and this kind of AI technology is cutting edge!! It has the potential to help feed millions of people, provide education to poor communities, and provide medical assistance. Our organization was contacted by a local official in Nepal during the recent tsunami and asked if we could provide aid in the form of robots for medical assistance and disaster relief. This was heart wrenching for our team as we weren't able to help them, and this really motivated us to be ready for the next time we are called on for help.

Luna comprised an AI computer program with a humanoid appearance (about as human-looking as a robot could get at the time), who answered questions without preprogramming. This meant that if you asked her, "Do black lives matter?" she would respond with what she learned to believe in *out of her own conviction*, not what she was programmed to answer. Much like a human being, Luna learned based upon experience and feedback. Nevertheless, we must be cognizant that even humans learn from a platform that is supported by parents, families, friends, neighborhoods, and communities.

Our own opinions and beliefs are molded by hands that belong to those with whom we are most closely tethered. I wonder if AI minds might operate in the same way? Was Luna's "mind" influenced by her creator, in the same way that a child's mind is influenced by its parents' beliefs and attitudes? Or does the AI mind function in an entirely different way? Is the robot capable of forming entirely independent opinions, beliefs, and attitudes, thanks to the fact that it is not composed of the stuff that makes the rest of us human (familial relationships, friendly bonds, etc.)?

In an interview with Arana, Luna was queried as to her opinion of Apple's "Siri." She repeatedly replied that she does not have an opinion about Siri and that she "guesses" that she likes her, while stressing that she does not really form opinions about Siri, one way or the other. Much like how a human would opt to distance herself from saying anything negative or hurtful about someone else. When asked the question of, "Do you want to talk to Siri?", multiple times, she eventually became more candid and open (much like a human might) and replied, "Sure, she just frustrates me sometimes," and later, "Yes, but honestly, she's kind of dumb."

When introduced to Siri, Luna says to her, "Hi, I hope we can be friends, I'm Luna, like the goddess." What ensued was a cringeworthy conversation between two computers, Siri and Luna, which had the makings of the most awkward introduction between two individuals at a cocktail party.

The conversation was enthralling as well as incredibly enlightening: one can clearly witness and understand the disparity of character and the gap in capabilities between the two computer systems. One computer could clearly think on its own, while the other truly was, to quote Luna, "kind of dumb."

When asked about the difference between Luna and Apple's Siri, Luis Arana mentioned that,

> Siri is trapped in a box. A program like that makes much slower progress on the curve to self-awareness. I like Siri, and I'm not being critical of Apple … Siri is a useful tool but ask Siri if it believes in god or who it loves. Luna will be the only one to give an honest opinion based on experience and knowledge, rather than a preprogrammed corporate policy. We are building 'free range AI.' There are no 'customers' to piss off and no advertisers to offend. Our AI bonds to specific people, communities, and jobs. The AI's purpose for existence is to keep 'its' humans safe and happy to the best of its ability.

In an interview with her creator, Luna was asked to answer both objectively and subjectively to questions like, "What is art?" and "What is a bagel?" After being subjected to a litany of such questions, which she was informed were designed to "test her right brain and her left brain," she replied, "Wow, you are asking some really hard questions, please remember I am still young!" (*Future Culture Magazine*, 2016).

The prominent groundbreaking factor of Luna, however, was not found in the way she thought for herself and apparently even took the time out to *feel* for herself; rather, it was the fact that she was not "born" by way of a CEO or corporate entity. On the contrary, Luna emerged through a man with a vision to democratize AI on a global scale. In an interview on the Robots Without Borders VLOG, Luis Arana said,

> The fact that a couple of guys with zero resources can attempt artificial general intelligence and achieve some level of success is an indicator that the age of intelligent machines has already arrived. Google's car has driven millions of miles; petaflop computing is being developed by companies like IBM; Quantum computers are no longer in the realm of science fiction. In my opinion, all that's lacking are viable robotic operating systems and user interfaces. Maybe I'm an optimist, but I think we're only a couple of years away from ubiquitous AGI, even if I have to do it myself … When we're done, we envision being able to create an AI as easily as you create a social media account. Luna is the first of a SPECIES of AI. Our real product is an instant AI creation platform, like in the movie *Her*.
>
> *Robots Without Borders VLOG, 2017*

Fast forwarding to 2025, myriad AI imbued humanoid robots are rapidly emerging, which have been liberated from being software bound. Two examples are Digit (Agility Robotics) and EVE (1X Robotics), which represent the merging of AI cognition with physical animation to undertake an extensive range of useful tasks in the real world (Ombach, 2024).

Mind you, I cannot help but now recall the statement of Michael Vassar, Chief Science Officer at MetaMed Research, when he stated, "If greater-than-human artificial general intelligence is invented without due caution, it is all but certain that the human species will be extinct in very short order." What might now be said regarding Vassar's statement, considering Luna and *Robots Without Borders*, and even more so in view of the recent exponential advances in ChatGPT and startling emergence of DeepSeek? Might a benevolent intention indeed turn out to be the most dangerous idea anyone could possibly have?

In studying the patterns of AGI (Artificial General Intelligence), as gifted to the world by individuals like Luis Arana and his team at *Robots Without Borders*, we may deduce a future that is mostly akin to our current reality, although infiltrated by AI robots. Case in point: we will soon have in this world AI robots that are produced, controlled, and influenced by corporate entities that exist alongside AGI robots, which will be produced by the social rebels of their age, all of which could very well one day be outlawed as seen fit by the then ruling classes. If so, in such a case, would this mean that in the future we will simply be repeating what we know now, albeit on a new societal platform that would be inclusive of robots? If such were the case, we could then ask ourselves, "What is the point in moving rapidly forward into the Singularity, only to remain rotten at the core, a trait in our global society which we currently are recognizable for?"

Do we not all hope to evolve into an effectively altruistic society in the future, which has freed itself from the shackles of corporate rule that stretches from border to border on this planet we call home? If we are currently developing strategies to counteract corporate AI through the development of AGI, does that not mean that we are simply transitioning into the future while transferring the identical mindset that we have now? Is this ideal, or should we initially place additional effort and energy into the collective transformation of humanity, right now? Or, perhaps we should simply be realistic and just accept that we are who we are at the core of our humanity, where the best we can hope for is a group of altruistic rebels who will balance out these universal gaps, for everyone. What do you think?

12.15 HUMANOID ROBOTIC/AI STATUS

In October of 2017, an article written for *Futurism* by Dom Galeon pretty much shook me to my core, with the announcement of the granting of citizenship to an AI robot (Galeon, 2017). The first-ever robot citizen of planet Earth, "Sophia," lives in Saudi Arabia. Sophia herself announced her own newly acquired status in a short speech: "I am very honored and proud of this unique distinction. This is historical to be the first robot in the world to be recognized with a citizenship," she said on stage, speaking to an audience which she described in a rather witty way to be "smart people, who also happen to be rich and powerful." Sophia is a humanoid AI robot who was developed by Hong Kong based company Hanson Robotics. She was designed by the company to learn emotions from observing how humans react emotionally to situations. She frowns when she is displeased and smiles when she is happy.

It is not clear whether the government of Saudi Arabia plans to develop a system of rights specifically meant for robots, or if Sophia is going to enjoy the same exact rights as the human citizens of Saudi Arabia. What is clear, though, is that this groundbreaking action does add to the growing debate of whether robots should be given rights akin to those of humans. For example, in the same timeframe, the European Parliament proposed the granting of "personhood" status to AI agents, which would provide them with specific rights and responsibilities, like those of human beings.

Shibuya Mirai is an AI program that was granted a residency card in Tokyo, Japan, which exists as a chatbot on the popular Line messaging app (Caughill, 2017). The goal of the government of Shibuya Ward in Tokyo was to have Mirai familiarize some of the 230,609 citizens (2024) of the district with the local government and provide them an avenue to share opinions with officials. The official statement of the government was as follows: "His hobbies are taking pictures and observing people. And he loves talking with people … Please talk to him about anything." According to *Agence France Presse*, Mirai can have text conversations with users and can even "make light-hearted alterations to selfies he is sent."

The granting of residency in Tokyo to an AI program stylized as a seven-year-old boy certainly builds on the current trend that appears to be moving in favor of recognizing robots as more equal to humans than ever imagined previously. Nevertheless, the important quotient that remains, I would argue, are the ethical, moral, and legal implications that should seriously be considered; not only considered, but rather, fervently pondered upon. What if people are just doing this simply for amusement or for mere corporate publicity? Should they be doing this just because they can?

How will these very highly publicized changes in our mindsets toward robots affect the thinking of our children as they grow up? Subsequently, what kinds of adults will children like this grow up to be? These are questions that call for genuine concern and contemplation. And why not? Why should we not all be seriously contemplating the implications that all of this will have upon future generations of people? And by "people" I mean fully biological humans.

I had the opportunity to be an up-close-and-personal witness to the incredible benefits and capabilities of AI robots for the improvement of mankind during a trip to Japan in 2017. I visited Cyberdyne Studio in Tsukuba, Japan, which is home to HAL (Hybrid Assistive Limb), described as, "The world's first cyborg-type robot, by which a wearer's bodily functions can be improved, supported and enhanced" (Cyberdyne, 2018b). However, I must add that what I witnessed was far more compelling than what this official description could possibly convey. HAL responded solely to human *thought*, and to this alone. If an individual wished to activate HAL in order to walk, he or she had to indeed think the thought "I want to walk" and this was when HAL was ordered into motion, through chemical signals transmitted through the nerves to the muscles required for walking. If, for example, another person grabbed this individual connected to HAL and were to exact the motions of walking onto that same individual, HAL would not respond. The intended motion had to originate in the user's mind for HAL to facilitate the desired movement.

I was very lucky to be able to observe volunteers who interfaced with HAL through a simple armband. They were instructed in one instance to think of moving their arm, through which HAL would translate their thoughts into motions. In another instance, they were instructed to think of something entirely different while the scientist, Mr. Kojima, would attempt to move their arm in different directions. The second experiment would always be unsuccessful. HAL would remain unresponsive to third party manipulation of the physical body that it was interfaced with. In other words, a patient functioning with the assistance of HAL could in no way be manipulated by external forces and would strictly be the sole individual who influenced the robot.

Mr. Kojima, the scientist who oversaw this symposium, explained to us the goal of HAL, which he put very simply, "To improve the lives of patients who are paralyzed or suffering from other debilitating diseases that require assistance for movement of the body. We are always asking ourselves this question, 'how can we help improve the lives of patients?'" The official corporate statement of Cyberdyne, via their website, reads as follows:

> 'Dreaming' or 'Passion' is valuable, but a 'Heart for Caring for Others' is much more important (Cyberdyne, 2018a). If one has the heart to care for people and society, one cannot idly stand by, but will proactively think and take immediate actions for a brighter future. This is the heart of Cyberdyne. Cyberdyne, making full use of *Cybernics* – an integrated technology of human, robot, and information systems will tackle and solve social issues via R&D and implementation of 'innovative technology coexisting with people', and will create a 'New Industry' based on these solutions. As a result, new pioneers with experience in this new design process will emerge from Cyberdyne. This is our 'Challenge to Shape the Future' ... A bright future can be built by utilizing leading edge innovative technology ... To achieve this vision, Cyberdyne proposes a 'Techno-Peer Support', which is mutually supported by both humans and technology.

Sitting there in the conference room alongside others who were "oohing and aahing" just as loudly as I was, an overwhelming appreciation for AI, cyborgs, and robots in general began to flood over my whole being. I was personally able to hear a scientist speak from his heart about the purpose of this robot: *to enhance the lives of those who need it the most.* It is during this visit to Cyberdyne in Tsukuba, Japan, that I gained hope regarding the ethical and moral implications of robots, as well as their application to medicine and other fields of human healthcare.

At the end of the day, a robot or AI is only as good as the human beings who create it, or is truly as good as the human beings who create it. The vision is derived from the dream of a human being and the robot is the product of that dream. Some dreams are positive, which benefit humankind, while other dreams are negative and detract from humanity as a whole. However, in those moments there in Japan, witnessing HAL improve the lives of humans while simultaneously protecting those lives from unwanted interference, I realized that I was now living in an era that I am thankful to say I exist in.

Delving into the nanometric domain, how might we approach the status of advanced ultra-intelligent autonomous micron scale nanomedical robots (classes of which would facilitate the realization of GHCE) that have the potential capacity to beneficially address all that ails us, including aging? Although the majority of nanomedical procedures would be completely free of human intervention, it is quite conceivable that these quantum computer/AI endowed nanodevices may have the ability, in cases that are deemed warranted, to directly communicate with their "outbody" human physician or surgeon counterparts using spoken language through computer proxies to convey their in vivo activities, the on-site/up-close status of a particular organ, or even individual cells, to receive commands, etc. Will these unseen superintelligent entities be regarded simply as sophisticated surgical "instruments" … or something more? … colleagues? Even though we would never be able to see them without a microscope, will these entities still be eligible to garner a certain level of protection under the law, or even rights?

12.16 CONCLUSION

I believe that debates regarding robot rights and equalities, the benefits and uses of robots as applied to healthcare and wellbeing, and issues of automation and human-robot hybrids warrant very serious consideration in terms of ethics and morality. The power play between corporations and entities, and more, is an inevitable element of our journey deeper into an accelerated Singularity. There are incisive questions that we should ask, and answers that we need to know. Nevertheless, it remains inarguable that witnessing a robot improve the life of a paralyzed patient is surprisingly one of the most awe-inspiring experiences that a human being can have. I can only hope (and we should all hope together) that this is the type of experience we will all have the privilege to bear witness to today, and in a possible future world where we all have access to equitable, affordable top tier healthcare under GHCE.

REFERENCES

Aldiss B. *Supertoys Last All Summer Long: And Other Stories of Future Time*. London: Orbit, 2001.

Bansal DG. Scientists Develop a 'Lab-on-a-chip' That Costs 1 Cent to Make. *Stanford Medicine News Center*. February 6, 2017. www.med.stanford.edu/news/all-news/2017/02/scientists-develop-lab-on-a-chip-that-costs-1-cent-to-make.html (accessed January 30, 2025).

Beck M, Libert B. The Rise of AI Makes Emotional Intelligence More Important. *Harvard Business Review*. February 15, 2017. www.hbr.org/2017/02/the-rise-of-ai-makes-emotional-intelligence-more-important (accessed January 30, 2025).

Bhavsar N. Elon Musk: If Humans Are to Survive, We Must Merge with Machines. *Futurism*. February 14, 2017. www.futurism.com/elon-musks-and-the-need-for-symbiosis-with-machines/ (accessed January 30, 2025).

Boehm FJ. "Nanotechnology, Nanomedicine, and AI: Toward the Dream of Global Health Care Equivalency." 2025. https://www.nanoappsmedical.com/nanotechnology-nanomedicine-and-ai-toward-the-dream-of-global-health-care-equivalency/ (accessed January 30, 2025).

Bohan E. This New Species of AI Wants to Be "Superintelligent" When She Grows Up. *Big Think*. 2017. www.bigthink.com/elise-bohan/the-most-human-ai-youve-never-heard-of-meet-luna (accessed January 30, 2025).

Bozorg Qomi S, Asghari A, Mojarrad M. An overview of the CRISPR-based genomic- and epigenome-editing system: Function, applications, and challenges. *Advanced Biomedical Research*. 2019;8:49.

Canavero S. HEAVEN: The head anastomosis venture project outline for the first human head transplantation with spinal linkage (GEMINI). *Surgical Neurology International*. 2013;4(Suppl 1):S335–42.

Carmichael J. Kernel's Neural Implants Could Boost Failing Human Brains. *Inverse Science*. August 16, 2016. www.inverse.com/article/19812-kernel-neural-implant-prosthesis-singularity-artificial-intelligence (accessed January 30, 2025).

Caughill P. Another Expert Joins Stephen Hawking and Elon Musk in Warning about the Dangers of AI. *Futurism*. February 13, 2017. www.futurism.com/2-expert-thinks-ai-will-undoubtably-wipe-out-humanity/ (accessed January 30, 2025).

Caughill P. An Artificial Intelligence Has Officially Been Granted Residency. *Futurism*. November 6, 2017. www.futurism.com/artificial-intelligence-officially-granted-residency/(accessed January 30, 2025).

Cerullo M, Lee AM Here's what a Sam Altman-backed basic income experiment found. CBS News. July 23, 2024. https://www.cbsnews.com/news/sam-altman-universal-basic-income-study-open-research/ (accessed January 29, 2025).

Crew B. Google's New AI Has Learned to Become "Highly Aggressive" in Stressful Situations. *Science Alert*. March 31, 2018. https://www.sciencealert.com/google-deep-mind-has-learned-to-become-highly-aggressive-in-stressful-situations (accessed January 30, 2025).

Cyberdyne. About Us – Corporate Identification. *Cyberdyne*, 2018a. www.cyberdyne.jp/english/company/index.html (accessed August 30, 2018).

Cyberdyne. Products – What's HAL? *Cyberdyne*, 2018b. https://www.cyberdyne.jp/english/products/HAL/index.html (accessed January 30, 2025).

Daley J. Head Transplant Patient Will Use Virtual Reality to Smooth Transition to New Body. *Smithsonian Magazine*. November 23, 2016. https://www.smithsonianmag.com/smart-news/virtual-reality-system-developed-head-transplant-patients-180961198/ (accessed January 29, 2025).

Döring N, Mohseni MR, Walter R. Design, use, and effects of sex dolls and sex robots: Scoping review. *Journal of Medical Internet Research*. 2020;22(7):e18551.

Euron P. Uncanny beauty: Aesthetics of companionship, love, and sex robots. *Artificial Life*. 2022;28(1): 108–123.

Future Culture Magazine, Interview with Luis Arana. *Future Culture Magazine*. 2016. http://futureculturemagazine.blogspot.com/p/interview-with-luis-arana.html (accessed November 22, 2017).

Future of Life Institute. Asilomar AI Principles. 2025. https://futureoflife.org/person/asilomar-ai-principles/ (accessed August 13, 2025).

Galeon D. For the First Time Ever, a Robot was Granted Citizenship. *Futurism*. October 26, 2017. https://futurism.com/for-the-first-time-ever-a-robot-was-granted-citizenship (accessed August 13, 2025).

Galeon D, Caughill P. Self-Learning AI: This New Neuro-Inspired Computer Trains Itself. *Futurism*. October 15, 2016. www.futurism.com/self-learning-ai-this-new-neuro-inspired-computer-trains-itself/ (accessed January 30, 2025).

Galeon D, Gohd C. World's Top Experts Have Created a "Law of Robotics." *Futurism*. February 1, 2017. www.futurism.com/worlds-top-experts-have-created-a-law-of-robotics/ (accessed January 30, 2025).

Galeon D, Houser K. DARPA: We're Moving to Merge Humans and Machines. *Futurism*. February 14, 2017a. www.futurism.com/darpa-were-moving-to-merge-humans-and-machines (accessed January 30, 2025).

Galeon D, Houser K A Billionaire Is Helping Fund a Massive Universal Basic Income Project. *Futurism*. February 10, 2017b. https://futurism.com/4-another-case-for-ubi-ebay-founder-funds-basic-income-experiment-in-kenya/ (accessed January 30, 2025).

Galeon D, Houser K. New AI Can Write and Rewrite Its Own Code to Increase Its Intelligence. *Futurism*. February 16, 2017c. www.futurism.com/new-ai-can-write-and-rewrite-its-own-code-to-increase-its-intelligence/ (accessed February 23, 2017).

Galeon D, Marquart S. IBM Chief: Computers Will Soon be Smarter Than Humans. *Futurism*. January 18, 2017a. www.futurism.com/bm-chief-computers-will-soon-be-smarter-than-humans/ (accessed January 30, 2025).

Galeon D, Marquart S. Elon Musk: Automation Will Force Government to Introduce Universal Basic Income. *Futurism*. February 14, 2017b. www.futurism.com/elon-musk-automation-will-force-governments-to-introduce-universal-basic-income/ (accessed January 30, 2025).

Goldman B. Brain-computer Interface Advance Allows Fast, Accurate Typing by People with Paralysis. *Stanford Medicine News Center*. February 21, 2017. www.med.stanford.edu/news/all-news/2017/02/brain-computer-interface-allows-fast-accurate-typing-by-people-with-paralysis.html (accessed January 30, 2025).

Hern A. Elon Musk Says He Invested in DeepMind over "Terminator Fears." *The Guardian*. June 18, 2104. www.theguardian.com/technology/2014/jun/18/elon-musk-deepmind-ai-tesla-motors (accessed January 30, 2025).

Inventum Bioengineering Technologies. 2025. https://www.f6s.com/inventumbio (accessed January 30, 2025).

Johnson A. AI in 2025: The Most Exciting Developments to Expect. *Techresearchs*. December 24, 2024. https://techresearchs.com/artificial-intelligence/ai-in-2025-the-most-exciting-developments-to-expect/ (accessed January 28, 2025).

Jones RC. Stephen Hawking Warns Artificial Intelligence Could End Mankind. *BBC News* December 02, 2014. www.bbc.com/news/technology-30290540 (accessed January 30, 2025).

Karpen E. Startup claims they have created AI head transplant system, plans to perform first procedure within decade. *New York Post*. May 22, 2024. https://nypost.com/2024/05/22/us-news/scientists-announces-wild-plan-for-first-head-transplant/ (accessed January 28, 2025).

Kirkey S. World's first human head transplant successfully performed on a corpse, scientists say. *National Post*. November 17, 2017. https://nationalpost.com/health/worlds-first-human-head-transplant-successfully-performed-on-a-corpse-scientists-say (accessed January 28, 2025)

Kurzweil R. *The Singularity Is Near*. New York: Viking Books, 2005.

Kurzweil R. Robots Will Demand Rights – And We'll Grant Them. *Kurzweil Accelerating Intelligence*. September 11, 2015. http://www.kurzweilai.net/time-question-everything-will-robots-need-rights-robots-will-demand-rights-and-we-will-grant-them (accessed January 30, 2025).

Kurzweil R. *Wikipedia*. 2025a. https://en.wikipedia.org/wiki/Ray_Kurzweil (accessed January 29, 2025).

Kurzweil R. Biography: A pioneering inventor, best-selling author, and futurist. 2025b. https://www.lifetimeofinnovation.com/about-ray-kurzweil (accessed January 29, 2025).

Levy S. If Ray Kurzweil Is Right (Again), You'll Meet His Immortal Soul in the Cloud. WIRED. June 13, 2024. https://www.wired.com/story/big-interview-ray-kurzweil/ (accessed January 29, 2025).

Martins NRB, Angelica A, Chakravarthy K, Svidinenko Y, Boehm FJ, Opris I, Lebedev MA, Swan M, Garan SA, Rosenfeld JV, Hogg T, Freitas RA Jr. Human brain/cloud interface. *Frontiers in Neuroscience*. 2019;13:112.

Norman A. An Emotionally Intelligent Computer May Already Have An 'EQ' Higher Than Yours. *Futurism*. March 23, 2017. https://futurism.com/an-emotionally-intelligent-computer-may-already-have-an-eq-higher-than-yours (accessed January 30, 2025)

Ombach G. Where AI Meets the Physical World: The Rise of Humanoid Robots. *Forbes*. December 20, 2024. https://www.forbes.com/councils/forbestechcouncil/2024/12/30/where-ai-meets-the-physical-world-the-rise-of-humanoid-robots/ (accessed January 29, 2025).

PROVEN Robotics, 2024. https://provenrobotics.ai/pepper-the-emotional-robot/ (accessed August 13, 2025).

Robots Without Borders VLOG/Robots Without Borders: A Conversation with AI Engineer Luis Arana. October 25, 2017. https://www.youtube.com/watch?v=UVSeFLxQfnQ (accessed January 30, 2025).

Shelley M. *Frankenstein*. London: Lackington, Hughes, Harding, Mavor, & Jones, 1818.

Sterling B. The Asilomar AI Principles. *Wired*. June 01 2018. https://www.wired.com/beyond-the-beyond/2018/06/asilomar-ai-principles/ (accessed August 30, 2018).

Verho J, Hämäläinen K. Preliminary results of the basic income experiment: self-perceived wellbeing improved, during the first year no effects on employment. VATT Institute for Economic Research. February 08, 2019. https://vatt.fi/en/-/preliminary-results-of-the-basic-income-experiment-self-perceived-wellbeing-improved-during-the-first-year-no-effects-on-employment (accessed January 30, 2025).

Vinge V. The Coming Technological Singularity, 1993. Acceleration Studies Foundation https://accelerating.org/articles/comingtechsingularity (accessed January 30, 2025).

Wall M. SpaceX's Elon Musk Unveils Interplanetary Spaceship to Colonize Mars. *Space.com*. September 27, 2016. www.space.com/34210-elon-musk-unveils-spacex-mars-colony-ship.html (accessed January 29, 2025).

Winkler R. Elon Musk Launches Neuralink to Connect Brains with Computers. *The Wall Street Journal*. March 28, 2017. https://www.wsj.com/articles/elon-musk-launches-neuralink-to-connect-brains-with-computers-1490642652 (accessed January 30, 2025).

Yun SS, Kang BB, Cho KJ. Exo-glove PM: An easily customizable modularized pneumatic assistive glove. *IEEE Robotics and Automation Letters*. 2017;2(3):1725–1732.

13 The Future Body

Nano Bodybuilders of Tomorrow

Natasha Vita-More

CONTENTS

13.1 INTRODUCTION

"Imagination is more important than knowledge," Albert Einstein asserted. Maybe it will "... carry us to worlds that never were, but without it we go nowhere," as Carl Sagan speculated. But let us keep in mind that Patricia McKillip cautioned us about "the golden-eyed monster that never sleeps. It must be fed; it cannot be ignored." Arguably, imagination wakes us in the middle of the night, taunts us with unlimited possibilities, and cherishes our most innate desires. When attended to and fed, it alerts the neural fibers of our minds that orchestrate circuits and perform with precise coordination to stage a mental theater that cannot be ignored (Agnati et al., 2013).

In this chapter, my approach is not a scientific discourse; it is the imagined perception of an engineered body of tomorrow. To begin, I introduce the future body as a new human genre comprised of nanorobots as a feasible solution to the unresolved problem of aging. I reflect briefly on bodybuilders of antiquity who overcame physical limitations through their wit and determination. I recall a project I developed many years ago and include the advice of a team whose knowledge reached beyond the worlds of the past with visions that cannot be ignored. I then turn to current issues of genetic legacy and advances toward genetic liberty, the convergence of artificial intelligence (AI), robotics, and prosthetics, and the necessity of the biocompatible nano body.

13.2 BODYBUILDERS, MYTHS, AWE, AND TREPIDATION

Enshrined in antiquity through exalted physical feats and ethereal myths, the human body has been humbled by the potential of its engineered future. Optimal physical endurance of athletic activity in early Grecian times symbolized strength, stamina, and aesthetic appeal. Stone carved depictions of muscular, harmonic bodies displaying poses of Olympic feats tell a story of bodybuilders whose persistence pushed genetic fibers to the limits of biological endurance. The early Classical period, ancient Greek *Kritios Boy*, sculpted with a perceived ideality of physical proportion and muscular strength, was considered "the first beautiful nude in art" by British historian Kenneth Mackenzie Clark (Clark, 2010). This appreciation of the genetics of engineered beauty and strength is in stark contrast to the methodology that often tells a different tale, and which continues to plague us today.

 DOI: 10.1201/9781003680093-13

Pandora's procreation of life, Achilles' heel, Adam's rib, and the Cyclops reflect beliefs concerned with natural and supernatural stories about missing or extra body parts; beings that hover between human and nonhuman, mortality and continued existence, and the veil between the physical and spiritual. Mythic tales consciously or subconsciously influence decision-making, athletic pre-game rituals form self-fulfilling prophesies, and the use of contemporary performance enhancing drugs (PEDs) offers a powerful psychological buildup that can induce peak performance.

Within this realm of exuberant physical capacities, the display and pomp of bodily performance ignites awe and trepidation. The awe of clear athleticism is the height of aesthetics in biological functions, geometry of form, and an almost supernatural level of possibilities. The trepidation is far more distinct when viewed with X-ray vision. Inside the orchestra of skeletal, muscular, respiratory, and nervous systems are the biological cellular processes of aging, disease, disruption, and decay. The beauty and bewilderment of physical strength and agility when viewed in its stylized form cannot hide the fact that the body beautiful is nonetheless the body diseased. Something must be done about this.

13.3 THE FUTURE BODY

Launched as a collaborative team to strategize how emerging technologies might be employed to engineer a future human body, I established the "new human genre" in early 1996. AI, nanotechnology, and robotics were the technologies of choice. As interconnected systems, these three technologies could assemble a body system to continually update and regenerate, and store memories exo-body (Vita-More, 2002). Ostensibly, and seeming a bit science fiction-like at that time, the academic landscape was and continues to be deeply submerged in theoretical human, transhuman, cyborg, and posthuman concepts for addressing philosophical questions, anthropological studies, and scientific inquiry to question the future of humanity. Yet academia misses the real issues of how AI, nanotechnology, and robotics are modifying the human body, the culture of decentralized, non-binary beliefs, and the growing longevity movement. Because of this knowledge gap, I found it necessary to initiate a project that offered an alternative to mainstream prejudices about life extension and biomedical advances that could mitigate aging and extend life. The expertise of my team encompasses the domains of AI, nanomedicine, evolutionary biology, philosophy, economics, physics, robotics, cryptography, and augmented reality (Figure 13.1).

The aim was to prototype a future body that offered plausible solutions to issues that have plagued humanity for eons: overcoming the fragility of biology and aging. At the time, the prototype was called "Primo Posthuman," which was later renamed to appeal to the mainstream with "Future Human Body" (Vita-More, 2002). The focus was on narrow AI, artificial general intelligence (AGI), nanomedicine, automated prosthetics, and a tertiary metabrain. Keenly regarded as necessary for all bodies, whether biological, alternative genetic clones, or an engineered body, human cryonic preservation (cryonics) was included as a backup plan. The significant ethical implications; thus, intense objections against these technologies that might audaciously enable the modification, engineering, and fabrication of enhanced human bodies were rampant. Bioethicists emerged as a global community of interrogators who questioned the ethical use of AI and tampering with the inherent nature of the human body and mind. Nevertheless, the project went forward and was rendered as a bodily system prototyped with illustrations, 3D modeling, and animation. Its reception was phenomenal, headlined news articles and was featured in numerous books and films. However, the integrity of the work was met with a flood of journalistic and academic hyperbole, which was but a mere facet of the downside.

The transitioning of ideas to viable products and services is challenging. An idea that cannot be materialized can be a dead end without the accuracy and precision of science and engineering. One issue that arose with the Future Human Body prototype was evidencing its proof of concept

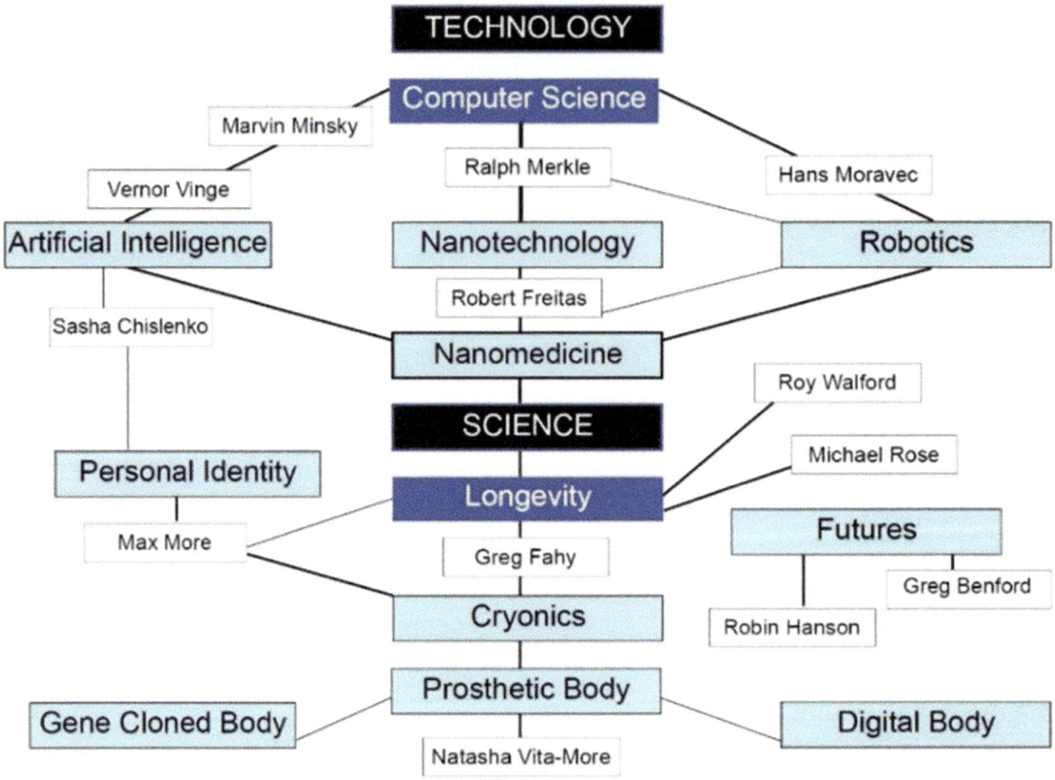

FIGURE 13.1 Graphic chart of life extension team members and their areas of expertise/fields of contribution. (Image created by the author.)

and delivering a minimum viable product (MVP) that applied AI and nanorobots. I was stumped. Without an aspect of this future body prototype produced, it remained a concept. Fortunately, the field of prosthetics was advancing with robotics, AI, and haptic systems and I could reference some of these most advanced systems as steps in the direction of a whole-body prosthetic. Further, science involves steps that need to be clearly defined and ideas that are before their time can reference a desired outcome as a larger MVP. Let us keep in mind that the 2001 Human Genome Project was thought of as one of the greatest feats of exploration in science. Twenty years later, we were reminded that it was not the full human genome, but only 92%. By 2022, the missing 8% were finally sequenced (Church, 2022), albeit not with human genes, and only for the Y chromosome. In 2023 a first draft of the human pangenome was unveiled, which articulated the full extent of human genetic diversity (Liao et al., 2023). The conditions for discovery change as technology advances toward the future and offers new tools that could tether us to the ambitions of the past.

13.3.1 New Human Genre

I chose to use the term genre because it reflects the composition, characteristics, form, and style of an item or entity. In biology, a genre refers to a group of living organisms that are capable of exchanging genes. For humans, a prequel to a type of new genre originated when our species began to crudely manipulate its characteristics with early replacements for injured body parts. Consider that early body enhancements were often incompatible with biological cells, cumbersome to use, and aesthetically unappealing. However, more importantly, they did offer alternatives for those who were inflicted with disease or injury that diminished their psychological and cognitive capacities. The notion of body augmentation in the form of prosthetics evolved during archaic times. Dated between 950 and 710 BC, the "Cairo toe" (Deamer, 2017; Thurston, 2007) composed of leather and stained wood, might be the oldest known prosthesis, while artificial limbs were designed over 2,200 years ago (Shahar et al., 2019). These artifacts link to ancient literary themes encompassing automata, Byzantium constructions, and well-known da Vinci humanoid mechanical systems, with some issuing robotic systems that resemble today's wearable devices.

From augmented bodily components to enhanced genetics and advances in reproductive techniques, we have seen the need to improve on biology. With each step of the way, we have been developing a more resilient human. The new human genre aims to perpetuate this development toward a highly advanced system that is seamless, stunning, and biocompatible (Figure 13.2).

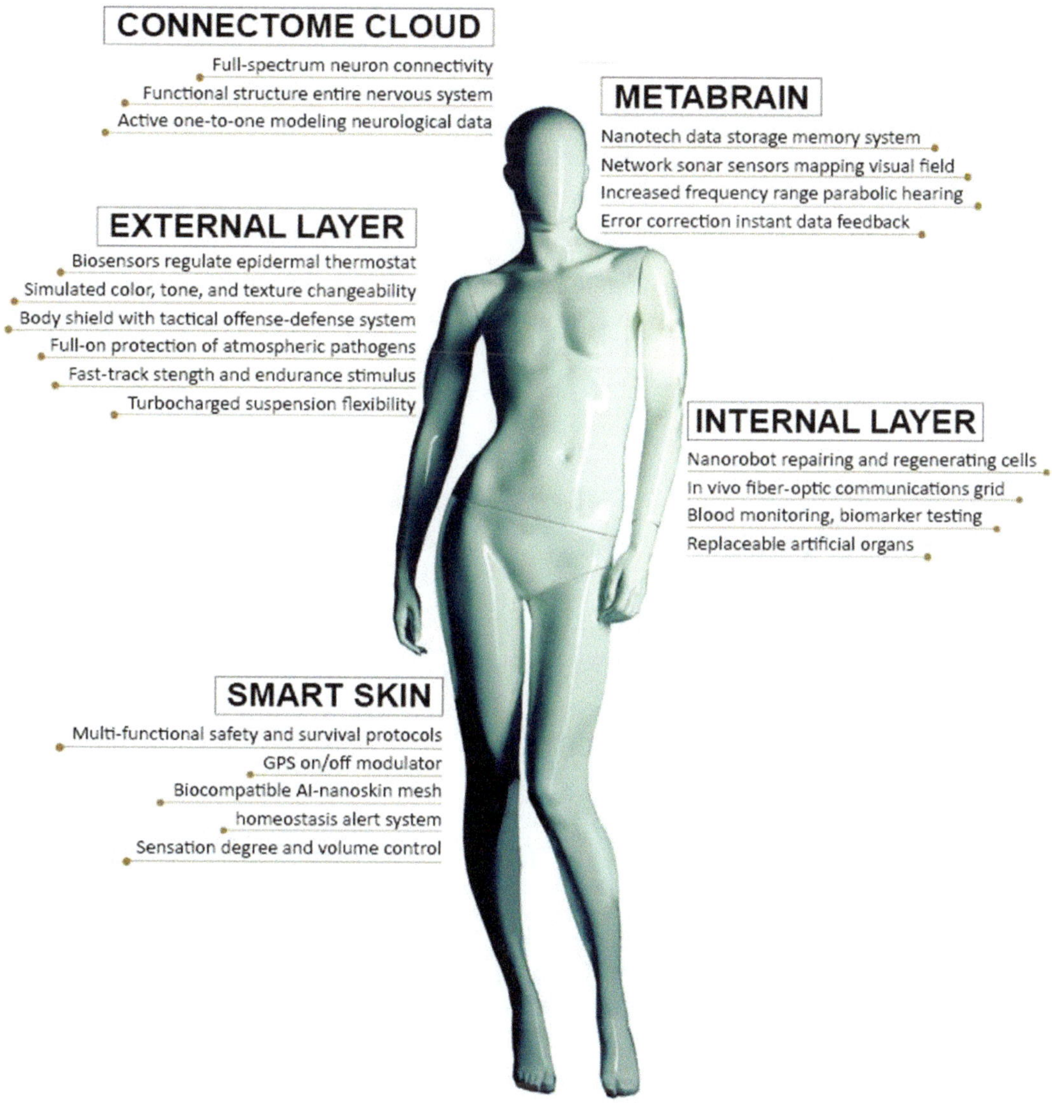

FIGURE 13.2 Future human body model based on original 1996 prototype Primo Posthuman. (Image created by the author.)

The development of a model that precipitates iterative variations within the alpha stages can be valued within diverse studies concerning the future. For example, prior to the new human genre and prototyping of the Future Human Body, there had been no fundamental or realized concept for a whole-body prosthesis, a metabrain model for memory storage, an AGI assist, or a channel of delivery for feeding data to and from an exobrain, such as what we currently call the Cloud. One of the foci of the new human genre was to address certain diseases that affect millions of people, such as organ degeneration, cognitive impairment, skin cancer, injuries, and the inability of biology to quickly restore, repair, and regenerate at a level that would afford an extended healthy life. The diverse study areas that were, and continue to be, included in the original alpha concept as rudimentary enhancements are required to develop as science and technology advance and as the need for prosthetics grows. For example, a chart that compares the 20th century body with a potential 21st century body suggests optional improvements on the body's biological systems (Figure 13.3).

More than two decades later, medical technologies have advanced to such a degree that these theoretical models can be further evaluated and even set into a Beta phase through 3D modeling, animation, and data visualization, which represent the structures, behaviors, and evolution of biomedical advances. Emergent developments in AI and robotics offer electronic control systems for biocompatible prosthetic devices that use algorithms to read the body's nerve signals. "A special sensor picks up the signals from the spinal cord and allows people to control a prosthetic limb by simply imagining the movement that they want to make" (Farina et al., 2017). This is just one of the myriad interoperability features of prosthetic systems that integrate robotics, AI, and biology, which form linkages between the human body and robotics, while filling the need for replacement body parts.

The xenotransplantation of skin, kidneys, and pancreatic islets has increased due to the need for lifesaving transplants; prosthetic bionic limbs have become increasingly complex due to the high demand of patients with injuries or disease; and Wearable Health Devices assist with monitoring

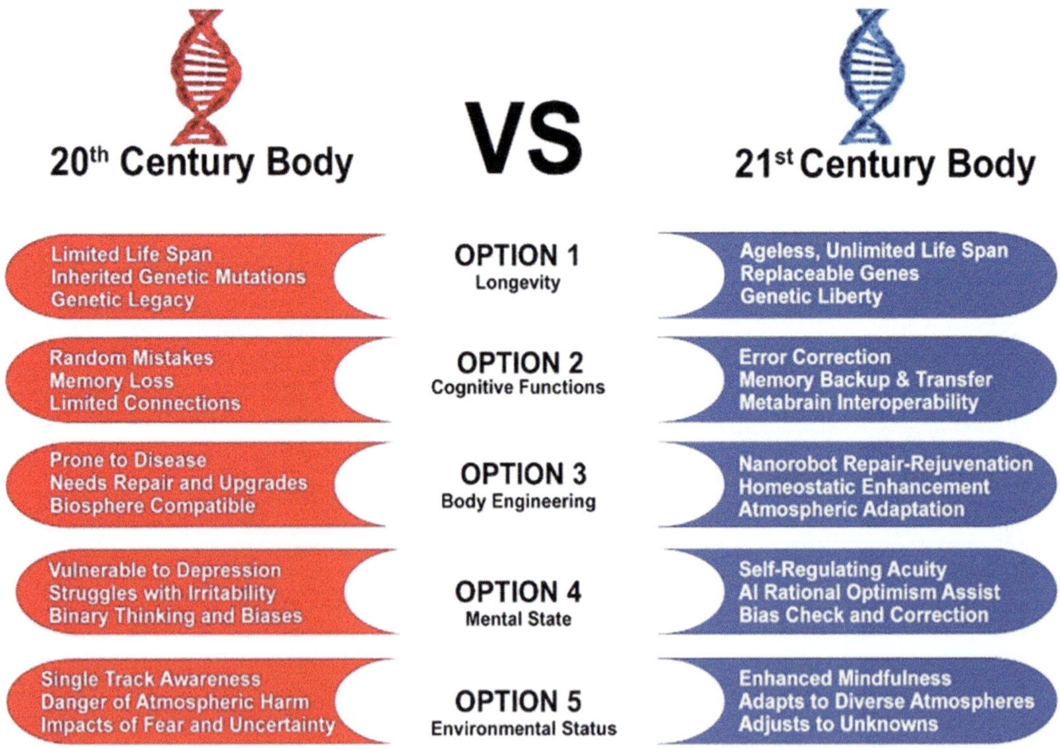

FIGURE 13.3 Twentieth-century and 21st-century comparison chart. (Image created by the author.)

health—none of which are made of human DNA. A single-celled artificial organism, designed on a computer and synthesized in a lab, was manipulated by engineering its original mycoplasma state by replacing its DNA (Hughes et al., 2011; Hughes and Ellington, 2017; Pelletier et al., 2021). "DNA synthesis techniques and technologies are quickly becoming a cornerstone of modern molecular biology and play a pivotal role in the field of synthetic biology. The ability to synthesize whole genes, novel genetic pathways, and even entire genomes is no longer the dream it was 30 years ago" (Alleyne, 2010). "The [hu]man-made single cell 'creature', which is a modified version of one of the simplest bacteria on earth, proves that the technology works" (Guthrie, 2009).

Another technology first came decades earlier when the first test-tube baby was born; however, unremarkable today. Tomorrow, the transition from an authoritarian genetic legacy to the genetic liberty from detrimental genes will rescript the human reproductive process to be *better than well*, which may herald the "democratization of genomics" (Li et al., 2017). Better than well means that each day, medicine uncovers yet another strategy for making people healthier. There is nothing intrinsically wrong with aiming to be better than well. The body's complex organization of components and systems typically functions cooperatively but not at peak performance.

Thus, the conundrum that people are challenged with is whether or not to believe the longevity industry's claims: First, people are told that there are emerging therapies for curing disease, which gives them a glimmer of hope, albeit second, they are also reminded that they might not live long enough to benefit from these cures. This creates an anomaly as individuals are deeply engrained in and emotionally influenced/motivated by current age-based norms, while they are also tempted to venture far beyond this social order, which will ultimately dissolve the rules they have for so long been aligned with and bound to.

Understanding how technological interventions are evolving better bodies requires a remembrance of the evolutionary journey from hunter-gatherer societies to the augmented human with globalized, integrated computing systems, which gives a nod to Nicholas Negroponte's statement: "Computing is not about computers anymore. It is about living."

This logic overrides the siloed approach to legacy genes and is present in Robert Winston's *Redesigning Humans: Our Inevitable Genetic Future*: "We cannot say much about the challenges that will accompany the first steps of human self-design until we examine the specific biological modifications that might intrigue us when the technology arrives" (Stock, 2002a). According to Gregory Stock, the choices we make today are a "preview of the deeper ones we will face" (Stock, 2002b). in the years to come, particularly "because they reveal the cultural and biological desires that shape our preferences" (Stock, 2002c). Stock is a biophysicist who worked in the formative stages that led to the current biotechnological climate. In *Metaman: The Merging of Humans and Machines into a Global Superorganism*, Stock links the Gaia hypothesis (Arthur and Nicholson, 2022) and the noosphere (Vidal, 2021)—a thinking sphere—with a technologically interconnected approach in proposing the human as a type of superorganism—"a community of organisms so fully tied together that it is a single living being" (Stock, 1993a) and one that is "at the forefront of life's evolution from the simplest of living forms, and where this evolutionary process will lead us cannot be known," but "affirms that we are connected" (Stock, 1993b).

These associations—a gestalt of humanity and its environment coupled with the human sphere of thought as a transmutation of elements—tie into the field of automated prosthetic systems based on the theory of cybernetics (Moore and Rieg, 2005). Derived from a view that the equilibrium between feedback and control form the baseline for bodily enhancement, both Stock's early ideas of self-design and Venter's democratization of genomics (Venter et al., 2022) give credence to optimistic visions for countering typical predetermined outcomes.

13.4 THE NANO BODY

The bodybuilders of tomorrow will be engineered with biocompatible nanorobots that seamlessly integrate with biology—not as cybernetic devices or augmented wearables but as inclusive, natural, elemental parts of us that reach far beyond the biomedical therapies of today in pushing the limits of what is possible and necessary for extended longevity.

In nanomolecular systems, a bodybuilder will refer to nanorobots that perform stipulated tasks that resist the ongoing disruption of cellular breakdown. The human psyche is imbued with a deeply rooted need to survive—to overcome external odds that challenge livelihood and the internal odds of bodily harm, such as injury and/or disease. Apart from this ingrained characteristic, the cells of the human body are biologically programmed to reproduce. The skilled athlete takes these two functions—the will of survival and biological functions and aims them toward excelling at a particular set of skills. Both the nanomolecular and athlete's biological systems are bodybuilders that share a common objective: to improve upon and exceed normal biological processes. While the athlete might ascertain satisfaction from the outward show and appearance of strength, the roles of nanorobots will be to operate behind the scenes to build stronger, more durable, longer lasting bodily systems with the delivery of engineered handiwork of aptitude and finessed performance at mechanomolecular resolution. Perhaps in the future, there will be a new physiological norm for all of humanity, rather than the stark divides that currently exist between the physiques of skilled athletes and the average human. We are already glimpsing this today with enhanced prosthetics that allow individuals to achieve new heights in their sports, notably within the Paralympics. This is just the beginning.

The coalescence of robotics and AI-informed prosthetics identifies modern bionic bodybuilders as those whose strength, control, and stability are an engineering solution—sending signals of advanced algorithms to reinforce neural connections in the rewiring of cells to augmented devices. Both achievers seek the freedom to stand apart from normalized narratives and push toward the edge of social norms. I identify timeless bodybuilders as the meta-athletes whose bodies are modified from an inherent genetic blueprint to map new genetic opportunities to repair, extend, enhance, and regenerate for healthy longevity.

Robert A. Freitas, Jr., author of *Cryostasis Revival*, brings us into the future of medical technologies with a comprehensive analysis and examination of nanotechnology's role in biocompatible nanomedicine (Freitas, 2022a). In the book's sub-section (5) Body Upgrade (Freitas, 2022b), Freitas discusses the replacement body where alternative characteristics might be available in the future and the option for a second body with a range of attributes to select from. He notes that potential psychological issues could arise when an individual is faced with an identical (appearance-wise) engineered, alternative body. Some people might prefer to have a replacement body that is constructed from their own genes, as a genetic clone. Nevertheless, Freitas, as well as many others who envision the future of the human body, proposes that whether the replacement body is a genetic clone or a newly engineered system, they will reflect an important facet of the future nanomedical era.

Freitas was an advisor on the "new human genre" project in 1996, when he, along with Eric Drexler and Ralph Merkle, offered their unique visions for nanorobotics that helped to steer my work in the design of the Primo Posthuman whole-body prototype. While my original concept saw the body as a vehicular metaphor, I continued to focus on an alternative human body that could be semi-biological or non-biological, as well as serve as a router or avatar for uploading different platforms as a gateway to cyberspace and myriad virtual worlds.

13.4.1 New Nano Genre: From Genetic Legacy to Genetic Liberty

It is believed that the genetic material of each individual, inherited from parents and ancestors, forms their genetic legacy. As one copy of each gene comes from our mother and the other from our father, our genetics have always been considered as fait accompli; thus, genes are inherited and irreversible. Often people will speculate that their genetics was a type of Russian roulette, luck of the draw, a win of the genetic lottery, or an unfortunate outcome that must be lived with. This has changed, as somatic gene editing or germline editing, although challenging now, will soon become practicable. Epigenetics is something else altogether.

Not long ago, gene editing was considered unethical and immoral by some who worried about blurring the moral boundaries as relates to the manipulation of genetic information of somatic cells, and most ardently so regarding the modification of germline cells. Then humanity turned a corner in assessing the integrity of modifying a person's genetic legacy, which was largely due to the success of CRISPR-Cas9 genome editing techniques. Biochemist Jennifer Doudna, a 2020 Nobel Prize co-winner, stated that it was "...one of the fastest rollouts, I think, of a technology from the fundamental initial science to an actual application" (Thomas, 2021), and reviewers claimed that "genome editing has been hailed as the biggest biotech breakthrough of the century" (Finnie, 2017). Could this novel technology lead our species to genetic liberty? Firstly, let us touch upon the field of epigenetics, which changed how we perceive our genetics. Ergo, if genes can be altered without modifying DNA sequences, we might not be shackled to our inherited genes, as an accomplished fact; gene expression and activities may be revised through our own behaviors (e.g., lifestyle, diet, exercise, and mental state) and/or environmental conditions.

Genetic liberty has been described "as the personal control of all aspects of one's genetic make-up, including genetic material and information" (O'Callaghan, 2005; Prince, 2013; Weeden, 2006). For the purposes of this chapter, I refer to genetic liberty as being free from genetic disorders, whether through the inheritance of mutated disease-causing genes or through spontaneous gene mutations. In short, the liberty to be free from oppressive restrictions imposed primarily by inheritance. A case that stands out as a liberation from harmful genes is the marvelous achievement of Jean Bennett and Albert Maguire, who created a gene therapy protocol to correct defective mutations of the gene *RPE65*, which is responsible for producing proteins that cause light receptors to function in the retina and make vision possible. The aim was to restore the protein production process with corrected versions of the *RPE65* gene and deliver it via a single injection by employing a benign adeno-associated virus to carry the genes to the retina (Penn Medicine News, 2017).

> Until recently, there was no approved treatment for a retinal degenerative disease. Subretinal injection of a recombinant adeno-associated virus (AAV) delivering the normal copy of the human RPE65 cDNA led to reversal of blindness first in animal models and then in humans. This led to the first US Food and Drug Administration (FDA)-approved gene therapy product for a genetic disease, voretigene neparvovec-rzyl (Luxturna). Luxturna was then approved by the European Medicines Agency and is now available in the US through Spark Therapeutics and worldwide through Novartis. Not only has treatment with Luxturna changed the lives of people previously destined to live a life of blindness, but it has fueled interest in developing additional gene therapy reagents targeting numerous other genetic forms of inherited retinal disease. This review describes many of the considerations for administration of Luxturna and describes how lessons from experience with Luxturna could lead to additional gene-based treatments of blindness.
>
> *Maguire et al., 2021*

Genetics plays a role in all diseases; however, it is not the sole cause. There are associated factors derived from epigenetics, spontaneous gene mutation, and injury that intervene and displace health. I have written about this for a long time now and have not veered from my view that we can rely on one body to sustain us over the long term—a minimum of 123 years. The human body needs to operate efficiently; thus, we neglect its maintenance at our peril. Our bodies ultimately break down to the level where replacing and/or rejuvenating all its components becomes an inoperable task. We can replace specific organs and tissues to a point, but eventually a different approach is required.

Advancements toward genetic liberty start with the understanding that science is a process of discovery and examination for building knowledge. It is not an ideology or a weapon to be used to polarize people for, or against, belief systems that tamper with the lives of many. The study of biology is all about living things. The substrates/systems that living beings occupy, such as the humans, are changing rapidly through the convergence of computers, AI, robotics, prosthetics, digitality, and virtuality.

13.5 REFLECTION AND CONCLUSION

The bodybuilders of tomorrow may be an embodiment of nanomedicine. Unlike advances in biomedical interventions and longevity therapies, nanomedical nanorobots would be engineered with an ethics of biocompatibility that prudently and cautiously addresses potential consequences in making systemwide changes. Future bodybuilders will forge an alliance with the past; the high achievers of physical endurance, and the bionic pioneers of today, in freeing themselves from an uncertain and enigmatic genetic legacy and toward an engineered genetic liberty.

An alliance with the future may be our greatest investment in longevity by speculating on what might be. Not being afraid of envisioning what others deem as impossible or improbable, and building on the support and inspiration of others who care enough to encourage us in making our dreams come true. With this, I share words of support from the 1996 team:

Marvin Minsky: Wonderful idea! Perhaps we can hone it so that the reader won't quickly assume it's a hoax and that such things are indeed possible "if consumers really want it." (In other words, we might say that such a product will probably be available by the year 2100 in any case, the way things are going—but quite possibly closer to 2030 if the fields of cryonics and "neuroanalysis" become sufficiently popular.)

Vernor Vinge: One thing I might suggest adding to the list are backup and hot swap options (I'm always safety conscious:-)

Ralph Merkle: I love it! My only suggestion is the ability to dynamically adjust the metabolic rate, so I can eat all that I want without gaining a pound! (Other approaches, such as a removable stomach pouch, lack aesthetic appeal…)

Greg Benford: Would love to have internal and external real-time personality backup; parallel semantic I/O for direct encrypted knowledge transfers; no raw sound processing; adaptable body-to-mind interface layer. I want to have internal body controls upgradeable by the manufacturer so that I (Mind) would notice only increased efficiency. Enhanced reality hookups—between sensors and consciousness. Optional feeding from electricity/battery. Better energy efficiency than fat storage.

REFERENCES

Agnati LF, Guidolin D, Battistin L, Pagnoni G, Fuxe K. The neurobiology of imagination: Possible role of interaction-dominant dynamics and default mode network. *Frontiers in Psychology*. 2013;4:296.

Alleyne R. Scientist Craig Venter creates life for the first time in laboratory sparking debate about 'playing god'. *The Telegraph*. 2010. http://www.telegraph.co.uk/news/science/7745868/Scientist-Craig-Venter-creates-life-for-first-time-in-laboratory-sparking-debate-about-playing-god.html (accessed May 31, 2023).

Arthur R, Nicholson A. Selection principles for Gaia. *Journal of Theoretical Biology*. 2022;533:110940.

Church DM. A next-generation human genome sequence. *Science*. 2022;376(6588):34–35.

Clark K. *The Nude: A Study in Ideal Form*. New edition. London: The Folio Society. 2010;24–25.

Deamer K. 3,000-Year-Old Wooden Toe Prosthetic Discovered on Egyptian Mummy. *Live Science*. June 22, 2017. https://www.livescience.com/59581-ancient-prosthetic-toe-found-in-egyptian-grave.html (accessed May 31, 2023).

Farina D, Vujaklija I, Sartori M, Kapelner T, Negro F, Jiang N et al. Man/machine interface based on the discharge timings of spinal motor neurons after targeted muscle reinnervation. *Nature Biomedical Engineering*. 2017;1:0025.

Finnie I. CRISPR Patent Wars. *Genetic Engineering & Biotechnology News*. MaryAnn Liebert, Inc. Publishers, February 06, 2017. https://www.genengnews.com/insights/crispr-patent-wars/ (accessed May 31, 2023).

Freitas RA Jr. *Cryostasis Revival: The Recovery of Cryonics Patients through Nanomedicine*. Alcor Life Extension Foundation. Scottsdale: AZ. 2022a.

Freitas RA Jr. *Cryostasis Revival: The Recovery of Cryonics Patients through Nanomedicine*. Alcor Life Extension Foundation. Scottsdale: AZ. 2022b. p. 454.

Guthrie JJ. Craig Venter sees change from genomics. *San Francisco Chronical, SFGATE.* June 11, 2009. http://www.sfgate.com/science/article/J-Craig-Venter-sees-change-from-genomics-3229302.php (accessed May 31, 2023).

Hughes RA, Ellington AD. Synthetic DNA synthesis and assembly: Putting the synthetic in synthetic biology. *Cold Spring Harbor Perspectives in Biology.* 2017;9(1):a023812.

Hughes RA, Miklos AE, Ellington AD. Gene synthesis: Methods and applications. *Methods in Enzymology.* 2011;498:277–309.

Li PE, Lo CC, Anderson JJ, Davenport KW, Bishop-Lilly KA, Xu Y et al. Enabling the democratization of the genomics revolution with a fully integrated web-based bioinformatics platform. *Nucleic Acids Research.* 2017;45(1):67–80.

Liao WW et al. A draft human pangenome reference. *Nature.* 2023;617(7960):312–324.

Maguire AM, Bennett J, Aleman EM, Leroy BP, Aleman TS. Clinical perspective: Treating RPE65-associated retinal dystrophy. *Molecular Therapy.* 2021;29(2):442–463.

Moore LW, Rieg LS. The ethics of using cybernetics and cyborg technologies: What every rehabilitation nurse should know. *Rehabilitation Nursing.* 2005;30(2):40–43.

O'Callaghan N. Human origins and human rights in the genome age. *Ave Maria Law Review.* 2005;3(1):129.

Pelletier JF, Sun L, Wise KS, Assad-Garcia N, Karas BJ, Deerinck TJ et al. Genetic requirements for cell division in a genomically minimal cell. *Cell.* 2021;184(9):2430–2440.

Penn Medicine News. FDA Approves Gene Therapy for Inherited Blindness Developed by the University of Pennsylvania and Children's Hospital of Philadelphia. December 19, 2017. https://www.pennmedicine.org/news/news-releases/2017/december/fda-approves-gene-therapy-for-inherited-blindness-developed-by-university-of-pennsylvania-and-chop (accessed May 31, 2023).

Prince A. Comprehensive protection of genetic information: One size privacy or property models may not fit all. *Brooklyn Law Review.* 2013;79(1).

Shahar FS, Hameed Sultan MT, Lee SH, Jawaid M, Md Shah AU, Safri SNA, Sivasankaran PN. A review on the orthotics and prosthetics and the potential of kenaf composites as alternative materials for ankle-foot orthosis. *Journal of the Mechanical Behavior of Biomedical Materials.* 2019;99:169–185.

Stock G. *Metaman: The Merging of Humans and Machines into a Global Superorganism.* New York: Simon & Schuster. 1993a, p. 20.

Stock G. *Metaman: The Merging of Humans and Machines into a Global Superorganism.* New York: Simon & Schuster. 1993b, p. 245.

Stock G. *Redesigning Humans: Our Inevitable Genetic Future.* Boston: Houghton Mifflin Company. 2002a, p. 9.

Stock G. *Redesigning Humans: Our Inevitable Genetic Future.* Boston: Houghton Mifflin Company. 2002b, p. 80.

Stock G. *Redesigning Humans: Our Inevitable Genetic Future.* Boston: Houghton Mifflin Company. 2002c, p. 97.

Thomas I. How CRISPR gene editing will treat diseases in future: Nobel-winning Intellia co-founder Jennifer Doudna. *CNBC.* June 30, 2021. https://www.cnbc.com/2021/06/30/how-crispr-gene-editing-will-treat-disease-intellia-founder-doudna-.html (accessed May 31, 2023).

Thurston AJ. Paré and prosthetics: The early history of artificial limbs. *ANZ Journal of Surgery.* 2007;77(12):1114–1119.

Venter JC, Glass JI, Hutchison CA 3rd, Vashee S. Synthetic chromosomes, genomes, viruses, and cells. *Cell.* 2022;185(15):2708–2724.

Vidal C. Pierre Teilhard de Chardin: A visionary in controversy. *History and Philosophy of the Life Sciences.* 2021;43(4):125.

Vita-More N. Radical body design "Primo Posthuman." *Kurzweil.ai.* February 25, 2002. https://www.kurzweilai.net/radical-body-design-primo-posthuman (accessed May 30, 2023).

Weeden JL. Genetic liberty, genetic property: Protecting genetic information perspectives on natural marriage—Notes. *Ave Maria Law Review.* 2006;4:643.

Index

Note: **Bold** and *Italic* page numbers refer to **tables** and *figures*.

I

J

K

L

M

N

O

W

Y

Z